Forensic Dentistry
Second Edition

Forensic Dentistry
Second Edition

Edited by
David R. Senn
Paul G. Stimson

CRC Press is an imprint of the
Taylor & Francis Group, an **informa** business

First published in paperback 2024

First published 2010
by CRC Press
2385 NW Executive Center Drive, Suite 320, Boca Raton FL 33431

and by CRC Press
4 Park Square, Milton Park, Abingdon, Oxon, OX14 4RN

CRC Press is an imprint of Taylor & Francis Group, LLC

© 2010, 2024 Taylor & Francis Group, LLC

This book contains information obtained from authentic and highly regarded sources. While all reasonable efforts have been made to publish reliable data and information, neither the author[s] nor the publisher can accept any legal responsibility or liability for any errors or omissions that may be made. The publishers wish to make clear that any views or opinions expressed in this book by individual editors, authors or contributors are personal to them and do not necessarily reflect the views/opinions of the publishers. The information or guidance contained in this book is intended for use by medical, scientific or health-care professionals and is provided strictly as a supplement to the medical or other professional's own judgement, their knowledge of the patient's medical history, relevant manufacturer's instructions and the appropriate best practice guidelines. Because of the rapid advances in medical science, any information or advice on dosages, procedures or diagnoses should be independently verified. The reader is strongly urged to consult the relevant national drug formulary and the drug companies' and device or material manufacturers' printed instructions, and their websites, before administering or utilizing any of the drugs, devices or materials mentioned in this book. This book does not indicate whether a particular treatment is appropriate or suitable for a particular individual. Ultimately it is the sole responsibility of the medical professional to make his or her own professional judgements, so as to advise and treat patients appropriately. The authors and publishers have also attempted to trace the copyright holders of all material reproduced in this publication and apologize to copyright holders if permission to publish in this form has not been obtained. If any copyright material has not been acknowledged please write and let us know so we may rectify in any future reprint.

Except as permitted under U.S. Copyright Law, no part of this book may be reprinted, reproduced, transmitted, or utilized in any form by any electronic, mechanical, or other means, now known or hereafter invented, including photocopying, microfilming, and recording, or in any information storage or retrieval system, without written permission from the publishers.

For permission to photocopy or use material electronically from this work, access www.copyright.com or contact the Copyright Clearance Center, Inc. (CCC), 222 Rosewood Drive, Danvers, MA 01923, 978-750-8400. For works that are not available on CCC please contact mpkbookspermissions@tandf.co.uk

Trademark notice: Product or corporate names may be trademarks or registered trademarks and are used only for identification and explanation without intent to infringe.

Publisher's Note
The publisher has gone to great lengths to ensure the quality of this reprint but points out that some imperfections in the original copies may be apparent.

Library of Congress Cataloging-in-Publication Data

Forensic dentistry / editors, David R. Senn, Paul G. Stimson. -- 2nd ed.
   p. cm.
 Includes bibliographical references and index.
 ISBN 978-1-4200-7836-7 (hardcover : alk. paper)
 1. Dental jurisprudence. I. Senn, David R. II. Stimson, Paul G.

RA1062.F67 2010
614'.18--dc22   2009043228

ISBN: 978-1-4200-7836-7 (hbk)
ISBN: 978-1-03-291956-0 (pbk)
ISBN: 978-0-429-29276-7 (ebk)

DOI: 10.4324/9780429292767

Visit the Taylor & Francis Web site at
http://www.taylorandfrancis.com

and the CRC Press Web site at
http://www.crcpress.com

Dedication

This dedication necessarily encompasses many people and a few memories.

First, we wish to remember our former editor of the first edition, Curtis A. Mertz, D.D.S. Dr. Mertz died at the end of 2007. He was one of the founding fathers of the American Board of Forensic Odontology (ABFO) and did his fair share to get the first edition completed and published.

We also want to remember and salute that small group of concerned odontologists who met on Fire Island, New York, after the impetus for the formation of various forensic boards was announced. This group consisted of Dr. Edward D. Woolridge, at whose home the meeting was held, and Drs. Lowell J. Levine, Robert B. J. Dorion, Arthur D. Goldman, Curtis A. Mertz, George T. Ward, and Manual M. Maslansky. They planned and developed the framework for the ABFO. Several others, including one editor and another contributor to this book, were invited to be included in the original group. The board was incorporated in the District of Columbia with the first certificates awarded on February 18, 1976. This board has grown and developed and now includes diplomates from many American states and Canadian provinces. Unfortunately, the board lost its only European diplomate, the late Dr. Michel Evenot of France. We are proud of the progress the board has made and its continuing support of educational and research efforts. The ABFO is the only forensic odontology board accredited by the Forensic Specialties Accreditation Board.

We want to especially dedicate this book to each of you who hold it in your hands. If you are a forensic odontologist, you must strive to constantly improve the science and the field, as did *your* mentors, with lectures, papers, and in person. In order for forensic odontology to progress to a specialty of dentistry there must be a consistent stream of new ideas and original and applied research. If you are not a forensic odontologist and are referring to this book, we welcome you to this challenging and fascinating field. It is our hope that the material presented in this book will be, in some way, helpful to you for your inquiry.

As coeditor of the first edition with Dr. Mertz, I welcome Dr. Senn as coeditor of the second edition. He was my student in pathology in dental school and has gone the extra mile for this second edition. His efforts are reflected in the high caliber of the chapters in the book before you. This project would not have been possible without his hard work and vigorous encouragement to our contributors. Thanks also to our publishers for their help and cooperation. Our joint wish to you all is happy forensics!

Paul Stimson

Table of Contents

Preface ix
The Editors xi
The Contributors xiii

1 Science, the Law, and Forensic Identification 1
CHRISTOPHER J. PLOURD

2 History of Forensic Dentistry 11
PAULA C. BRUMIT AND PAUL G. STIMSON

3 Scope of Forensic Odontology 25
BRUCE A. SCHRADER AND DAVID R. SENN

4 Death Investigation Systems 31
RANDALL E. FROST

5 Forensic Medicine and Human Identification 61
D. KIMBERLEY MOLINA

6 Fingerprints and Human Identification 79
AARON J. UHLE

7 DNA and DNA Evidence 103
BRION C. SMITH AND DAVID SWEET O.C.

8 Forensic Anthropology 137
HARRELL GILL-KING

9 Forensic Dental Identification 163
MICHAEL P. TABOR AND BRUCE A. SCHRADER

10	**Forensic Dental Radiography**	**187**
	RICHARD A. WEEMS	
11	**Forensic Dental Photography**	**203**
	FRANKLIN D. WRIGHT AND GREGORY S. GOLDEN	
12	**Dental Identification in Multiple Fatality Incidents**	**245**
	BRYAN CHRZ	
13	**Age Estimation from Oral and Dental Structures**	**263**
	EDWARD F. HARRIS, HARRY H. MINCER, KENNETH M. ANDERSON, AND DAVID R. SENN	
14	**Bitemarks**	**305**
	DAVID R. SENN AND RICHARD R. SOUVIRON	
15	**Abuse: The Role of Forensic Dentists**	**369**
	JOHN D. MCDOWELL	
16	**Jurisprudence and Legal Issues**	**379**
	ROBERT E. BARSLEY, THOMAS J. DAVID, AND HASKELL M. PITLUCK	
17	**Evidence Management**	**395**
	SCOTT HAHN	
18	**Future of Forensic Dentistry**	**405**
	DAVID R. SENN AND PAUL G. STIMSON	

Appendix: U.S. Federal and State Court Cases of Interest in Forensic Odontology   **411**
COMPILED BY HASKELL M. PITLUCK AND ROBERT E. BARSLEY

Index   **423**

Preface

Since the publication of the first edition of *Forensic Dentistry* in 1997 the discipline of forensic odontology has experienced considerable growth. Like all forensic specialties, forensic dentistry or forensic odontology has enjoyed (some may say suffered) a great increase in public interest during this period.

Forensic dentists assist medical examiners, coroners, police, other law enforcement agencies, and judicial officials to understand the significance of dental evidence in a variety of criminal and civil case types. Prosecution, plaintiff, and defense attorneys rely on forensic odontologists to analyze, report, and explain dental findings that impact their cases.

The growth and evolution of forensic odontology has not taken place without significant growing pains. The editors and contributors have chosen not to attempt to rationalize those problems but to report them, analyze the causes, and offer alternate courses to minimize the probability of similar difficulties in the future.

The editors did not intend for this book to include comprehensive, step-by-step instructions on how to practice each phase of forensic odontology. Instead, the editors and contributors have endeavored to look objectively and philosophically at the development, current state, and future of forensic dentistry and other closely associated forensic disciplines. We are of the mind that if sound scientific principles are applied from the beginning, and continued throughout, then the specific steps taken will follow that same model and will have the best opportunity to meet success.

The editors are confident that the assembled contributors are outstanding. They have produced thoughtful and sometimes provocative chapters that offer substance, fact, and ideas suitable for experienced forensic investigators or those who are just embarking on forensic careers.

The editors want to offer particular thanks to our families and especially to our wives, who not only gave us gracious support, but endured, mostly graciously, our extended physical, emotional, and mental absence. We owe them much in retribution.

Finally, we thank the publishers for their patience and support.

David R. Senn, D.D.S.
San Antonio, Texas

Paul G. Stimson, D.D.S., M. S.
Sugar Land, Texas

The Editors

David R. Senn, D.D.S., attended the University of Texas in Austin and received his dental degree from the University of Texas Dental Branch at Houston. He practiced general dentistry from 1969 until 1992 and has practiced and taught forensic odontology exclusively since 1992. He is board certified by the American Board of Forensic Odontology.

Dr. Senn is clinical assistant professor in the Department of Dental Diagnostic Science at the University of Texas Health Science Center at San Antonio (UTHSCSA) Dental School. He is director, Center for Education and Research in Forensics; director, Fellowship in Forensic Odontology; and director, Southwest Symposium on Forensic Dentistry. He has authored book chapters and articles in refereed journals on forensic odontology topics.

As a forensic odontologist for DMORT (Disaster Mortuary Operational Response Team), he worked in victim recovery and identification in New York following the World Trade Center attacks, in East Texas after the Shuttle *Columbia* crash, and in Louisiana following Hurricanes Katrina and Rita. He is a forensic odontology consultant and chief forensic odontologist for the Bexar County (Texas) Medical Examiner's Office.

He serves on the board of editors for the *American Journal of Forensic Medicine and Pathology* and is an editorial consultant for *Forensic Science International*. He has served on the board of governors for the American Society of Forensic Odontology, currently serves on the board of directors of the Forensic Specialties Accreditation Board, and is the president (2009–2010) of the American Board of Forensic Odontology.

Paul G. Stimson, D.D.S., M.S., is a graduate of Loyola University Dental School and has an M.S. degree in general pathology from the University of Chicago. He is an emeritus professor in the Department of Oral and Maxillofacial Pathology at the University of Texas Dental Branch in Houston. He began his teaching career there after completing graduate school in 1965, retiring in 1997. He taught oral and general pathology and forensic odontology and was an oral pathologist affiliated with M. D. Anderson Cancer Hospital and the Veterans Hospital Dental Department. In 1968, he became the forensic odontologist for the Harris County Medical Examiner. He is presently the chief consultant in forensic odontology for the medical examiner. He has taught forensic odontology for over forty years, and has written refereed journal articles, book chapters, and edited books on this subject. He taught in the forensic odontology course at the Armed Forces Institute of Pathology from 1968 until 1998. He has lectured on this topic in Mexico, Europe, Canada, and the United States. He was one of the founding fathers of the American Society of Forensic Odontology and has held every office in that organization in the earlier years of the society. He is one of the thirteen original members that represented the founding of the American Board of Forensic Odontology (ABFO). He has held every office in that organization and has served on various study groups and committees. He is a registered emeritus diplomate of the American Board of Oral and Maxillofacial Pathology and an active diplomate of the American Board of Forensic Odontology.

The Contributors

Kenneth M. Anderson
Division of Oral and Maxillofacial
    Pathology
College of Dentistry
University of Tennessee
Memphis, Tennessee

Robert E. Barsley
Professor and Director
Dental Health Resources
LSUHSC School of Dentistry
Chief Forensic Odontologist
New Orleans Forensic Center
Orleans Parish Coroner
New Orleans, Louisiana

Paula C. Brumit
Forensic Odontology Consultant
Dallas County Medical Examiner
Dallas, Texas
and
Bexar County Medical Examiner's Office
San Antonio, Texas
and
Clinical Instructor-Fellowship in Forensic
    Odontology
University of Texas Health Science Center at
    San Antonio
San Antonio, Texas

Bryan Chrz
Consultant to the Office of the Chief
    Medical Examiner
State of Oklahoma
Perry, Oklahoma

Thomas J. David
Forensic Odontology Consultant
Georgia Bureau of Investigation
Division of Forensic Sciences
Atlanta, Georgia

Randall E. Frost
Chief Medical Examiner
Bexar County Medical Examiner's Office
San Antonio, Texas

Harrell Gill-King
Center for Human Identification
University of North Texas Health
    Science Center
Graduate School of Biomedical Sciences
Fort Worth, Texas

Gregory S. Golden
Deputy Coroner
Chief Forensic Odontologist
County of San Bernardino
San Bernardino, California

Scott Hahn
Special Agent
Federal Bureau of Investigation
Coordinator
Evidence Response Team
CAPT, DC, USNR
Miami, Florida

Edward F. Harris
Professor
Department of Orthodontics and
    Department of Pediatric Dentistry
College of Dentistry
University of Tennessee
Memphis, Tennessee

John D. McDowell
Professor
Department of Diagnostic and
    Biological Sciences
Director Oral Medicine and Forensic Sciences
University of Colorado School of
    Dental Medicine
Aurora, Colorado

Harry H. Mincer
Professor
Department of Biologic and Diagnostic
    Sciences
College of Dentistry
University of Tennessee
and
Odontology Consultant to the Shelby
    County Medical Examiner
Memphis, Tennessee

D. Kimberley Molina
Deputy Chief Medical Examiner
Bexar County Medical Examiner's Office
San Antonio, Texas

Haskell M. Pitluck
Circuit Court Judge, Retired
19th Judicial Circuit
State of Illinois
Crystal Lake, Illinois

Christopher J. Plourd
Certified Criminal Law Specialist
Forensic Evidence Consultant
San Diego, California

Bruce A. Schrader
Forensic Odontology Consultant
Travis County Medical Examiner
Austin, Texas
and
Bexar County Medical Examiner's Office
San Antonio, Texas
and
Clinical Instructor-Fellowship in Forensic
    Odontology
University of Texas Health Science Center at
    San Antonio
San Antonio, Texas

David R. Senn
Director
Center for Education and Research in
    Forensics
University of Texas Health Science Center at
    San Antonio
San Antonio, Texas
and
Chief Forensic Odontologist
Bexar County Medical Examiner's Office
San Antonio, Texas

Brion C. Smith
Deputy Director for Forensic Services
American Registry of Pathology
Armed Forces DNA Identification
    Laboratory
Rockville, Maryland

Richard R. Souviron
Chief Forensic Odontologist
Miami-Dade Medical Examiner
    Department
Miami, Florida

Paul G. Stimson
Professor Emeritus
University of Texas Dental Branch
Houston, Texas
and
Forensic Dental Consultant, Chief
    Odontologist
Harris County Medical Examiner
Houston, Texas

David Sweet O.C.
Director
BOLD Forensic Laboratory
Professor
Faculty of Dentistry
University of British Columbia
Vancouver, British Columbia, Canada

Michael P. Tabor
Chief Forensic Odontologist
Davidson County
State of Tennessee
Nashville, Tennessee

Aaron J. Uhle
Major Incident Program Manager
Latent Print Support Unit
Federal Bureau of Investigation Laboratory
Quantico, Virginia

Richard A. Weems
University of Alabama School of Dentistry
Forensic Odontologist Consultant
Alabama Department of Forensic Sciences
and
Jefferson County Chief Medical Examiner
Birmingham, Alabama

The Contributors

Franklin D. Wright
Forensic Odontology Consultant
Hamilton County Coroner's Office
Cincinnati, Ohio

Science, the Law, and Forensic Identification

1

CHRISTOPHER J. PLOURD

Contents

1.1	Introduction	1
1.2	Science	2
1.3	The Law	3
1.4	Forensic Identification and Forensic Dentistry	4
1.5	Conclusion	8
References		9

1.1 Introduction

Forensic science is simply defined as the application of science to the law or legal matters. In today's *CSI* and *Forensic Files* world, this area of science is much more widely known to the general public. However, it is also misunderstood due to Hollywood's resolve to complete every case within the context of a one-hour, commercials included, pseudo-real-life crime drama. When the actual real-life judicial system needs science to resolve a question, the person who is called upon to bring science into the courtroom is often a forensic scientist. The law and science are strange bedfellows. Science is an empirical method of learning, anchored to the principles of observation and discovery as to how the natural world works. Scientific knowledge increases human understanding by developing experiments that provide the scientist with an objective answer to the question presented. Through the scientific method of study, a scientist systematically observes physical evidence and methodically records the data that support the scientific process. The law, on the other hand, starts out with at least two competing parties with markedly different views who use the courthouse as a battleground to resolve factual issues within the context of constitutional, statutory, and decisional law.

1.2 Science

The essence of any scientific study involves developing an alternative hypothesis, devising an experiment or series of experiments to test the accuracy of the hypothesis (question presented), and finally, carrying out the scientific experiment so as to yield an unbiased result. Science meets the law only to the extent that the legal system must look to science to help resolve a legal dispute. Scientists in today's world no longer maintain the fiction that all science is equal. This inequality is often played out in courtrooms throughout the United States. The fundamental paradigm of the judicial system in America is that science is an open process, collegial in nature, unlike the legal system, which is adversarial in nature and legal strategies are developed in secret. The overriding objective of the parties in a legal dispute is to win. With a scientist, the objective of the scientific endeavor is to reach a correct result that will withstand scrutiny from fellow scientists who can review the methodology and examine the data. Science is premised upon observable phenomena, logical deductions, and inferences that are transparent and open to scrutiny. The inherently conflicting underpinnings between science and the law frequently make forensic science controversial and the courthouse an open arena in which forensic scientists are used as pawns in the resolution of legal disputes. To complicate the legal process, each of the nonscientist parties has an interest in the outcome, be it significant sums of money, personal freedom, or even life itself in cases involving the death penalty. At the center of legal cases there sits a person who wears a long black robe to whom we refer as a judge. The judge's job, usually with the help of a jury, is to keep the adversarial parties at bay long enough to accomplish the orderly resolution of the factual questions raised by the warring litigants using applicable law. The logic of the legal system is further complicated for the forensic scientist because often conflicting forensic scientific evidence that is generated by the opposing parties is ultimately submitted to the review and decision of twelve citizens, known as a trial jury. Those jurors are selected on the basis of each juror not having any knowledge or understanding of forensic or real-world science other than that occasional episode of *CSI* or *Forensic Files*.

The most common question asked by the legal system of a forensic scientist is a request to provide proof of identity of an item or person, which is a component of criminalistics. This area of forensic science involves the association of an evidentiary item that is typically related to a crime. A forensic identification has two essential steps: The first step is a comparison between an unknown evidentiary item and a known item and having the forensic scientist render a judgment as to whether there is a sufficient concordance to say there is a "match." Examples of these comparative sciences include latent prints located at a crime scene thereafter compared to the known prints of a person, and bullet(s) collected from a body at autopsy compared to test bullets fired from suspected weapons. The second part to the identification analysis should give some meaning to the concordance (match) by providing a scientific statement that would allow the trier of fact, a judge or jury, to weigh the significance of the matching association and answer a simple question for the benefit of the trier of fact: What does "match" mean?

A forensic investigation requires a skillful blend of science using both proven techniques and common sense. The ultimate effectiveness of the scientific investigation depends upon the ability of the forensic scientist to apply the scientific method to reach a valid, reliable, and supportable conclusion about a question in controversy. Overall, science and the law must coexist within the framework of our judicial system, although each discipline may and often does have conflicting and competing interests. Any expert who is interested in the practice of a forensic science specialty must have a clear understanding not only of the fundamental principles of science, and presumably his or her chosen field, but also of the applicable legal standards relating to that area of forensic science; they must know quite a lot about that area of the law.

1.3 The Law

Expert testimony is a common and essential component in both civil and criminal trials. Every forensic scientist who is called into court to give the results of his or her study must first be qualified as an expert witness. Courts allow expert testimony out of necessity to assist the fact finder. A witness qualifies as an expert by reason of "knowledge, skill, experience, training, or education."[1] The trial judge determines if a witness is qualified as an expert and in what field of areas of science the expert may testify.[2] The forensic scientist may qualify as an expert on the basis of education, background, or study.[3] Evidence being offered by a qualified forensic expert is subject to admissibility standards for the specific scientific evidence being presented. A judge must determine admissibility of that scientific evidence. Before a judge can make that determination, the proffered scientific evidence must first pass a simple test of relevancy. Relevant evidence is defined by the Federal Rules of Evidence and most state court jurisdictions as "evidence having any tendency to make the existence of any fact that is of consequence to the determination of the action more probable or less probable than it would be without the evidence."[4] Once a court determines that the proffered scientific evidence is relevant, there are two different legal standards that courts apply in determining the admissibility of evidence: the *Frye*[5] general acceptance standard and the *Daubert*[6] scientific reliability standard. The original scientific admissibility test developed in the case of *Frye v. United States*[7] held that, to be admissible, scientific evidence must be "sufficiently established to have gained general acceptance in the particular field in which it belongs."[8] After the development of the *Frye* general acceptance standard, federal and state courts attempted to apply the rule to a wide variety of scientific evidentiary issues with mixed results. Courts often struggled with the *Frye* standard because the inquiry did not focus on the reliability of the particular scientific evidence; instead, the *Frye* test focused upon the general reliability of the scientific testing as a whole and its acceptance by others in the field. Another problem was that it was difficult to identify the appropriate expert community to answer the question of general acceptance. Some courts became concerned with the correctness of the *Frye* standard because the standard unfairly discredited new tests and accepted scientific principles. In 1993, the Supreme Court developed a new standard for scientific evidence in *Daubert v. Merrell Dow Pharmaceuticals*.[9] In *Daubert*, the Supreme Court concluded that in order for scientific evidence to be admissible, it must be shown to be scientifically valid and relevant to at least one issue in the case.[10] The Supreme Court offered numerous factors to aid federal judges in making the determination of scientific admissibility. These factors included whether the technique has been or can be tested, whether the technique has been subjected to peer review or publication, the known or potential rate of error, whether the technique is generally accepted in the community, and whether the technique was created outside of the litigation process. The *Daubert* test still allows courts to consider the issues addressed in the *Frye* standard because the "generally accepted" prong is one of many factors—instead of the sole factor in the analysis. By replacing *Frye* with *Daubert*, the U.S. Supreme Court made the trial judge a "gatekeeper" for the admissibility of any scientific evidence[11] (see Chapter 16).

1.4 Forensic Identification and Forensic Dentistry

The field of forensic dentistry or the more professional term, forensic odontology, is the application of dentistry to the law. Forensic dentistry now has been an integral part of the American judicial system for well over three decades. Overall, forensic dentistry includes multiple areas of scientific study, where the legal system and dentistry coincide. This specialized area of dentistry includes the gathering and interpretation of dental and related evidence within the overall field of criminalistics. Forensic dental evidence ranges from the identification of persons using dental records (Chapter 9) to the identification and analysis of bitemarks on an object such as a food item, or a bitemark on a victim compared to a suspect, or on a suspect compared to a victim (Chapter 14), to the estimation of a person's age based upon dental development or other characteristics (Chapter 13).

The forensic dentist is often an expert witness in civil disputes where dental injuries are at issue or there is a question of dental malpractice. Legal liability cases relating to injuries to the teeth, mouth, or jaw may involve the expertise of a forensic dentist (odontologist). A qualified dental expert can provide opinion testimony on issues relating to the loss or damage to teeth and the effect of the loss or damage to an injured individual. For example, if a person was involved in an automobile accident or an altercation where legal liability is in question, the forensic dentist may explain to the jury how the accident or assault caused the dental injury to occur. In criminal cases, the forensic dentist will assist the judge or jury by relating expert testimony concerning a dental identification examination or by identifying bitemarks and giving an opinion as to who may have made the bitemark (Chapter 16).

Dental identification of a person from dental records by a qualified forensic dentist has long been established and accepted by courts as a means to prove the identity of an individual (Chapter 9). A question as to the identification of a person may arise from a mass disaster, such as an airplane crash, natural disaster, or a situation where multiple people died in a fire and the bodies are not otherwise recognizable (Chapter 12). Dental identifications relying on x-rays and dental records universally have been considered to be a reliable identification method and rarely has a legal challenge been raised in court. Age estimation using dental evidence is necessary when a question arises as to a person's correct age as it relates to court proceedings. Typically, if a person is accused of a crime, it may be significant to determine if the individual is a minor and therefore subject to the juvenile court jurisdiction or whether the person has reached adulthood, where he or she would be prosecuted as an adult (Chapter 13). Each of these subdisciplines of forensic dentistry is discussed in one or more of the chapters of this book.

One area of forensic dentistry merits additional discussion. Forensic bitemark evidence to determine identity has become controversial over the last decade and has undergone a fundamental challenge by the greater scientific community. The catalyst for this change was the development and acceptance of DNA identification genetic testing, which is now considered to be the gold standard of biological human identification (Chapter 7). Genetic DNA identification began to be used in the late 1980s and, in cases where the traditional fingerprint or dental identification cannot be done, has dominated the field of human identification.

DNA profiling over the past decade is the most significant advance in forensic science since the development of fingerprinting in the 1900s. DNA analysis has now set a high standard against which other forensic sciences are being judged. A working knowledge and understanding of the development and use of forensic DNA identification sciences is therefore essential to all scientists who practice in other areas of the forensic sciences. Not only has DNA identity testing redefined the standard of acceptability of other scientific evidence, but it has also fostered an awareness among juries that non-DNA-based identification techniques are less supported scientifically and, in some cases, should be less accepted than DNA profiling as a method of scientific investigation.

Understanding all of the identification sciences, including DNA typing, how each developed, and how they are applied to specific casework, is essential to the forensic dentist. They are discussed in the following chapters.

Forensic DNA typing evolved from medical diagnostic techniques. Medical diagnostic DNA typing involves clean samples from known sources. In contrast, forensic DNA typing involves samples that are often degraded, contaminated, and may originate from multiple, unknown sources. Forensic DNA analysis also involves matching of samples from a wide range of alternatives present in the population. Except in cases where the DNA evidence excludes a suspected donor, assessing the significance of an apparent match requires a statistical analysis of population frequencies using a scientifically reliable database.

There are different types of DNA that are of interest to forensic scientists. They include nuclear DNA, mitochondrial DNA, and Y chromosome DNA. The DNA sequence, or order, of the base pairs is the same for every cell in a person's body that has a nucleus, with the exception of reproductive cells (ova and sperm), each of which contains only one-half of that person's DNA.[12] Approximately 99.9% of the sequence of the 3.3 billion bases is identical for all humans and performs the same function. However, approximately 1/1,000 of the sequence of the DNA molecule is different among all individuals, with the exception of identical multiple birth siblings (twins, triplets, etc.). The fact that people vary to this extent allows forensic scientists to determine whether DNA from a particular evidence sample could or could not have originated from a known person. DNA profiling is a catchall term for a wide range of methods for studying genetic variations. DNA technology for human identity purposes was designed for detection of variation (polymorphism) in specific DNA sequences. Forensic scientists have identified multiple small segments, or loci, where the DNA strand varies among groups of people. Highly variable loci are called polymorphic and are useful to identify biological material as unique (discussed further in Chapter 7).

Mitochondrial DNA (mtDNA) is a small genome that is found multiple times in the cytoplasm of each cell surrounding the nucleus. Mitochondrial DNA is passed from a mother to each of her children. A man's mtDNA is inherited from his mother, but he does not pass it on to his children. This maternal inheritance pattern has two important implications in forensic testing. The first implication is advantageous; the mtDNA of only a single maternal relative, even distantly related, can be compared to the mtDNA of another individual, for instance, the skeletal remains of an unidentified body, and help to solve both a missing person case and an unidentified body case. The second implication is disadvantageous; mtDNA is not a unique identifier.

Because maternal relatives share the same mtDNA type, the individual source of a biological sample can never be conclusively identified with mtDNA.

In a similar manner to how mtDNA is inherited from the maternal parent, the Y chromosome is inherited (only by males) from the male parent. All members from the same paternal lineage will therefore have the same Y-STR (short tandem repeat) profile. The STR genetic markers present on the Y chromosome may be used to obtain the genetic profile of the male donor(s) in mixtures of body fluids from males and females. Y-STR analysis will only target the Y chromosome; the DNA from the female contributor will be ignored.

Other mixture cases in which Y-STR analysis may be useful include sexual assaults involving saliva/saliva and saliva/vaginal secretion mixtures and instances in which the postcoital interval between the incident and the collection of intimate samples from the victim is greater than two days. DNA and DNA profiling are discussed in detail in Chapter 7.

In order to understand the present status of forensic dentistry as a forensic identification science within the overall forensic science community, it is helpful to understand and trace the history of the development of forensic dentistry. As with many changes in our American society, forensic dentistry emerged as the result of landmark events (cases) that established and shaped forensic dentistry as a useful scientific tool within the greater forensic science legal community. The issue of the scientific admissibility of bitemark evidence was established in 1976 in a landmark case in California. The use of bitemark evidence after that case grew dramatically and bitemark evidence became a sought-after identification technique by law enforcement and prosecutorial agencies. Additional new bitemark identification methods were developed and used in thousands of cases throughout the United States and around the world (see Marx in Chapter 14).

In a noteworthy case from the state of Florida, a clean-cut serial killer, originally from Washington state, was convicted and eventually sentenced to death based upon bitemark evidence. The bitemarks identified at autopsy were ultimately pivotal evidence against him. The significance of this case sent a clear message to law enforcement in the United States and elsewhere that bitemark evidence could be a critical link in establishing proof of identity and obtaining a conviction. The case received widespread media attention, which resulted in public acknowledgment and acceptance of bitemark evidence (see Bundy in Chapter 14).

Beginning in the later half of the 1990s, the forensic science community was shaken by numerous instances where errors occurred in cases and individuals were exonerated after a determination was made that they were wrongfully convicted. The problem of innocent people being convicted and unjustly imprisoned for crimes they did not commit became a growing national concern that received public acknowledgment by politicians and caught the attention of the general public, with more cases arising in which DNA identity testing technology exonerated factually innocent people. A number of DNA exoneration cases involve forensic science errors relating to evaluation of trace and biological evidence such as hair comparison and serology evidence. DNA exonerations also occurred where the person was convicted by forensic dentistry using expert bitemark identification analysis.

In the discipline of forensic dentistry, a milestone case of a wrongful conviction was the case of Ray Krone, convicted and sentenced to death for a capital murder. He was the hundredth person in the United States who had been sentenced to death to walk free from prison since the reinstatement of the death penalty in the United States in 1977. The bitemark evidence was evaluated independently for the prosecutors by two forensic dentists, one of which was an American Board of Forensic Odontology (ABFO) board-certified forensic dentist who said positively, "better than a fingerprint," the bitemark matched the suspect. "The bite marks on the victim were critical to the State's case. Without them, there likely would have been no Jury submissable case against Krone."[13] Again, this case and its unusual and provocative outcome will be examined in the bitemark chapter (see Krone in Chapter 14).

Another bitemark conviction followed by a DNA exoneration will also be discussed. The suspect was sentenced to death for the murder of his girlfriend's three-year-old daughter. Even though other forensic dentists concluded that the marks were not even bitemarks, the jury found him guilty. The case demonstrated again that DNA collected from a crime victim can prove actual innocence in cases even where seemingly reliable evidence persuaded a jury to convict a person and sentence that person to death (see Brewer in Chapter 14).

1.5 Conclusion

The investigation of bitemark cases by forensic dentists has necessarily evolved as the result of deficiencies uncovered after convictions that relied on bitemark evidence were overturned by DNA evidence. Improved technology and an increasing awareness of previously untested assumptions by forensic dentists have developed. This is the result of a concerted effort by some forensic dentists to build a solid scientific foundation and reliable protocols for bitemark comparisons. As a direct result of past mistakes there is now a better understanding by forensic dentists of the inherent variability and resulting distortion of marks left by human teeth in human skin. Although much work remains ahead, progress has been made. There is an increasing acceptance by forensic dentists that there is rarely, if ever, a scientific basis to justify an opinion that a specific person in an open population made a bitemark on human skin with scientific certainty, be it total or reasonable, based solely on the analysis of the pattern information. Therefore, a "positive match" in these cases is not scientifically supportable.

Those forensic dentists who have accepted the lessons of DNA exoneration cases have promoted an emphasis on conducting objective empirically based scientific research that will support bitemark opinion evidence and hold that evidence to a higher, more reliable scientific standard. One suggested approach being discussed by some forensic dentists is to unify the bitemark pattern analysis to the DNA profile testing as part of a single scientific study rather than independent scientific investigations.[14] This proposed method would avoid situations were the DNA and bitemark analysis are not in agreement. Scientific studies being performed by forensic dentists are expected to demonstrate that there are reliable methods and approaches to comparing bitemark evidence that minimize the potential for subjective bias and other factors that have, in the past, led to errors. As these studies are examined and other studies are undertaken by the forensic dental community they are expected to improve this troubled area of forensic science.

References

1. Federal Rule of Evidence 702.
2. Federal Rule of Evidence 104(a).
3. J. Wigmore, Evidence §556 at 751 (Chadbourn RN Rev. 1979).
4. Federal Rule of Evidence 401.
5. *Frye v. United States* (D.C. Cir. 1923) 293 F. 1013.
6. *Daubert v. Merrell Dow Pharmaceuticals, Inc.* (1993) 509 U.S. 579.
7. *Frye v. United States* (D.C. Cir. 1923) 293 F. 1013.
8. *Frye v. United States* (D.C. Cir. 1923) 293 F. 1013.
9. *Daubert v. Merrell Dow Pharmaceuticals, Inc.* (1993) 509 U.S. 579.
10. *Daubert v. Merrell Dow Pharmaceuticals, Inc.* (1993) 509 U.S. 579.
11. *Kumho Tire Co., Ltd. v. Carmichael, et al.* (1999) 526 U.S. 137.
12. U.S. Congress, Office of Technology Assessment, *Genetic Witness: Forensic Uses of DNA Tests*, OTA-BA-438 (Washington, DC: U.S. Government Printing Office, July 1990), 3–4.
13. *State of Arizona v. Ray Milton Krone* (1995) 182 Ariz. 319, at pp. 322, 897, P.2d 621, at p. 624.
14. Vale, G., "Coordinating the DNA Pattern Analysis Studies in Bite Mark Cases," in *Proceedings of the American Academy of Forensic Sciences*, Vol. XIII, February 2007.

History of Forensic Dentistry

2

PAULA C. BRUMIT
PAUL G. STIMSON

Contents

2.1	The Garden of Eden	12
2.2	Aggripina the Younger—Lollia Paulina	12
2.3	Jai Chand, Last Raja of Kanauji	12
2.4	The Earl of Shrewsbury	12
2.5	Charles the Bold, Duke of Burgundy	13
2.6	Peter Halket	13
2.7	Dr. Joseph Warren—Paul Revere	13
2.8	Janet McAlister—Dr. Pattison	13
2.9	Guerin	14
2.10	Caroline Walsh	14
2.11	Louis XVII	14
2.12	Dr. John Webster—Dr. George Parkman	15
2.13	William I, the Conqueror	15
2.14	Misidentification Corrected	16
2.15	A. I. Robinson—His Mistress	16
2.16	Winfield Goss—Mr. Udderzook	16
2.17	John Wilkes Booth, 1865 and Again in 1893	17
2.18	Dr. Oscar Amoëdo—The Bazar de la Charite, 1898	17
2.19	Strenuous Cross-Examination, 1898	18
2.20	Iroquois Theatre—Chicago, 1903	19
2.21	Bites in Cheese, 1905 and 1906	19
2.22	Chilean-German Discord Averted, 1911	19
2.23	Tooth Numbering Systems and Denture Marking	20
2.24	John Haig—Denture-Aided Identification, England	21
2.25	Denture Tooth-Aided Identification—Australia	21
2.26	Facial Reconstruction—Kollman and Buchley, Then Gatliff	21
2.27	Adolf Hitler	21
2.28	*Texas v. Doyle, Doyle v. Texas*, 1954	22
2.29	Lee Harvey Oswald, 1963 and Again in 1981	22
2.30	Other Cases	22

2.31 Summary	22
References	23

2.1 The Garden of Eden

Vale wrote in 2005, "It is always tempting to suggest that the history of bitemark evidence [and hence forensic dentistry] began with the eating of forbidden fruit in the Garden of Eden."[1] Temptation now, as then, is genuine. However, forensic odontologists and court reporters were very rare at that time; there is no dependable record of the event, analysis, comparisons, or testimony. Moreover, there were a limited number of suspects in this closed-population case and the suspects reportedly confessed.

2.2 Aggripina the Younger—Lollia Paulina

A later, but still early, and better-documented reference to the use of teeth for identification occurred during the first century CE. Agrippina the Younger, fourth wife of Emperor Claudius I and the ambitious mother by a previous marriage of Nero, contracted for the death of Lollia Paulina. To ensure that the contract was accurately concluded, Agrippina had Paulina's head brought to her. The confirmation of identification was made based on dental misalignments and other peculiarities.[2]

2.3 Jai Chand, Last Raja of Kanauji

In 1193, a great Indian monarchy was destroyed when Muhammad's army established the seat of his empire at Delhi. A significant battle during the invasion of the sacred city of Kanauji involved the sacking of the holy shrines of Muttra, the birthplace of Krishna, an important site in the Hindu religion. During the siege, Jai Chand, the Raja of Kanauji, was murdered after being taken prisoner and was identified by his false teeth when he was found among those slain.[3]

2.4 The Earl of Shrewsbury

The Earl of Shrewsbury was killed in the battle of Castillon in 1453. His herald was able to identify him by his teeth.[4]

2.5 Charles the Bold, Duke of Burgundy

After inheriting additional lands, Charles the Bold, Duke of Burgundy, decided to create an independent state between France and Germany. He was killed in the battle of Nancy in 1477 while trying to accomplish the task. The duke's page was able to identify him according to his dentition, as he had lost some teeth in a fall years previously.[5]

2.6 Peter Halket

During the French and Indian Wars Peter Halket was killed in a battle near Fort Duquesne in 1758. The fort was later captured by British General Forbes, who arranged to have the dead buried prior to leaving for Philadelphia. Three years later, a Native American who had fought in the battle remembered Officer Halket and was able to lead Halket's son to the area where he was killed during the battle. The son was able to recognize his father's skeleton by an artificial tooth.[6]

2.7 Dr. Joseph Warren—Paul Revere

In Boston in 1776, at the battle for Breed's Hill (often misidentified as Bunker Hill), Dr. Joseph Warren was killed. His face was unrecognizable as he suffered a fatal head wound, a rifle ball to the left side of his face. Paul Revere, silversmith and dentist, identified the decaying body of Dr. Warren by the small denture that he had fabricated for him. The denture was carved in ivory and was held in place by silver wires. The identification made it possible to bury Dr. Warren with full military honors on April 8, 1776.[7,8]

2.8 Janet McAlister—Dr. Pattison

The earliest known use of a dentist as an expert witness in court occurred in 1814 in the case of a Janet McAlister in Scotland. In *His Majesty's Advocate vs. Pattison et al.*, the High Court in Edinburgh charged a lecturer of anatomy and two of his students for the violation of Mrs. McAlister's grave. Mrs. McAlister had died at the age of forty years. The night after her burial, the trio was alleged to have moved her body to the nearby College Street Medical School. Mrs. McAlister's husband gave artificial teeth worn by his wife to a dentist, Dr. James Alexander, who was able to fit the dentures into the skull. The presence of a "pivot tooth" was helpful in defining his opinion. The defense testimony stated the dentures could be "fitted to any skull" and, therefore, did not fit just this skull. The jury returned a verdict of not guilty.[9]

2.9 Guerin

Three years after the disappearance in 1829 of a Mr. Guerin, a new tenant discovered a human skeleton buried in the basement. Guerin's identification was accomplished by the abrasions caused by clay pipes he had a habit of using when smoking. The abrasive marks in the dentition were unique and were similarly described by multiple witnesses.[10]

2.10 Caroline Walsh

An elderly Caroline Walsh moved in with a young Irish married couple in 1831. She was never seen again. Later, the son of the married couple accused them of murder, stating that he saw his mother leave the home with something heavy and large in a bag. A woman fitting the description of the missing woman was found on the streets in a "squalid" condition and stated her name was Caroline Walsh. She was hospitalized and subsequently died. It was pointed out in the trial that the missing Caroline Walsh had perfect teeth. This Caroline Walsh had lost her front teeth many years previously. The remains of the missing Mrs. Walsh were never found, but the accused were convicted.[11]

2.11 Louis XVII

Louis XVII died in prison in Paris in 1795 at the age of ten years two months from advanced tuberculosis of the lymph nodes (scrofula). In 1816, a plan to erect a monument to the young prince generated rumors that he was still alive, now thirty-one years of age, and that another child had been buried in his place. The story did not end there. In 1846, during the reconstruction of a church, a lead coffin containing the skeleton of a child was found near a side entrance. Dr. Milicent, a physician, examined the bones and concluded the child had died of bad health and neglect. Another physician, Dr. Recamier, examined the bones and said they were those of an individual, fifteen or sixteen years of age. All twenty-eight teeth were present and the third molars could be seen. Dr. Recamier's age assessment was accepted and the body was reinterred in an unmarked place. The quest for the Dauphin continued and in 1897, a relative of Louis XVII gained permission to again search for the coffin.

A coffin was found that contained the skeleton of a young male. Based on tooth development, three experts aged the remains at between sixteen years plus and eighteen years plus. It was concluded the remains were not those of the Dauphin. These cases represent, perhaps, the first cases of forensic dental age estimation.[12]

2.12 Dr. John Webster—Dr. George Parkman

Dr. George Parkman, a respected professor at Harvard University, failed to return from dinner on November 23, 1849. Dr. Parkman was a physician, but also a real estate speculator and moneylender. He was sixty-four years of age and a man of very regular habits. When he failed to appear as expected, suspicion of foul play fell on his colleague, John White Webster, a professor of chemistry at the same university. Dr. Webster had been behaving somewhat irregularly of late, and it was known that he owed Dr. Parkman a considerable sum of money. His laboratory was searched and, in a tea chest, human remains were found. In a nearby assay furnace fragments of a lower jawbone, three blocks of artificial teeth in porcelain, and melted gold were also found. At Webster's trial for murder, Dr. Nathan Cooley Keep, a dentist, identified the teeth as part of an upper and lower denture he had made for Dr. Parkman three years earlier. He recalled the circumstances of the denture's construction in exact detail, as Parkman had been anxious about having the dentures ready for the opening of a new medical college at which he was to give a speech. The day before the event, when some of the bottom teeth collapsed during the baking process, Dr. Keep and his assistant worked through the night and fitted the denture some thirty minutes before the ceremony. Dr. Parkman returned in a short time and complained that the lower cramped his tongue. An adjustment was made by grinding away portions of the inside of the lower denture. Dr. Keep fit portions of the lower denture to models he had retained in the production of it and showed the court where he had done the grinding adjustment of the lower denture. The dental evidence was overwhelming and Webster was found guilty and hanged. The Parkman–Webster case represents the first case of a dentist giving expert testimony in courts in the United States.[13]

2.13 William I, the Conqueror

Struck by a stray arrow in France in 1089, William the Conqueror fell from his horse and died at the age of forty-four. In 1868, his tomb was opened. All who were present stated the bones and teeth were in "good condition as if the King had died only yesterday, instead of 768 years ago."[14] The durability and longevity of teeth enable forensic dentists to make identifications even when bodies are severely damaged or long buried.

2.14 Misidentification Corrected

In the United States, in 1869, two women victims of a boat fire on the Ohio River were subsequently returned to Philadelphia, where one of the bodies was misidentified. The family dentist later examined the bodies and was able to correctly identify them.[15]

2.15 A. I. Robinson—His Mistress

Although well respected within the community, in 1870, a Mr. A. I. Robinson was suspected of murdering his mistress. Five distinct bitemarks were found on her arm, which clearly showed individual tooth marks. An investigating dentist actually bit the arm of the deceased and later had Robinson bite his (the dentist's) arm to make comparisons. The bitemark on the body showed that five teeth in the maxillary arch caused the mark. One suspect had a full complement of teeth and was excluded. Mr. Robinson had five maxillary front teeth but at trial was found not guilty.[16]

2.16 Winfield Goss—Mr. Udderzook

In 1873 outside of Baltimore, Maryland, a body was found in the ashes of a burned cottage. The body was tentatively identified as Winfield S. Gross, who was known to have used the cottage for his chemistry experiments. His widow and ten witnesses were certain that the body was that of Gross. Mr. Gross had insured himself for $25,000 eight days prior to the fire. The insurance companies refused to pay the widow's claim. A dental consultation was then requested. Mrs. Gross stated that "there were no artificial teeth to her knowledge and he never complained of pain or decayed teeth. No dentist saw him during the time we lived together." The remains were examined at the Baltimore College of Dental Surgery, where Dr. F. J. S. Gorgas gave a full and detailed description of the jaws and the remaining teeth. There were two teeth in the upper jaw and some misalignment in the lower jaw. These statements were at variance from those of Mrs. Gross and other witnesses. The insurance company thus claimed at trial that the remains were not those of Mr. Gross. The verdict of the jury, however, was in favor of Mrs. Gross. The insurance companies appealed the verdict. Within a month, the body of a murdered man was discovered in Pennsylvania. Mrs. Gross's brother-in-law, a Mr. Udderzook, had been seen traveling in Pennsylvania with an unnamed friend. When the body was examined, the height and other characteristics were similar to Mr. Gross's. The teeth were in good shape and were well preserved. Ultimately, Udderzook was charged and prosecuted for the murder of Gross. He was found guilty and executed in 1874.[6] We do not know the fate of Mrs. Gross.

2.17 John Wilkes Booth, 1865 and Again in 1893

After shooting President Lincoln on April 14, 1865, John Wilkes Booth escaped and took final refuge in a barn on a farm in Virginia. The U.S. Calvary located him there on April 26. They surrounded the barn and set it on fire. Booth exited, was shot, and died at the scene. In later years, it was rumored that he had somehow escaped, was alive, and living abroad. Because of this rumor, his body was disinterred and examined in 1893. The family could not visually identify the body, but the family's dentist was able to recognize his work as well as a peculiar "formation" of the jaw that he had noted in his records during a dental visit for the placement of a filling.[17]

2.18 Dr. Oscar Amoëdo—The Bazar de la Charite, 1898

Considered by many to be the father of forensic odontology, Dr. Oscar Amoedo was born in Matanzas, Cuba, in 1863. He began his studies at the University of Cuba, continued at New York Dental College, and then returned to Cuba in 1888. He was sent as a delegate to the International Dental Congress in Paris in 1889. Paris was very appealing to him and he decided to stay. He became a dental instructor and teacher at the Ecole Odontotechnique de Paris in 1890 and rose to the rank of professor, writing 120 scientific articles on many topics (Figure 2.1). A tragic fire at a charity event, the Bazar de la Charité, stimulated his interest in dental identification and the field of forensic odontology. Amoedo was not involved in the postfire identifications, but knew and interviewed many who were. His thesis to the faculty of medicine, entitled *L'Art Dentaire en Medicine Legale*, earned him a doctorate and served as the basis for his book by the same name, the first comprehensive text on forensic odontology (Figure 2.2).[12] He lectured and worked in the field until 1936, finally stopping at the age of seventy-three. His accounts of the identifications following the Bazar de la Charite were given in a paper at the Dental Section of the International Medical Congress of Moscow and published in English in 1897, one year before the book was published. In that paper he revealed that neither a dentist nor physician generated the idea of dental identification: "It was then that M. Albert Hans, the

Figure 2.1 Dr. Oscar Amoedo.

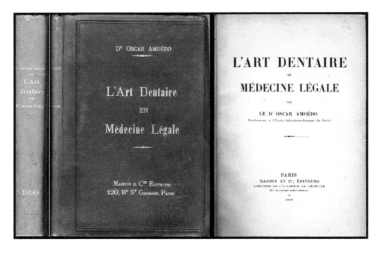

Figure 2.2 Amoedo textbook cover and title page. (Courtesy of Dr. Adam Freeman.)

Paraguay Consul, conceived the idea of calling the dentists who had given their services to the victims. His counsel was followed, and with excellent results. In the face of the powerlessness of the legal doctors, since all ordinary signs of identification had disappeared, our *confreres* were appealed to ... Drs. Burt, Brault, Davenport, Ducourneau, Godon, and some others."[18]

2.19 Strenuous Cross-Examination, 1898

In 1898, a girl was found dead. A local dentist described the state of her mouth and teeth. A missing girl's family dentist in another town was located.

The body was disinterred and the dentist was able to identify his work. The dentist complained about how strenuously and thoroughly the defense attorney grilled him while he was on the witness stand.[19] Many dentists still today dread having to go into courts of law and give sworn testimony.

2.20 Iroquois Theatre—Chicago, 1903

In 1903, the Iroquois Theatre in Chicago burned and 602 of the 1,842 patrons in the theatre died. The stairways had been closed and chained to prevent the "lower-class ticket holders" from coming downstairs. Also, the outside doors opened inward, a popular design of the day, but one that proved disastrous when frightened throngs pushed others against the doors, preventing their opening. Although no records of the identifications can be found today, Dr. Cigrand stated in his article that "hundreds" were "unmistakably identified" from their dental records.[20]

2.21 Bites in Cheese, 1905 and 1906

In 1905 and 1906, two cases were reported concerning tooth marks left in cheese. In the 1905 case in Germany, a robber bit into the cheese then left it on a windowsill. Plaster casts of the cheese were later interpreted to be from a pipe smoker. Just such a man was found among the suspects.[21] The 1906 British case involved a store break-in. The dentition of a store worker fit "exactly" a cast of the cheese. The store worker was arrested, but requested in court that his mouth be examined again, revealing that he had a broken tooth, the crown was missing, leaving only the root. In spite of this apparently attempted subterfuge, he was found guilty.[22]

2.22 Chilean-German Discord Averted, 1911

In the early 1900s, forensic odontology can be credited for the help of establishing a dental school in Chile. Residents of the small village of Caleu mistook a group of German tourists for bandits and, fearing an attack, fired upon them. In the ensuing disagreements with German officials, the German consulate in Valparaiso was set on fire. Shortly after this fire, the German litigation building in Santiago burned to the ground. A body was found in the rubble. It was first identified as the secretary to the litigation, a Mr. William Becker, according to clothing, a wedding ring (with his wife's initials in it), a watch, and glasses. An autopsy also identified the body as Becker's. The German minister, however, was not satisfied. Two German physicians, members of the faculty of Santiago University, performed a second autopsy. The anterior teeth were severely burned, but the posterior portion of the remaining dentition was described and charted. A stab wound to the heart was discovered. During this time, news was given that a considerable amount of money was missing from the consulate. The immediate suspect was a servant, Mr. Ezekel Tapia. A Chilean dentist was then asked by a judge in the case to examine the body and any pertinent records. As a result, the body was found to be Tapia's, and it was believed that Mr. Becker may have murdered him, dressed him with his own clothes and personal effects, and burned the anterior portion of his face to hide the fact that the secretary had gold bridgework. A witness claimed to have seen Mr. Becker during the night after the fire in Santiago. The judge in the case asked a Chilean dentist, Dr. Guillermo Valenzuela Basterra, to review the dental facts of the case. Mr. Becker's dentist, Dr. Dennis Lay, had placed anterior gold and platinum fillings for Mr. Becker, and removed five posterior teeth. He shared these records with Dr. Valenzuela. The findings were inconsistent with those of the remains found in the fire. Law enforcement officials were alerted and the secretary was captured at a border crossing, trying to escape into Argentina. It is ironic that Mr. Becker was able to travel from Santiago into the mountains by wearing dark glasses and a handkerchief, hiding his identity by simulating a toothache. Mr. Becker was found guilty of multiple crimes and executed on July 5, 1910. This eased the problems between Chile and Germany, and the relationship between the two nations was repaired. To show its gratitude, the government of Chile asked Dr. Valenzuela what he most desired as a form of reward. Dr. Valenzuela asked to see the long-planned dental school building completed. The wish was granted and the school was built two years later.[23]

2.23 Tooth Numbering Systems and Denture Marking

Dr. Zsigmondy published a method of numbering teeth in 1861. He numbered permanent teeth from one to eight from the anterior midline and distinguished the quadrants by placing the numbers in segments of a cross.[24] Deciduous teeth were designated with Roman numerals. Palmer later made similar proposals in 1891.[25] In 1883, Dr. Cunningham proposed numbering all teeth from one to thirty-two. Numbering the teeth in this manner, starting with the upper-right third molar (1) and ending with the lower-right third molar (32), is commonly known as the universal system and is widely used in the United States. In this system the deciduous teeth are lettered from A to T in the same pattern. Most of the rest of the world uses the Federation Dentaire Internationale (FDI) numbering system, which is similar to the system proposed by Dr. Zsigmondy. Denture marking to assist in identification was first proposed by Cunningham.[26]

History of Forensic Dentistry 21

2.24 John Haig—Denture-Aided Identification, England

Dr. Keith Simpson describes a most interesting case in which dentures were useful for the identification of a body placed in an acid bath. A wealthy widow, living in a hotel in England, went out for an afternoon with a John Haig, who lived in the same hotel. She was never seen again. Investigation of Mr. Haig showed he had a police record and led to a two-story shed he used for what he called "experiments." Some interesting things were found: two carboys of sulfuric acid, papers relating to five other individuals who had disappeared, a pistol, and blood spatter on a wall from a possible shooting. During his interrogation, Haig admitted killing the widow and said he destroyed her body in acid. After a fourth sifting of a pile of black slush found behind the shed, a set of upper and lower dentures was found. The dentures were identified by the widow's dentist. It was fortunate in the case that Mr. Haig admitted to the murder, as the dentures were made totally of acrylic resin and would have dissolved completely, given enough time.[27]

2.25 Denture Tooth-Aided Identification—Australia

In the Carron murder case in Australia, the victim was thoroughly incinerated, but artificial denture teeth of a type known as diatoric were found. A dentist was able to identify the individual by the use of this particular type of denture teeth.[28]

2.26 Facial Reconstruction—Kollman and Buchley, Then Gatliff

Kollman and Buchley did the first scientific work in facial reconstruction. They proposed twenty-three points of skin thickness measurements, which they provided in the form of a table. Soft materials were then used to sculpt the face, a technique that has been widely used and is still used with modifications today.[29] Although computerized methods are becoming more common, Betty Pat Gatliff of Oklahoma has taught many forensic artists and a significant number of forensic dentists facial reconstruction techniques.[30] She also contributed chapters to the excellent and comprehensive text on forensic art published in 2001 by Karen Taylor.[31]

2.27 Adolf Hitler

After the end of World War II, rumors were rampant that Adolf Hitler had escaped with his wife, Eva Braun. They had in fact died together in 1945, but their bodies had been burned and then buried in secret by Russian soldiers. Due to a lack of antemortem and postmortem records, it was a challenge to dispel the rumors. Finally, pieces of Hitler's jaw were found that showed remnants of a bridge, as well as unusual forms of reconstruction, and evidence of periodontal disease. Hitler's identity was confirmed when the dental work matched the records kept by Hitler's dentist, Hugo Blaschke.[32]

2.28 *Texas v. Doyle, Doyle v. Texas,* 1954

Although bitemark evidence had been used earlier, the *Doyle v. State* case in Texas in 1954 marked the first time that this type of dental evidence was used in court in the United States.[33] Like in some earlier cases, Doyle, in the process of committing a burglary, allegedly left the imprint of his dentition in a partially eaten piece of cheese. The analysis of the evidence was made by having the suspect bite into another piece of cheese for the comparison. Dr. William J. Kemp, a dentist and longtime dental examiner for the State of Texas, testified that the bites in both pieces of cheese matched.[34]

2.29 Lee Harvey Oswald, 1963 and Again in 1981

Several years after the assassination of John F. Kennedy, an English author named Michael Eddowes raised suspicion concerning the identification of Lee Harvey Oswald. It was his belief that the body buried in 1963 in Oswald's grave was really that of a Russian spy. To set the record straight, the body was exhumed and a positive identification of Oswald was made on October 4, 1981, with the aid of military antemortem dental records.[35]

2.30 Other Cases

Other significant dental identification cases in recent years include those concerning the Symbionese Liberation Army (1973–1975), the Los Angeles police shootout (1974), Jonestown in Guyana (1978), the terrorist attacks on the World Trade Center and the Pentagon (2001), and Hurricanes Katrina and Rita (2005).

2.31 Summary

There is a defining constant found in the historical cases discussed above: forensic odontologists were involved in helping to resolve difficult questions and bring closure to the families of the victims. Forensic odontologists will continue to make these types of valuable contributions to society and forensic science.

There are several historical cases of interest in the area of bitemark analysis. That historical information will be discussed in more detail in Chapter 14.

References

1. Dorion, R.B.J. 2005. *Bitemark evidence*. New York: Marcel Dekker.
2. Cassius Dio, C., Earnest, F., Baldwin, H. 1914. *Dio's Roman history*. London: W. Heinemann.
3. Hunter, W.W. 1885. *The imperial gazetteer of India*. 2nd ed. London: Trübner & Co.
4. Barr, E.A. 1967. Forensic dentistry [Letter to the Editor]. *Br. Dent. J.* 122:84.
5. de Troyes, J. 1620. *Histoire de Loys XI, Roi de France, et des choses memorables aduentuës de son regne, depuis l'an 1460*. Paris: Escrite par vn Greffier de l'Hostel de ville de Praris Imprimèefur sur le vray Original.
6. Grady, R. 1884. Personal identity established by the teeth: The dentist as a scientific expert. *Am. J. Dent. Sci.* 17:384–405.
7. Forbes, E. 1943. *Paul Revere and the world he lived in*. Boston: Houghton Mifflin Co.
8. Ring, M.E. 1976. Paul Revere—Dentist, and our country's symbol of freedom. *N.Y. State Dent. J.* 42:598–601.
9. Campbell, J.M. 1963. *Dentistry then and now*. Glasgow: Pickering & Inglis, Ltd.
10. Orfilla, C. 1829. Lesuere, Guerin case. *Ann. Hyg. Publ.* 1:464.
11. Smith, F.J., ed. 1905. *The late A. S. Taylor's manual*, 139–41. 5th ed. London: J&M Churchill.
12. Amoëdo, O. 1898. *L'Art Dentaire en Medicine Legale*. Paris: Masson et Cie.
13. Dilnot, G. 1928. *The trial of Professor Webster*. Famous Trial Series.
14. Mackenzie, R.S. 1869. Disinterment of the remains of William Rufus. *Dental Cosmos* 11:13–16.
15. McGrath, J.M. 1869. Identification of human remains by the teeth. *Dental Cosmos* 11:77–78.
16. Hill, I.R., ed. 1984. *Forensic odontology*. Bichester, UK: The Old Swan.
17. Marco, B.B. 1898. A system to assist in the identification of criminals and others by means of their teeth. *Dental Cosmos* 40:113–16.
18. Amoedo, O. 1897. The role of the dentists in the identification of the victims of the catastrophe of the "Bazar de la Charite," Paris, 4th of May, 1897. *Dental Cosmos* 39:905–912.
19. Rosenbluth, E.S. 1902. A legal identification. *Dental Cosmos* 44:1029–34.
20. Cigrand, B.J. 1910. Dental identification—A public service. *Am. Dent. J.* 9: 356–63.
21. Prinz, H. 1915. A contribution to the tooth in its relation to forensic medicine. *Br. Dent. J.* 36:383–86.
22. Anon. 1906. Identification by teeth. *Br. Med. J.* 12354:343.
23. Valenzuela, J. 1916. Identification of the dead by means of the teeth. *Oral. Hyg.* 6:333–34.

24. Zsigmondy, A. 1861. Grundzuege einer praktischen Methode zur raschen und genauen Vermarkung der azhnaerztlichen Beobachtungen und Operationen. *Dtsch. Vierteljahresschr. Zahnheilk.* 1:209.
25. Palmer, C. 1891. Palmer's dental notation. *Dental Cosmos* 33:194–98.
26. Cunningham, G. 1883. On a system of dental notation, being a code of symbols from the use of dentists in recording surgery work. *J. Br. Dent. Assoc.* 4:456.
27. Simpson, K. 1951. Dental evidence in the reconstruction of crime. *Br. Dent. J.* 91:229–37.
28. Woodforde, J. 1968. *The strange story of false teeth*, 137. London: Routledge & K. Paul.
29. Kollman, A., Buchley, W. 1898. Die Persistenz der Bassen und die Rekonstruktion der Physiognomie prahistorischer Schadel. *Arch. F. Anth.* 25.
30. Gatliff, B.P. 2008. Forensic artist. Available from http://www.skullpturelab.com/about.php.
31. Taylor, K.T. 2001. *Forensic art and illustration.* Boca Raton, FL: CRC Press.
32. Highfield, R. 1999. Dental detective work gets to the root of Hitler mystery. *Daily Telegraph*, London. October 26, 1999.
33. *Doyle v. State.* 1954. 159 Tex. C.R. 310, 263 S.W.2d 779.
34. Pierce, L. 1991. Early history of bitemarks. In *Manual of forensic odontology*, ed. D. Averill. 2nd ed. Colorado Springs, CO: American Society of Forensic Odontology.
35. Norton, L.E., Cottone, J.A., Sopher, I.M., DiMaio, V.J.M. 1984. The exhumation of identification of Lee Harvey Oswald. *J. Forensic Sci.* 29:20.

Scope of Forensic Odontology

3

BRUCE A. SCHRADER
DAVID R. SENN

Contents

3.1	Introduction	25
3.2	Dental Identification	26
3.3	Multiple Fatality Incident Management	27
3.4	Bitemark Evidence Collection and Analysis	28
3.5	Abuse	28
3.6	Age Estimation	29
3.7	Expert Testimony in Criminal and Civil Litigation	29
3.8	Summary	30
References		30

3.1 Introduction

When the subject of forensic dentistry arises, the first reaction of many people tends to be toward one of two extremes: either very cool or decidedly gross. The public assumption seems to be that forensic dentistry deals with "the dead." This view is not totally inaccurate. Although the majority of dental identification cases do involve the dead, there is much more involved, including cases dealing with the living, in this interesting field of art and science. With training, ongoing continuing education, and experience, the forensic odontologist will find the application of this knowledge to be personally rewarding. If dentists are interested but do not wish to pursue the areas of forensic dentistry that are associated with "wet work," they will find that they can practice "dry fingered" forensic dentistry in their own offices by accurately recording their patient's oral information on an ongoing basis. Forensic dentistry or forensic odontology involves several areas that will be discussed generally in this chapter and explained in more detail in later chapters. The general definition of this discipline is that forensic odontology is the combination of the science and art of dentistry and the legal system, a crossroads of dental science and law. The general topics to be discussed include the subdisciplines of forensic odontology, dental identification, multiple fatality incident management, bitemarks, abuse, age estimation, and expert testimony in criminal and civil litigation.

3.2 Dental Identification

When considering the many processes that are involved in forensic dentistry, most laypersons are familiar with identification of a deceased individual through the comparison of dental radiographs. Identification by dental means is a fast and reliable method. Dental identification is most often accomplished by comparing postmortem dental radiographs from the unidentified person with antemortem radiographs of a known individual. This process of dental forensics is often interpreted on currently popular forensic television series by the actor-dentist holding a dental radiograph backlit by the room lights with the film overhead while standing in the elevator lobby. This generally occurs following a brief evaluation of a body in the morgue. But, of course, the actor-dentist is certain that the radiographs he was just handed for evaluation are from the decedent. The positive identification is completed and without further discussion the district attorney's case theory is confirmed and the suspect is incarcerated.

In real forensic cases the process of using dental radiographs and dental charting can be an accurate and efficient method for making a positive identification or exclusion. But, the comparison must be completed in a controlled and methodical manner, with attention to the details of the dental structures and restorations that may be seen in the radiographic comparison. A comparison of an antemortem radiograph with a body in the morgue occurs only in the virtual reality of television and film world.

In a dental identification, the initial goal of the forensic dentist is to obtain a set of postmortem photographs, radiographs, and accurate dental charting on the unidentified person. This can be a straightforward or difficult process, depending on the condition of the postmortem specimen and the physical resources available to the dentist. The problems most often involve limited available resources in the morgue setting.

Procuring antemortem records can also be a challenge. Often, but not always, there will be some information on the unidentified person, a clue to his or her identity. Once a putative identity is known, the process of procuring antemortem dental records begins. Many dentists are concerned that their original records must remain in their possession and resist the release of their records. Although it is true that the dentist is expected to maintain the original record, this hurdle is easily cleared by discussion with the dentist concerning the necessity to use the record for comparison of a possible patient and the possible consequences of their interference in a medicolegal death investigation. Also, with the current ability to digitize a paper record by using a flatbed scanner or to take digital photographs of a dental chart and analog radiographs by placing them on an x-ray view box, the problem of resistance from a dental office can be reduced or eliminated. Dental records are readily available from any number of dental facilities that could have previously collected dental information on a patient as part of their examination.

Any dental charting of the teeth, financial records for treatment rendered, insurance claim forms, photographs, and radiographs that would be part of a dental examination are important items to collect as part of the antemortem reconstruction. These items could be part of the dental record created during an examination in a dental or medical facility. These items could be found as part of a dental-medical record in a private dental practice, dental teaching facility, military in-processing facility, hospital-based dental program, dental in-processing examination as part of incarceration, or medical records of an emergency room. An emergency room could potentially have radiographs of the head/neck region that include dental structures that are found on dental radiographs. The dental radiographs that are most often seen in a dental comparison are dental bitewing x-rays, as these are generally taken during regular dental checkup visits and are the most recent radiographs available. After the postmortem charting and radiography is complete and the antemortem records are procured, the comparison process can begin. The detailed reconstruction of the dental records and the comparisons that result in positive identifications are rewarding parts of the work. The forensic odontologist is able to aid in the closure process for a grieving family (see Chapter 9).

3.3 Multiple Fatality Incident Management

A multiple fatality incident (MFI) develops when the number of fatalities in the incident exceeds the number the medical examiner or coroner's facilities were designed to handle. The process of collection of dental information on victims in a mass disaster is identical to the processes that are used in the identification of a single fatality. The major difference in this process is the potential magnitude of the event and the unique set of circumstances that can surround the event. These may include the location, climate, and coverage area of the event, for example, a plane crash in mountainous terrain, a tsunami in a tropical area, the collapse of multistory structures in a major city, or a hurricane in a coastal area. Each of these incidents has unique issues that must be addressed with regard to recovery, processing, and storage of remains. Each potential MFI will have its unique problems to overcome, but accurately collecting and comparing the data is the common process in all of these situations.

With each MFI, there will be the need for personnel with different levels of experience to work together to accomplish the common goal of identifying all of the victims of the disaster. Personnel in all areas of the operation should have the ability and desire to be detail oriented, as errors can lead to missed or misidentifications. A mass disaster team should be organized and trained in coordination with the local or state government to allow the most expeditious deployment of a dental team when its services are needed. These areas will be discussed fully in Chapter 12.

3.4 Bitemark Evidence Collection and Analysis

Bitemark analysis is the most complex and controversial area of forensic odontology. Consequently, some forensic dentists are reluctant to enter into this arena. Bitemarks can occur in a wide variety of substrates, although the most common of these is, unfortunately, human skin. The proper documentation of a bitemark is not overly complex, and the techniques for collecting evidence are manageable by most forensic dentists with practice and attention to detail. The bite site can be evaluated in the third dimension by using a very accurate dental impression material and dental stones or resins to create a solid model for viewing under magnification, light microscopy, or with scanning electron microscopy. This three-dimensional model of the bitten area can then be compared to suspects' dental casts. Technique shortcomings exist and include that solid models of bitemarks on skin are nonelastic. The problems associated with bitemark analysis will be discussed more fully in Chapter 14.

3.5 Abuse

Identification and reporting of abuse is a complex and emotional area. Healthcare practitioners are required by law in most jurisdictions to report suspected cases of abuse. The head and neck area is a common target in abuse. Extraoral injuries consistent in shape and appearance to a hand or object are identifiable. Intraoral trauma can occur as the result of strikes to the face, causing torn frena and fractured, mobile, or avulsed teeth. Intraoral soft tissue pathology may be noted following forced feeding or forced fellatio. Some cases may require the consideration of whether extensive or rampant caries are a result of the caregivers' lack of knowledge or stem from neglect or abuse. In areas where access to dental care is an issue there will likely be a higher caries incidence that could further exacerbate the determination of whether reporting of abuse may be necessary. Deciding to report suspected abuse requires sound judgment, especially considering that the parent or

3.6 Age Estimation

Researchers have studied the processes of human aging by many different methods. These include developmental, histological, biochemical, and anthropological techniques. Anthropologists analyze the fusion of the cranial sutures of the skull, the development of the long bones, features of the pelvic girdle, and along with forensic dentists, features of the teeth. These techniques can be valuable when creating a profile for an unidentified person, whether living or deceased. Estimating an individual's age can also be helpful in assisting law enforcement agencies in determining the attainment of the year of majority of a living individual that will ultimately affect the individual's treatment in the legal system as either a child or an adult.

The methods of age estimation using teeth include analyzing tooth development and eruption, studying tooth degradation, and measuring biochemical and trace element changes in dental structures. Each of these methods has its advantages and limitations in accuracy and in the ease of use. Some can be performed through the analysis of dental or other radiographs or with clinical examination; others require laboratory testing or tooth destruction. The individual jurisdiction's requirements and the odontologist's skill and knowledge will help to establish the appropriate techniques for each case (see Chapter 13).

3.7 Expert Testimony in Criminal and Civil Litigation

Forensic odontologists are frequently called to give sworn testimony in depositions and courtrooms. The testimony may involve the previously mentioned areas of dental identification, bitemark analysis, or age estimation. Dentists participating in forensic casework should expect that at some point they will be required to provide sworn testimony.

Forensic dentists also may be called to provide an opinion in standard of care, personal injury, dental fraud, or other civil cases. These cases, as with other forensic cases, require the evaluation of material and the development of an opinion concerning the case. Dental experts are *not* hired guns, or advocates for one point of view. Dental experts must be advocates for the truth and endeavor to find that truth by the application of their special knowledge and skills. The unwavering goal of the forensic dental expert must be impartiality, thoroughness, and accuracy (see Chapter 16).

3.8 Summary

Forensic dentistry is a multifaceted, interesting, and rewarding blend of dentistry and the law. For most who participate in the field of forensic odontology there is not great financial reward, but the satisfaction of performing difficult and challenging tasks well is immensely rewarding. A forensic odontologist's work can have great impact on the lives of individuals and families. Their opinions may influence judges and juries in cases that can and have involved exoneration, the loss of liberty, and even the loss of life. This is an awesome and sobering responsibility that should not be casually undertaken. "The majority of those who fail and come to grief do so through neglecting the apparently insignificant details."[1]

References

1. Allen, J. 1909. *The mastery of destiny*, vii. New York: G. P. Putnam's Sons.

Death Investigation Systems

4

RANDALL E. FROST

Contents

4.1	Introduction	31
4.2	Early Death Investigation	31
4.3	The Coroner System	35
4.4	Modern American Death Investigation Systems	38
4.5	Death Certification	42
	4.5.1 Cause of Death	45
	4.5.2 Manner of Death	47
4.6	Facets of a Modern Death Investigation Office	49
4.7	Quality Assurance	56
4.8	Summary	58
References		58

4.1 Introduction

Throughout human history, the inevitability of death has inspired not only a sense of fear, but also a paradoxical sense of fascination and curiosity. It is no surprise then that the investigation of death has a long and varied history, intimately involved with the rise and governance of human populations. The sociologist Stefan Timmermans[1] has noted that death is not an individual event, but a social one, and every developed society has had an interest in the phenomenon, be it from a legal or public health viewpoint in modern populations, or as part of a mythic or superstitious worldview in earlier societies. Beliefs about the phenomenon of death have also been inexorably linked to religious systems throughout history.[2-4]

4.2 Early Death Investigation

The most primitive societies likely had a well-developed sense of the causative relationship of trauma, old age, and illness to death, and early "investigations" occurred even in tribal societies to determine why a member of the family group had died, though they were more likely to invoke superstition or magical thinking instead of the "rational" methods employed by modern societies.[2-4] Early Mesopotamian civilizations, and those of ancient Egypt, Greece, and India, had well-developed legal codes (the earliest being the Code of Hammurabi, 2200 B.C.), and these laws often referenced medical issues, such as duties of physicians, allowable fees, the viability of the fetus, and discussions of injuries. These cultures also had well-developed medical systems, but there is little or no reference in their extant writings to suggest that medical practitioners were regularly involved in the investigation of death. Rather, common sense and experience were applied by various officials, magistrates, or priests in an attempt to explain why and how individuals died. Some of the earliest death investigations probably involved deaths due to suicide, which most societies have considered to be an unacceptable act for religious or superstitious reasons.[5] Taking one's own life might result in denial of funeral rites, reprisals against the decedent's family, or other penalties, so a rudimentary death investigation was necessary in such cases to determine if a death was self-inflicted.

The earliest written documentation specifically related to formal death investigation has been discovered in archaeological excavations in China.[6] Here, bamboo strips unearthed and dated from the period of the Ch'in Dynasty (221–207 B.C.) have been found inscribed with writings giving instructions to civil servants charged with the examination of corpses who died under suspicious circumstances.

Death investigation in ancient Athens was largely a private matter instead of a concern of the state.[7] As such, investigations by a governing body were not consistent, and Greek physicians were apparently not involved in certification or investigation of death, though there are reported instances of their testimony in legal proceedings involving injury.

In early Roman legal writings,[4,8] such as the Numa Pompelius (approximately 600 B.C.), the Twelve Tables (approximately 450 B.C.), and the Lex Aquila (572 B.C.), certain medical regulations were put forth, and questions of a medical nature were posed. Roman courts could also call various "experts" as witnesses in court cases, including physicians. They could be retained by the disputants, or the magistrate or judge could call on them to advise the court. This latter group formed a class of witnesses known as *amicus curiae*,[3] or "friends of the court," and they were appointed by the court to provide expert advice in a nonpartisan manner.

The actual investigation of death, however, does not appear to have involved the medical community in most cases, though there is apparently some disagreement on this point among historians of the period.[8] A well-known exception is the death of Julius Caesar.[8,9] The writings of Suetonius, a secretary and historian of the Emperor Hadrian, give an account of the examination of the body of Caesar by the physician Antistius. According to the historian, the physician examined Caesar after his assassination and opined that only one of his multiple stab wounds was fatal (though one wonders exactly how this was determined without an autopsy). Even so, there is no indication as to how the physician came to be involved in the case of Caesar's assassination. It is not known if he was appointed by a magistrate or was merely called into service by the associates of Caesar. If the former, there is little evidence that this was a common procedure in Rome, for in the majority of extant writings about early Roman court cases, there is a notable absence of physician involvement in death investigations.

In the later Roman period, during the reign of the emperor Justinian (483–565 A.D.), the laws of Rome were brought together and codified. In this corpus, often referred to as the Justinian Code, reference is made to physicians as expert witnesses, though again their involvement in the investigation of death is not directly discussed.[2]

Postmortem dissection or autopsy examination was forbidden in most ancient cultures,[4] but Egypt was something of a remarkable exception due to the fact that the elaborate funerary preparations of that culture provided some rudimentary degree of anatomic knowledge.[10] In contrast to Athens and Rome, early Egypt also appears to have had a formal system of death investigation that did involve physicians.[11] After the Ptolemeic era, and during the period of Roman rule in Egypt, there were physicians known as the *demosioi iatroi*, or "public physicians." The extant legal writings of the time are apparently replete with reports by these individuals, known as *prosphoneseis*, regarding their examinations of deceased persons. However, the actual investigation of death was under the control of an administrative official known as a *strategos*. This individual had the authority to dispatch an assistant, known as a *hyperete*, and a physician to conduct an examination of a deceased person and render a report on their findings. It seems that the relationship of the physician to the hyperete was of a subservient and secondary nature, similar to that found in the modern coroner system.

Somewhat ironically, greater progress in the *medical* investigation of death may have occurred after the barbarian invasions of Rome in the fifth century A.D., during the so-called dark ages. These invading tribes introduced the concept of the *weregeld*, a type of compensation or "blood money" paid to the victim of a crime, or his family, by the assailant.[2] Implicit in this payment was that a full evaluation of the extent of injury suffered by the victim must be undertaken by the courts. Medical experts were utilized to assist the court in conducting these examinations, as documented in the legal codes of the era. Later, in Charlemagne's *Capitularies*, the requirement for medical testimony in certain types of traumatic injury was required, but after the death of Charlemagne it seems that progress in death investigation languished in the West for some time.[2]

Much of the understanding of death in earlier times was based on superstition or myth, and nowhere is this more demonstrable than in the vampire legends of medieval Europe.[12] In an effort to explain epidemics, bad luck, or other untoward occurrences in a village, some individuals readily placed the blame for these events on the recently dead. In an effort to ameliorate the village's problems and confirm the culpability of a particular corpse, the grave of the recently deceased was sometimes opened. On exhuming the body, typical changes of decomposition would be noted as expected, though they were misinterpreted as indicating something much more menacing than the normal dissolution of the body. For instance, purging of the body is often seen in decomposing corpses. This is the issue of a dark bloody fluid from the oral and nasal passages due to autolysis, putrefaction, and liquefaction of the internal viscera. The resultant fluid is pushed out of the nose and mouth through the airways and esophagus by decompositional gas formation, which causes increased pressure within the thoracic and abdominal cavities. Though now recognized as a common postmortem artifact, this purging process was taken to represent blood soiling of the mouth due to recent feeding on the blood of living victims. Further, with decomposition, the epidermis of the skin separates from the underlying dermis, resulting in so-called skin slip. Rather than being recognized as a decompositional change, this slippage of the skin was attributed to the growth of new skin, and decompositional bloating and red discoloration of the body were described as a healthy, ruddy complexion, compared with the sallow appearance of the deceased at the time of burial. As decomposition progresses, rigor mortis (or stiffening of the extremities) disappears, but this suppleness of the limbs was considered a sure sign of vampirism. All of these factors were taken to indicate continued life beyond the grave, as well as nocturnal feasting on the blood of the living occupants of the village. Detailed methods of investigation were developed in order to confirm the identity of the "vampire," and there were also prescribed procedures for warding off the revenant, and for putting it to rest permanently. These included such treatments as decapitation, "staking" the vampire through the heart, removal of the heart, cremation, tying the mouth shut, and reburial face down, presumably to confuse the undead when he or she attempted to rise from the grave.

It seems difficult to believe that such misconceptions could occur, particularly as the phenomenon of postmortem decomposition should have been well known (refrigeration of decedents not being available in that era), but such is the case. And even though the beliefs in vampirism were manifestly erroneous by today's standards, they do indicate a depth of concern about the process of death. They also show the development of a detailed investigative and empiric method, and the development of an internally coherent and systematic way of explaining observations and understanding death and its relationship to other occurrences. Such internally consistent and systematic misinterpretations based on the best learning of the day should serve to give us pause when we become too certain of the validity of our own current positions.

Another example of early death investigation, from an Eastern perspective, can be found in the book *Hsi Yuan Chi Lu*[13] [*The Washing Away of Wrongs*] (from China, circa 1247 A.D.) as written by Sung Tz'u. This text gives detailed instructions on death investigation, and is probably the oldest extant full text on the topic. It includes discussions of decomposition, determination of time since death, homicidal violence, self-inflicted injuries, various accidental deaths, and deaths due to natural causes. In spite of its antiquity, the similarities between the investigational methods taught in the book and those utilized today are often striking.

4.3 The Coroner System

Though the cause of death was undoubtedly investigated in ancient times, it was likely an ancillary duty of tribal elders, magistrates, priests, or other authorities. The first instance of an official office charged with the investigation of death, as we know it today, was probably the English *coroner*.[14–18] Though officials with this responsibility are reported as far back as 871 A.D., in the time of Alfred the Great, the beginning of the coroner system in England is generally taken to have been in 1194, with the publication of the *Articles of Eyre*, the Eyre being a system of roving "circuit" justices in England in the twelfth century. These itenerant judges traversed the land to hear cases and dispense justice, but due to the long intervals between their visits (an average of seven years), it was necessary to have local officials perform careful investigations and keep records of offenses so that the cases could effectively be brought before the justices when they finally did arrive. Without proper records, many cases were never tried. This would not do, as many of them involved production of revenues for the monarch, at that time Richard the Lionhearted. Richard, a Norman king, was an absentee ruler with a penchant for expensive foreign wars that placed a heavy strain on the royal coffers. In addition to his travels and the need to equip large numbers of troops, he also managed to become captured and imprisoned in Germany during his return from the Holy Land in 1192. A huge ransom was required to secure his release and return to England in 1194. These expenses created an acute need for revenues, so no stone was left unturned in a desire to collect all taxes and other dues to which the Crown was entitled under the law. There were many such assessments, creatively applied in the name of law and order, to enrich the king at the expense of his subjects.

It was customary at that time to seize the property of felons, and because suicide was considered a crime against God, the property of those taking their own lives was also forfeited to the Crown. Furthermore, villages were penalized with a fine, or *amercement*, whenever a murder or other legal infraction occurred in their jurisdictions, a punishment for allowing civil disturbances to occur or for not properly following the complex system of laws in the realm. Sometimes the victims of these murders were members of the conquering Norman class who were unfortunate enough to find themselves in the midst of local Saxons bent on revenge. To prevent having large numbers of its Norman noblemen dispatched by the indigenous population, the Crown levied a fine known as the *lex murdrorum* on the lord of any village or territory in which a Norman was killed. And naturally this fine was passed on to the populace in the form of a tax. Even the object actually causing a death (referred to as a *deodand*) was subject to presentation to the Eyre, and it could be confiscated in the name of the Crown because of its culpability in the death or injury of a person. So if a person were injured or killed by a cart, animal, or farm implement, this item would likely be appropriated by the court, possibly depriving a farmer of the means of his livelihood.

It is obvious that the king had a vested interest in making certain that all of these types of cases were properly investigated and documented to ensure that all potential revenues were discovered. The law enforcement officers in each English county (shire) were the sheriffs (from *shire reeve*), but many of these officers were Saxons. Not only were they less than enthusiastic about supporting the Norman king, but they also had a well-earned reputation for embezzlement, to the detriment of the Crown's accounts, so another investigative authority was needed to counter their authority in favor of the king. To this end, the twentieth Article of Eyre established the office of *Custos Placitorum Coronae*, or "keepers of the pleas of the Crown," to represent the king's interests in locales throughout the country. The title *coroner* was a derivation of the Latin *coronae* for "crown," or perhaps of the term *crowner*, one who represented the interests of the king. Originally these men were knights, men of some wealth and means (presumably to lessen their propensity to embezzle funds), and their concern with death investigation was based entirely on the king's financial interests. In addition to death investigation, they were responsible for investigating almost any aspect of life that could conceivably yield revenue for the king, including confiscation of buried treasure and shipwrecks ("treasure troves"). In their pursuit of funds for the king (and themselves) coroners developed a reputation for greed and corruption that approached that of the sheriffs,[17] so needless to say, they were not particularly popular with their local constituents. In later years, other officials, such as justices of the peace, took over much of the original investigative functions, while the coroners' duties became focused exclusively on death investigation. They were empowered to hold public trials or "inquests," in which they questioned witnesses and empanelled juries to hear evidence regarding deaths and to make determinations as to how they came about. The coroner's inquest continues to persist today, and the office of the British coroner represents one of the oldest continuous judicial agencies in existence.

During its colonial period, England exported much of its culture and legal system throughout the world, including the American colonies. Not surprisingly, the office of coroner was a part of this export.[2,5] A search of the archives of the Plymouth Colony[19] reveals multiple references to coroners' inquests. The governor of Maryland appointed a sheriff-coroner in 1637, and the duties of the coroner are recorded in the state archives of that period.[5] Another early American coroner was reported to have been appointed by William Penn. As the nation grew and developed, the office became an integral component of local governments, responsible for investigation of death in a particular jurisdiction, though the incumbents were traditionally not physicians. Unfortunately, some coroners developed reputations for bribery, embezzlement, and lack of integrity as part of the political "spoils" system, resulting in a relatively low public opinion of the field.[10] As in the British system, the American coroner was a quasi-judicial figure, and not a medical professional, so in the absence of adequate medical training and experience, early coroners applied lay knowledge and common sense to the problem of death investigation.

Early in the development of the coroner system, the lack of medical involvement in death investigation was of no consequence. Medical knowledge and science were rudimentary at best, and even physicians viewed disease and death through superstitious and magical lenses. As no professional had a better grasp of the causative factors in a death than any other, no particular educational requirements were necessary or appropriate for the coroner. Through the centuries this nonmedical coroner system of death investigation changed little in British jurisdictions and their progeny. But in Europe, developing scientific and medical expertise was more readily brought to bear in death investigation as advances were made in the sciences,[2-5,10,16] with the first formal lectures in forensic medicine given at the University of Leipzig in the middle seventeeth century.[2,10] In this way, the continent was far ahead of the Anglo-Saxon model. There are numerous examples of cooperation between the medical profession and the law in Europe, and one such case in point is the *Constitutio Criminalis Carolina*, a code of law promulgated in 1553 in Germany. In it, expert medical testimony was required in cases of murder or traumatic deaths. Even the practice of autopsy examination (more properly known as the *necropsy*) was more readily accepted in Europe than in Britain.[10] The first significant move toward formal studies in forensic medicine in the British Isles was in Scotland, where a chair of "medical jurisprudence and police" was established at the University of Edinburgh in 1806.[17]

4.4 Modern American Death Investigation Systems

As will be seen subsequently, the systems of death investigation in the United States are widely varied. However, they generally share a number of mission components in common. To paraphrase DiMaio and DiMaio,[21] the various components of a death investigation system are as follows:

1. Determine the cause of death, and how the death came about
2. Identify the decedent
3. Determine the time of death and injury
4. Collect evidence from the body that may be useful in the police investigation
5. Document injuries that are present, or their absence
6. Deduce how injuries occurred
7. Document any natural disease present
8. Document or exclude any causative or contributory factors in the death
9. Provide testimony in court as needed

In spite of the widespread distribution and long history of the office of the coroner, as medical science and the understanding of death have become more complex, shortcomings of the system have became apparent. Most obvious is the fact that most coroner jurisdictions do not require a coroner to be a physician. Obviously, a nonphysician is in a poor position to render a medical opinion on the cause of death, a determination that is manifestly the practice of medicine. The likelihood of omissions or misinterpretations is increased in such jurisdictions where the responsibility for this duty is vested in someone other than a trained physician. Second, coroners are elected officials in the United States. The necessity of maintaining public electoral support has long been considered a potential impediment to the development and maintenance of absolute impartiality in death investigation by coroners. Death investigators must often make very hard and controversial decisions that are difficult for families, local political forces, and others to accept. In theory, at least, an appointed official, as opposed to an elected one, can make such decisions without the fear of alienating a voting block necessary for his or her continued employment in the position.

Proponents of the coroner system respond that the nonmedical coroner functions as an administrator and quasi-judicial agent, and can employ physicians to perform examinations and make medical decisions under his or her directions. However, since the coroner retains ultimate authority for the investigation and determination of the circumstances and cause of death, the possibility that such an individual will exercise that authority to veto or influence the medical decision of an employed physician still remains. Some coroner proponents also point to the greater political authority of an elected official over an appointed one. And in fact, many coroners have utilized their political acumen and the public inquest to call attention to issues of public safety. But the final decisions in a death investigation should ideally be made by an appropriately trained and experienced physician, and no one else. Only he or she has the expertise in the cause and interpretation of disease and injury to make such medical decisions.

Other concerns by opponents of the coroner system relate to potential conflicts of interest for the lay coroner.[20,21] In many coroner settings, particularly in small towns or rural areas, the office of coroner will not be a full-time position, and the person holding the office will have other employment. The coroner may be a law enforcement officer, a funeral director, occasionally a practicing physician, or hold other employment completely unrelated to death investigation. In one state, the county sheriff serves as the *ex officio* coroner of the county. The potential conflict of interest in such an instance is obvious, with the sheriff charged with the investigation of deaths that may be the result of an interaction of his or her officers with a decedent, such as an officer-involved shooting. Equally disturbing is the situation in which an inmate in a jail or detention facility operated by the sheriff is found dead. No matter how impartial and honest the sheriff, any such investigation will always have at least the potential for impropriety or lack of impartiality. If the coroner is a funeral director, there may be a potential conflict if the coroner stands to earn money by also arranging the funeral service of the deceased. In such cases there might be a tendency to avoid making any cause or manner of death determination that would alienate the next of kin. Some jurisdictions do require that the coroner be a physician. However, this is often little improvement over an untrained lay coroner. Most physicians have had practically no exposure to forensic pathology and death investigation, leaving them exceptionally ill-prepared to make determinations about cause and manner of death. Medical schools provide essentially no training in this area, and even most pathologists receive only a superficial introduction to the topic during their period of residency training. Most rarely or never perform autopsies in their practice. Without training and experience, such physicians are in a very poor position to properly examine complex cases or to identify subtle findings that may indicate foul play in what initially appears to be a natural death. In other cases, a physician might be the director or an employee of a medical facility accused of being responsible for the death of a patient. The death would then be investigated by the physician-coroner, with the attendant potential conflict of interest. Independence from other agencies must be considered an absolute requirement for optimal death investigation.

As early as 1928 the National Research Council stated bluntly that the office of coroner is anachronistic and "has conclusively demonstrated its incapacity to perform the functions customarily required of it."[22] The council went on to recommend the abolition of coroners' offices, replacing them with pathologist *medical examiners*. A recent report by the National Academy of Science also discusses inherent problems with the coroner system.

Partially because of these shortcomings with the coroner system, the medical examiner system came into being. The title of *medical examiner* is not a medical designation, but a governmental title, and it is defined differently by various jurisdictions (usually at the state level). But in the most basic sense, the medical examiner (ME) is a properly trained physician charged with the responsibility and authority to investigate deaths and to determine cause and manner of death in a particular jurisdiction.[21]

One of the first attempts at bringing medicolegal death investigation under the authority of physicians came in Maryland in 1860, with laws requiring the coroner to obtain the assistance of a physician in the investigation of violent death.[5,10] Later, in 1890, physicians were appointed in Baltimore as "medical examiners" to perform autopsies requested by the coroner or other officials. In 1877, the Massachusetts General Laws established physician medical examiners in Suffolk County and the remainder of the commonwealth, appointed by the governor for seven-year terms. These physicians assumed the duties of the coroner, but only in the case of deaths known to have come about by violent means. These physicians also were not able to perform autopsies on their own authority, but required an order from another official, such as the district attorney, to authorize this procedure.

The first true medical examiner system was formed in New York City in 1918.[3,5,10,23] This change was driven largely by a report of corruption in the coroner system at that time by the city's commissioner of accounts. In the reformed system, the medical examiner took on all the investigative duties of the coroner, whose office was abolished. The ME was required to be a physician appointed by the city and had the responsibility and authority to investigate any death occurring by violent means, or suddenly and unexpectedly. Deaths of prisoners or those not attended by physicians were also included in the medical examiner's jurisdiction. Other landmark facets of the New York system included civil service protection of the ME and the sole authority to decide what type of examination or investigation was necessary to certify the death. In the early years of this office, a toxicology laboratory was also added to the ME's office, establishing laboratory analysis as a critical component of death investigation.

When discussing the office of medical examiner, it is necessary to define the term *forensic pathology*. The term *forensic* derives from the Latin word *forensis* for "before the forum," or relating to argument and discussion. Pathology is the study of disease, or a medical specialty devoted to the diagnosis of disease by laboratory means. Forensic pathology can be defined as the subspecialty of medicine devoted to the medical investigation of death. It is one of many components of the modern forensic sciences, and is a subspecialty of the medical specialty of pathology. In the United States, training in forensic pathology requires completion of a course of medical study, culminating in a doctor of medicine or doctor of osteopathy degree, completion of three or more years of residency training in anatomic pathology, completion of an accredited year of fellowship training in forensic pathology in an accredited training program (usually a large medical examiner's office), and passage of national certification tests in both anatomic and forensic pathology, administered by the American Board of Pathology. The physician can then use the title of *forensic pathologist* and is considered board certified in that field.[24] While not all medical examiner positions require the incumbent to be a forensic pathologist, many do, and it is highly desirable that all physicians involved in medicolegal death investigation be board-certified forensic pathologists.

The formation of an academic specialty of forensic pathology owes much to early chairs of forensic medicine established in Europe and Scotland, but the first endowed chair of legal medicine in the United States was established at the Harvard Medical School in 1937.[10,25] In this program, a training program for physicians was established, as well as programs for law enforcement officers and other investigators. Forensic pathology was first recognized as a medical subspecialty in 1959, when examinations were administered and the first cadre of physicians was certified as forensic pathologists.

Over the years, the medical examiner system has been refined somewhat to include some fairly standard elements. Ideally the medical examiner is:

1. A physician trained in the field of forensic pathology. This ensures that the highest possible level of expertise is brought to bear on the cases that come before the ME. Forensic pathologists working under the direction of a coroner should not be referred to as medical examiners, since the coroner is the actual official imbued with the authority to investigate and certify death, not a physician.
2. Appointed, not elected. In this way, the office of medical examiner becomes a professional position, not a political one, and the office holder is not concerned with currying favor with an electorate and periodically campaigning for reelection.
3. Protected by civil service laws. Ideally the ME cannot be dismissed without good cause (such as incompetence or dereliction), and then only after appropriate due process procedures have been followed. This provides some protection for the ME, who must make decisions that might prove to be politically unpopular.
4. The sole authority for investigation and certification of deaths in his or her jurisdictional area, and is independent of law enforcement, prosecutorial, or judicial agencies. This ensures that the ME is independent and is an impartial witness for the truth in our adversarial legal system. It also gives the ME full authority to perform any type of examination or investigation necessary to determine cause and manner of

death, without seeking permission of families or other authorities. Though many medical examiners retain the authority to hold public inquests, such proceedings are rare in a modern ME system. Rather, the ME makes determinations based on a thorough investigation of the circumstances of death, along with findings of a scientific examination by autopsy, toxicology, or other disciplines. Only a trained and experienced physician has the knowledge to obtain and analyze such data and to synthesize a rational cause and manner of death conclusion from it.

During the first half of the twentieth century, medical examiner systems progressively replaced coroner jurisdictions throughout the United States. However, this replacement trend began to plateau in the mid-1980s, and new ME jurisdictions are now formed infrequently.[26] Currently about half of the population of the United States is served by a medical examiner, with the rest under the jurisdiction of a coroner. However, coroner offices outnumber ME offices by a factor of five in the United States. The disparity is due to the fact that ME offices tend to be clustered in highly populated counties, such that a single office will serve a much greater population than the average coroner's office.

The accompanying table (Table 4.1) shows a general distribution of ME versus coroner systems within the United States.[26] The country is served by a patchwork of different types of death investigation systems,[26,27] and the type of system is generally determined on a state-by-state basis. At this time, the populations of twenty-two states are served solely by medical examiners, eleven by coroners, and seventeen by a combination of medical examiner and coroner systems. Even within these broad groups, there are many variations. Some medical examiner states are under the authority of a statewide medical examiner. Others have authority vested at the regional or county level. In some states, the coroner function is subsumed by other office holders. For instance, in Texas, justices of the peace perform coroner duties in counties without a medical examiner. In many California counties, the sheriff has coronial authority. For these reasons, it is difficult to effectively categorize medicolegal death investigation systems in the United States, as the various systems often have little resemblance to each other.

4.5 Death Certification

A major common denominator in any modern death investigation system is the documentation of death and the determination of its cause and manner, also referred to as death certification. In early times, records of birth and death were kept inconsistently, if at all, but in 1538, clergy in England were required to keep a ledger of births, deaths, and marriages in their parishes.

Table 4.1 Distribution of Types of Death Investigation Systems in the United States

States Served by ME Systems	States Served by Coroner Systems	States Served by Mixed Systems
(all state MEs except where otherwise noted)		
Alaska	Colorado	Alabama
Arizona[a]	Idaho	Arkansas[c]
Connecticut	Indiana	California
Delaware	Kansas	Georgia
Florida[b]	Louisiana	Hawaii
Iowa	Nebraska	Illinois
Maine	Nevada	Kentucky[c]
Maryland	North Dakota	Minnesota
Massachusetts	South Carolina	Mississippi[c]
Michigan[a]	South Dakota	Missouri
New Hampshire	Wyoming	Montana[c]
New Jersey		New York
New Mexico		Ohio
North Carolina		Pennsylvania
Oklahoma		Texas
Oregon		Washington
Rhode Island		Wisconsin
Tennessee		
Utah		
Vermont		
Virginia		
West Virginia		

Source: Adapted from Hanzlick.[26]
[a] Medical examiner in each county.
[b] District medical examiners.
[c] Mixed-system states with state medical examiner.

This custom of registration persisted for many years, but gradually became a function of governments instead of the church. This change was given further impetus during infectious epidemics of the nineteenth century, when it came to be appreciated that it would be worthwhile to keep track of the numbers of deaths occurring as an infectious contagion progressed.[18] Subsequently, governments around the world began to require registration and certification of births and deaths. Modern death certification is a function of state governments, and all jurisdictions in the United States have a common requirement that the death of a person be officially documented, with attestation of the cause and manner of death by a physician, medical examiner, coroner, or other official. The document serving this purpose is referred to as a death certificate, and requirements regarding its use and filing are set forth

Table 4.2 State of Texas, Certificate of Death

by a state department of health, vital records, or equivalent.[28] State death certificates tend to be fairly similar, as most are based on the U.S. Standard Certificate of Death, which is in turn based on World Health Organization recommendations. A typical state death certificate is shown in Table 4.2. Many states are also moving toward a standardized digital death registration process that promises to make gathering of demographic and epidemiologic data much simpler and more effective.

Death Investigation Systems

The reasons for death certification and investigation are several. First, as has been noted above, a developed society has an interest in documenting the birth and death of its citizenry in order to provide for transfer of estates, administration of societal programs, payment of insurance settlements, etc. The tracking of deaths from an epidemiologic viewpoint allows for better public health surveillance in a society, be it related to epidemic diseases or public safety issues. As all purveyors of television crime dramas are well aware, adequate death investigation and certification is required for the criminal prosecution of deaths due to the action or inaction of another person or institution. And finally, knowledge of the cause and manner of death is often of importance in allowing appropriate grieving and closure for the family and loved ones of a decedent.

In most jurisdictions, deaths occurring solely by natural means may be certified by attending physicians. However, deaths due to trauma, intoxication, or unknown means usually fall under the jurisdiction of the medical examiner or coroner, and must be investigated and certified by that office. In addition to the demographic documentation related to the decedent, such offices must also attempt to determine the *cause and manner of death*.

4.5.1 Cause of Death

Cause of death is a concept applied somewhat more strictly and precisely by the medical examiner than by other physicians or the lay public.[21] A number of definitions of this term have been devised, but a simple one favored by the author is "that disease, injury or event, but for which death would not have occurred at the time it did." In death certification, the first insult that begins a cascading series of events leading to death is of primary interest. This is referred to as the *cause of death*, or as the *proximate cause of death*. This is in distinction to the subsequent resultant physiologic derangements caused by this event. These derangements are often referred to as *mechanisms of death* or the *immediate causes of death*.

For example, suppose an individual receives a gunshot wound that injures the spinal cord and renders the victim quadriplegic. If, years later, he or she succumbs to a urinary tract infection related to the paralytic bladder caused by the spinal cord injury, the cause of death should be appropriately certified as a "gunshot wound of the back," or "urosepsis complicating quadriplegia due to gunshot wound of back." Many physicians, however, would inappropriately list the cause of death as "urinary tract infection" (a mechanism of death) without referencing the true cause of the condition, the gunshot wound. The reasons for this are readily apparent, as most clinical physicians are concerned with diagnosing and treating acute conditions that can be ameliorated by medical or surgical therapy. The medical examiner, however, recognizes that the purpose of death certification is to provide statistical information on primary causes of death, and that the lapse of time between injury and death is of no importance in this documentation.

Often the cause of death cannot be arrived at by examination of the remains of the decedent. Just as a clinical physician must take a medical history prior to performing a physical examination, the medical examiner must have investigative information regarding the circumstances of death prior to reaching a conclusion. Review of past medical history, consideration of the presentation of the decedent at the time of death (sudden collapse, complaints of symptoms), and other factors are of equal importance to the autopsy and other examination techniques. It is for this reason that an adequate investigative team is required to assist the medical examiner in gathering initial and follow-up information.

Forensic pathology, like any field in medicine, is not an exact science. There are degrees of uncertainty in any cause of death determination, and the degree of likelihood necessary to make a cause of death statement varies from case to case. It is a matter requiring considerable professional judgment and experience, and is very difficult to quantify in most cases. The phrase "beyond a reasonable doubt" is a legal term referring to conclusions by a criminal trial judge or jury, but it has no place in the lexicon of the forensic pathologist. Instead, medical examiners are often asked to render their opinions to a "reasonable degree of medical probability." Though even this phrase is somewhat nebulous, it does at least recognize that medical determinations invariably involve some degree of uncertainty. It is important to realize that cause of death statements by a medical examiner are *opinions*, resulting from consideration of myriad different factors and observations, generation of a differential list of potentially fatal conditions or injuries, and selection of the most likely candidate(s) for cause of death from that list. When explaining this opinion to attorneys, families, juries, or any other group, the forensic pathologist must make every effort to convey any degree of uncertainty, to acknowledge other possible opinions, and to explain his or her rationale for selecting one over another. To simply state an opinion dogmatically, leaving no room for competing theories or argument, is incompatible with honest forensic medical practice. This is perhaps best summarized in a well-known statement by Dr. Paul Brouardel, a French physician of the late nineteenth century: "If the law has made you a witness, remain a man of science. You have no victim to avenge, no guilty or innocent person to ruin or save. You must bear witness within the limits of science." These "limits of science," and the application of various types of decision-making tools to forensic sciences, would form the basis for a long discussion far beyond the scope of this chapter. However, it is incumbent on every practitioner to understand the limitations, degrees of uncertainty, and sometimes ambiguity of medicolegal opinions and to readily acknowledge them when appropriate.

Sometimes a cause of death cannot be determined to a reasonable degree of probability. This may reflect the fact that multiple possible causes of death are present, and one cannot readily be chosen over another. It may also reflect the fact that not every fatal condition has accompanying anatomic changes that can be discovered on autopsy examination. The human body is in fact an electrochemical mechanism, and many fatal physiological processes are not associated with demonstrable anatomic alterations. When these processes cannot be inferred from historical or investigative information, the cause of death may remain undetermined. It is the mark of a good forensic pathologist that this conclusion is invoked whenever appropriate, without the attempt to form an unsupportable cause of death conclusion.

4.5.2 Manner of Death

Assignment of a manner of death is also a required part of the death certification process. Ostensibly, this is an attempt to classify the death as to the circumstances by which death came about; unfortunately, this classification is often problematic. There are five classical manner of death categories.[21]

- Homicide: Death caused by the intentional actions of another person.
- Suicide: Connotes a death due to one's own intentional acts.
- Natural: Death due to natural disease processes only, with no contribution from traumatic or external factors.
- Accident: Death due to unforeseen traumatic or external factors.
- Undetermined: The manner of death is not known or could not be determined.

Some jurisdictions also add additional categories, such as unclassified, therapeutic misadventure, etc., but these are not universal. The idea of a manner of death classification is an American invention,[31] and the manner of death categories available for use in death certification are promulgated by state vital records departments. Physicians, medical examiners, and coroners are bound to and limited by these available choices.

The problem with manner of death classification is that the "pigeonholing" of complex and disparate deaths into one of five (actually four) categories is fraught with problems. One difficulty in reproducibly assigning an appropriate manner of death category is the lack of agreement on definitions for the classification terms. The brief definitions listed above are quite rudimentary and broad, and are subject to considerable and substantive variation in various jurisdictions. For example, a homicide is generally considered to be a death at the hands of another person, whereas suicide is death at one's own hands. Yet a death due to a motor vehicle crash is generally classified as an accident, regardless of whether one or both of the drivers were at fault or caused the crash. If a hunter fires at a target he believes to be a deer, but inadvertently kills another hunter, many MEs will classify this death as a homicide, since the rifle was purposely fired at a target, which was struck and killed. Others would consider it to be an accidental death since the hunter did not intend to kill a human being. Some jurisdictions require that intent to cause one's own death be a factor in classifying a death as suicide. Others require only that the act leading to death be intentional, regardless of whether or not death was anticipated. For example, if a person is playing Russian roulette with a partially loaded revolver and dies of a gunshot wound of the head, many MEs would classify this as a suicide, without consideration of whether or not the player actually intended to die, or just was engaging in thoughtless, extremely risky behavior. Others would agree that the death represents a suicide, but largely because of the high inherent risk of the activity. Deaths due to acute intoxication by ethanol or other drugs are usually classified as accidents unless suicidal intent is evident. But deaths due to the chronic effects of the same drugs (cirrhosis, endocarditis, etc.) are typically classified as natural deaths. These are but a few of the inconsistencies and disagreements that may plague a manner of death determination.

As a demonstration of this point, a survey consisting of twenty-three separate and varied medical examiner case scenarios was sent to members of the National Association of Medical Examiners (NAME) in 1995.[29,30] The survey presented the cases and asked the respondents to answer a variety of questions, particularly relating to manner of death determination. Subsequently, a panel discussion was convened at a meeting of the National Association of Medical Examiners in 1996 in which five well-known and respected forensic pathologists discussed their opinions of the cases. The degree of discordance was striking in both the survey and the subsequent panel discussion. Complete agreement was reached in only one case, and in some cases, the level of disagreement was such that no majority opinion was identified. It is a fact that highly trained and experienced MEs frequently disagree on manner of death classification. Though it may be assumed that the use of a particular manner of death classification schema is based on consistent scientific principles, it is in fact based more on local tradition and habit, and national consistency has proven to be an elusive goal.

Another factor causing difficulty in producing consistent manner of death classifications is the exceptional complexity of various death processes. Each case is quite unique, with many facets of the death process being unknown, partially known, or merely inferred. In such cases, it is extraordinarily difficult to distill this complexity into one of four or five categories, and if it can be done, some would consider it to be of little practical use. After all, categorizing disparate and dissimilar types of deaths together into overly broad and artificial categories seems to be of questionable value.

In an attempt to better define the various manner of death categories, NAME published a booklet giving guidelines for death classification in 2003.[31] Its methodology has not, however, become universally adopted at this time, and variations in manner of death classification continue to abound.

It is often thought by uninformed individuals that a medical examiner's assignment of manner of death forms the basis for prosecution of crimes, insurance settlements, and other legal matters. In fact, however, the manner of death classification put forth by the medical examiner is not a legal opinion or criminal charge, nor is it binding to parties in civil disputes. It is, instead, a classification scheme for use by state vital records agencies charged with gleaning epidemiologic data from death certificates. By way of illustration, consider that if a driver causes a collision by virtue of his recklessness or intoxication, he may be prosecuted for manslaughter even though the medical examiner classifies the death as an accident. Conversely, if a police officer shoots and kills a weapon-wielding felon, it will generally be classified as a homicide by the medical examiner. But if the shooting is considered justifiable, no charges of murder or manslaughter may be brought against the officer. The belief that the medical examiner assigns criminal culpability when classifying manner of death is incorrect, but it is often a source of misunderstanding.

Because of these inconsistencies in the classical manner of death classification, some experts would prefer to delete this system in favor of a greatly expanded and more useful list of manner of death choices, or a more detailed and flexible narrative statement giving sufficient information so that the events of a death are clear to those perusing the death certificate. However, since death certification is driven by agencies other than medical examiners or coroners, the statutorily prescribed manner of death classification scheme must be utilized. Its shortcomings can be somewhat ameliorated if the forensic pathologist provides sufficient narrative detail in the autopsy or investigation reports to clearly define the circumstances of death to the extent that they are known, in spite of the limitations of the concept of manner of death.

4.6 Facets of a Modern Death Investigation Office

The majority of ME's offices are funded and chartered by government entities, such as counties, cities, or states. These organizations are established by statute, and function as agencies of that government. Typically, a chief medical examiner is appointed by the local city, county, or state executive, and he or she then appoints deputy medical examiners and other personnel as needed in order to meet the mission and statutory mandate of the office. The personnel of the office are employees of the jurisdiction, and the office is funded by the county, city, or state. In some other jurisdictions, private forensic pathology medical groups are appointed as medical examiners by governments on a contract basis. In coroner jurisdictions, forensic pathologists may be employed by government coroners' offices, or may be hired on a contractual basis to provide medical autopsy and examination services for the lay coroner. Small coroners' offices may send decedents to a large medical examiner's office or other medical facility for autopsy examination. A key feature of any type of system is that the examinations performed by the medical examiner or coroner's pathologist are done under the authority of the state, and as such, are not subject to approval of the decedent's next of kin, as are diagnostic hospital autopsies. A corollary of this authority is that there can be no room in a medicolegal examination for objections to the forensic pathologist's examination on personal or religious grounds. If the ME or coroner has the need to conduct an examination in order to adequately investigate the death, then it should be done regardless of any objections, though the examination techniques may be modified at the discretion of the autopsy physician to attempt to accommodate family beliefs. Any attempt to infringe upon this prerogative compromises the system of investigation significantly.

A medical examiner or coroner's office must first determine whether a case reported to the office falls under its jurisdiction. Jurisdictional criteria vary according to law from state to state, but in general, deaths due to trauma or intoxication, natural deaths occurring suddenly and unexpectedly, or those due to unknown causes fall under the jurisdiction of the ME or coroner. Other types of death may not fall under ME or coroner jurisdiction, and need not be examined by those offices.

Sometimes there is a need for autopsy examinations in cases that do not fall under ME jurisdiction. At one time in the United States, the autopsy rate of individuals dying in a hospital setting approached 50%. The autopsy was viewed as a valuable diagnostic and quality assurance, and teaching tool, and permission was sought from the patient's next of kin to perform an autopsy in most death cases. In cases not falling under a medical examiner or coroner jurisdiction, permission is required of the next of kin to perform an autopsy. In recent decades, however, the autopsy rate in this country has plummeted, and now autopsies are performed infrequently in most hospitals, even in teaching institutions. This is due to a multitude of causes.

First, there is an overreliance on modern diagnostic imaging techniques and a belief that computed tomography and magnetic resonance imaging scans will have discovered everything the autopsy might find. This is proving to be a very erroneous belief, as most autopsy physicians can attest. Imaging studies, in spite of their clinical utility, are poor substitutes for an adequate postmortem examination. Molina et al. have documented significant discrepancies between antemortem imaging findings and the autopsy,[32] which remains the gold standard of medical diagnosis. Second, the Joint Commission on Accreditation of Health Care Organizations deleted the autopsy requirement for hospital accreditation in 1971. This closely coincided with the precipitous drop in autopsy rates nationwide, as hospitals are no longer required to show a particular rate of autopsies in their institutions in order to be accredited.[33] The autopsy is not covered by third-party payers, or the payment is lumped under a general remuneration to a hospital without mandating actual autopsy rates, so there is no financial incentive to perform the examinations. Pathologists also tend to gravitate to other areas of practice that are not as time-consuming, less messy, and generate greater income.

As such, the autopsy today is practiced primarily in forensic pathology settings, such as medical examiners' offices. Yet these offices have strictly defined limits on the types of cases they may take under jurisdiction, leaving a host of "medical" cases unexamined each year. These represent a true treasure trove of diagnostic and research data that go untapped in the United States.

Medical examiners' offices vary in their organization, but it is possible to describe the organization and function of a "generic" office. A typical ME's office will be divided into some or all of the following sections:

- Investigations
- Autopsy section
- Toxicology laboratory
- Clerical section
- Administration

Investigations is in many ways the most important section of the medical examiner's office. It is typically an investigator who takes initial reports of a death, and makes a determination as to whether or not the case falls under medical examiner jurisdiction. If it is determined that the case will be investigated by the ME, the investigator must then obtain more information about the circumstances of the case. The investigator will also take a leading role in helping to establish positive identification of the decedent. Identification techniques are discussed at length in a subsequent chapter, and are one of the most critical functions of any medicolegal examination. Just as a physician must take a medical history before examining or treating a patient, so must the forensic pathologist obtain background information on a death before examining a decedent. As the subject of the examination obviously cannot be interviewed, it is up to the investigator to gather this information from whatever source is available. This may involve visitation of the scene of death to photograph and describe findings (Figure 4.1a and b). Often the position of the decedent may give critical information about the factors that lead to death. Conditions at the scene may implicate environmental factors in the death, or the finding of medications or intoxicants may result in suspicion of a drug-related demise. Indeed, interpretation of drug levels found in the body on toxicologic testing often relies heavily on scene or historical information about the decedent's prior drug use. Scene findings, correlated with autopsy

a

b

Figure 4.1 Two typical scenes requiring investigation: a motor vehicle crash and a homicidal death. Medical investigators will examine the body of the decedent at the scene, obtain information from police investigators and witnesses, examine the environmental factors and characteristics of the scene in an attempt to help the forensic pathologist interpret the autopsy findings, and take charge of the body for transportation to the medical examiner's office.

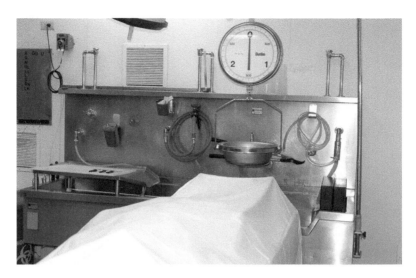

Figure 4.2 The body is prepared for examination in the autopsy room. Note the well-lit area for examination, with an area for dissection of organs by the physician.

findings, often provide information about how a death occurred (manner of death), in addition to what caused the death. Apart from the scene investigation, investigators obtain other information regarding the medical history of the decedent by interviewing family members or acquaintances, or by obtaining medical records from hospitals or physicians. They may be responsible for transporting decedents from death scenes or hospitals to the ME office, fingerprinting bodies, inventorying personal effects and medications, and admitting and releasing bodies from the morgue. Alternatively, some or all of these duties may be shared with other sections of the office. In large offices, the investigation section is usually composed of a number of full-time employees. In smaller offices, much of the investigative functions may be performed by the forensic pathologist or other staff.

The medical or autopsy section includes forensic pathologists and the technicians who assist them in performing examinations of decedents (Figure 4.2). These examinations may take the form of full or limited autopsy examinations, or be limited to external examination of the body. The extent of the examination will be determined by the medical examiner after reviewing the decedent's medical history and circumstances of death. In elderly individuals or those with extensive and potentially fatal medical histories, found dead under circumstances that indicate a death due to natural causes, examination may be limited to external inspection of the body to exclude any evidence of trauma. On the other hand, in cases of acute traumatic death, particularly in the case of apparent homicide, a full autopsy

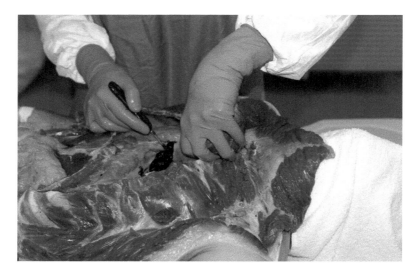

Figure 4.3 Beginning evisceration (removal of body organs) by the autopsy technician.

will generally be indicated. In most cases, blood and other body fluids or tissues will be drawn at the time of examination for submission to the toxicology laboratory at the discretion of the forensic pathologist. Trace evidence will be collected from the body as appropriate during the examination, and clothing and other salient materials will be preserved for evidentiary purposes as needed. Photographs of injuries and of the decedent for identification and documentation purposes will also be obtained. The degree to which autopsy technicians assist forensic pathologists will vary from office to office, depending on staffing, tradition, and local philosophy. In some offices, assistants may perform eviscerations of bodies under supervision of forensic pathologists (Figure 4.3), though the actual dissection of organs is almost always performed by the pathologist. In others, only pathologists eviscerate bodies. Either system is acceptable as long as all of these activities are under the direct supervision of a forensic pathologist. Assistants may also perform clerical duties, take radiographs, draw blood or other toxicology specimens, suture bodies closed, clean the body and examination area, assist in removal of clothing, and assist with inventory and preservation of clothing and evidence.

As a part of the medical or autopsy section, a modern medical examiner's office will maintain relationships with expert consultants to assist the medical examiner in specialized areas. One of the most important and commonly utilized consultants is the forensic odontologist. A trained and certified dental practitioner provides invaluable aid in helping to establish identification of decedents by dental comparison and in the evaluation of bitemark evidence,

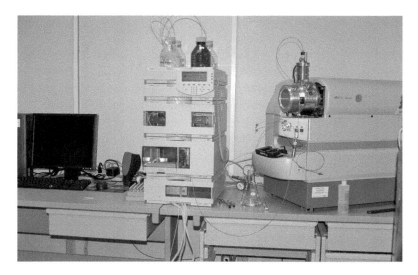

Figure 4.4 A view of one of the many types of analytical instruments (liquid chromatography–mass spectrometry analyzer) in the toxicology laboratory.

which may be invaluable in linking an assailant to a homicide victim. Anthropologists assist the pathologist in evaluation of skeletal remains, again, in an effort to establish identification by narrowing the age, race, and sex of the decedent. They may also assist in evaluating traumatic or other changes in the skeleton that may show acute or remote injury. Neuropathologists, pediatric pathologists, radiologists, and other medical specialists may provide valuable input into the investigation of specialized medical aspects of a case, and nonmedical specialists, such as engineers, electricians, entomologists, meteorologists, and geologists, may assist in the evaluation of the external or environmental factors in a death.

The toxicology section is integral to the function of any modern medical examiner's office. Proper analysis of the blood and other body fluids is not only required to confirm cause of death in cases of apparent drug overdose, but is routinely performed in cases of deaths due to unknown causes, deaths due to trauma, and some deaths due to apparent natural disease. Information on intoxication is critical to adjudication of many traumatic deaths, and often drug intoxication is found in deaths initially thought to be solely due to natural disease. Though many large medical examiners' offices will have an on-site dedicated toxicology laboratory (Figure 4.4), other, smaller offices may find that it is more cost-effective to utilize the services of a large off-site commercial laboratory. In either case, it is imperative that the laboratory be appropriately accredited and supervised, and that protocols for specimen collection, retention, and analysis be well established and meet the requirements of the chief medical examiner or supervising forensic pathologist. It is also necessary that the forensic toxicologist be available for ready consultation in difficult or problematic cases.

The clerical section of the medical examiner's office is responsible for providing a critical interface of the office with the public (reception), transcription of physician's dictated reports, preparation of death certificates, coordination of court appearances, and maintenance of documents, records, and data. This section will also respond to subpoenas, public information queries, and the myriad other requests for information that are submitted to the medical examiner's office each day.

In a medical examiner system, office administration must be under the auspices of the chief medical examiner. It is imperative that this individual have full authority over operational, budgetary, and personnel matters, though he or she may employ administrative, fiscal, and other assistants, and delegate duties to them as needed. It is the chief medical examiner who will determine the procedures and policies of the office, and final authority over the office must reside with him or her, as well as responsibility for the performance of all aspects of the office. This includes responsibility for all investigations and examinations, and for maintaining the quality of the office's work products. Adequate civil service protection is optimal for all medical examiners, to help ensure that these physicians are not intimidated or punished for their honest professional opinions, which form the true work product of the organization.

Coroner systems will have many organizational facets in common with medical examiners' offices, with the exception that final authority of the medical investigative agency does not reside with a trained physician (with the exception of rare cases in which the coroner is also a board-certified forensic pathologist). It is imperative, then, that the coroner's chief forensic pathologist has adequate authority over all operational matters to ensure that good forensic pathology procedures are followed in all areas. This will require adequate input into budgetary and personnel matters, as well as medical matters.

4.7 Quality Assurance

The single most important quality control or assurance (QC/QA) mechanism in the ME's office is the appointment of qualified and certified forensic pathologists, particularly in the position of chief medical examiner. In modern medical practice, board certification of physicians is expected and usually required for the full exercise of the practice privileges in a medical specialty.[34] Similarly, such certification is necessary in the field of forensic pathology to indicate that a practitioner has met the minimum standards of training and knowledge in the field. In the United States, the only path to certification in forensic pathology accepted by NAME[34,35] is through the American Board of Pathology (a member board of the American Board of Medical Specialties). This organization sets requirements and standards for training, examination, and certification in the field of pathology and its subspecialties, in this case, forensic pathology.[24] This requires completion of accredited (by the American Council of Graduate Medical Education) residency and fellowship training programs in both anatomic and forensic pathology, with subsequent successful completion of qualifying examinations given by the American Board of Pathology in both areas. If these requirements are met, the physician will receive certification in anatomic pathology, and special qualification in the field of forensic pathology, also referred to as board certification. At this point NAME recognizes the individual as a forensic pathologist.[35] While this certification does not guarantee excellent practice, it does show that the practitioner has met a minimum level of training and performance in the field.[34] While some physicians utilize the term *board eligible* to indicate that they have taken the requisite training in pathology or its subspecialties but have not passed the board certification examinations, this designation is not recognized by the American Board of Pathology, and should not be used or accepted.

In addition to board certification, practitioners may now be held to professional standards of practice in the field of forensic pathology. The National Association of Medical Examiners has published such standards to provide guidance and objective criteria for the assessment of the practice of forensic pathology.[35] In 2003, NAME formed a committee to investigate the advisability of formally adopting standards for medicolegal autopsy practice.[36] After much debate and discussion, including surveying the membership about proposed standards, the "Forensic Autopsy Performance Standards" were approved by the membership of NAME and published in 2006 in the *American Journal of Forensic Medicine and Pathology*. These set forth standards for practice of forensic autopsy pathology, provide some definitions related to the field, and discuss standards for associated techniques, such as toxicological analysis, radiography, histology, and written reports. While broad and relatively basic, such standards do require a certain level of practice by physicians and can serve as an objective guide in assessing an individual's level of practice.

While certification and standards refer to the practitioner and his or her practice, accreditation refers to the assessment of a death investigation system or office as an organization, without assessing the performance of any individual practitioner. Currently the only organization accrediting medical examiners' offices on a national basis is NAME.[37] In 1975 the organization established a voluntary peer review system using criteria developed for inspection and accreditation of death investigation systems. Accreditation of a medical examiner or coroner's office involves a thorough inspection of the office by an outside trained forensic pathologist, utilizing a checklist devised by the Standards and Accreditation Committee of NAME. Criteria within the checklist are divided into two categories: Phase I requirements, which are desirable, but the lack of which will not significantly impact on the function of the system, and Phase II requirements, which are considered essential for the system to function adequately. Upon successful completion of this inspection, with no Phase II deficiencies and no more than fifteen Phase I deficiencies, the office is issued a certificate of accreditation for a period of five years. If the inspection is not successful, the office management will be counseled regarding deficiencies and methods of correcting them. Provisional accreditation for a brief period and reinspection are available to assist offices in meeting this goal.

4.8 Summary

The field of death investigation is highly variable and there is often a lack of consistency in how it is practiced within the United States. Each state sets its own legal and governmental framework for the type and extent of medicolegal investigation, resulting in a patchwork of systems throughout the country that tends to confound any classification scheme. However, there are basic consistencies in the goals and practice of good quality death investigation, and thanks to organizations such as the National Association of Medical Examiners, uniform medical practice standards and accreditation criteria are now published, serving as a benchmark for the nationwide evaluation of forensic pathology and death investigation practice. Adherence to these criteria will help to ensure that medicolegal investigation in this country meets minimal approved standards and best serves the needs of the citizens of the various jurisdictions. Beyond this, recruiting and maintaining practitioners who adhere to the goals of truth, impartiality, high quality, and integrity in their investigations will provide the best insurance for escalating quality in the field of forensic death investigation. After all, the goal of any system of death investigation is ultimately to serve the needs of the society in which it operates. As a reflection of this goal, many medical examiners' offices prominently display a well-known adage regarding their work: "*hic locus est ubi mors gaudet succurrere vitae*," or "this is the place where death delights to help the living."

References

1. Timmermans, S. 2006. *Postmortem. How medical examiners explain suspicious deaths.* Chicago: University of Chicago Press.

2. Camps, F. E., et al. 1976. Historical and general law. In *Gradwohl's legal medicine*, section I. Chicago: John Wright & Sons, Ltd.
3. Curran, W. J., et al. 1980. History and development. In *Modern legal medicine, psychiatry, and forensic science*, chap. 1. Philadelphia: F. A. Davis Company.
4. Payne-James, J., et al. 2003. *Forensic medicine: Clinical and pathologic aspects*, 5–12. San Francisco: Greenwich Medical Media.
5. Fisher, R. S. 2006. History of forensic pathology and related laboratory sciences. In *Medicolegal investigation of death*, ed. W. U. Spitz, rev. M. S. Platt, 3–21. 4th ed. Springfield, IL: Charles C. Thomas.
6. Stark, M. 2000. *A physician's guide to clinical forensic medicine*. Totowa, NJ: Humana Press.
7. Amundsen, D., and G. Ferngren. 1977. The Physician as an Expert Witness in Athenian Law. *Bulletin of the History of Medicine* 51:202–213.
8. Amundsen, D., and G. Ferngren. 1979. The forensic role of physicians in Roman law. *Bulletin of the History of Medicine* 53:39–56.
9. Rolfe, J. C. (trans.). 1920. *Suetonius*. Vol I. Cambridge, MA: Harvard University Press.
10. Spitz, D. J. 2006. History and development of forensic medicine and pathology. In *Spitz and Fisher's medicolegal investigation of death*, chap. 1. 4th ed. Springfield, IL: Charles C. Thomas.
11. Amundsen, D., and G. Ferngren. 1978. The forensic role of physicians in Ptolemaic and Roman Egypt. *Bulletin of the History of Medicine* 52:336–53.
12. Barber, P. 1988. *Vampires, burial, and death*. New Haven, CT: Yale University Press.
13. Sung Tz'u. 1981. *The washing away of wrongs (Hsi Yuan Lu): Forensic medicine in thirteenth-century China*, trans. B. McKnight. Ann Arbor: Center for Chinese Studies, University of Michigan.
14. Knight, B. 2007. CROWNER: Origins of the office of coroner. http://www.britannia.com/history/coroner1.html (accessed December 12, 2007).
15. Holdsworth, W. 1903. *A history of English law*. London: Methuen & Co.
16. Garland, A. N. 1987. Forensic medicine in Great Britain. I. The beginning. *American Journal of Forensic Medicine and Pathology* 8:269–72.
17. Mant, A. K. 1987. Forensic medicine in Great Britain. II. The origins of the British medicolegal system and some historic cases. *American Journal of Forensic Medicine and Pathology* 8:354–61.
18. Davis, G. 1997. Mind your manners. Part I. History of death certification and manner of death classification. *American Journal of Forensic Medicine and Pathology* 18:219–23.
19. Delaney, T. 1999. The Plymouth Colony Archive Project. Deacon Philip Walker, Sr., of Rehoboth, Plymoth Colony. http://www.histarch.uiuc.edu/plymouth/Walker.html (accessed June 17, 2008).
20. Standing Bear, Z. G. 2005. *Funeral Ethics Organization*, Winter Newsletter 2005, pp. 1–3. http://www.funeralethics.org/winter05.pdf (accessed April 18, 2008).
21. DiMaio, V., and D. DiMaio. 2001. *Forensic pathology*. 2nd ed. Boca Raton, FL: CRC Press.
22. National Academy of Sciences. 2003. *Medicolegal death investigation system: Workshop summary*. Washington, DC: National Academies Press.

23. Eckert, W. 1987. Charles Norris (1868–1935) and Thomas A. Gonzales (1878–1956). New York's forensic pioneers. *American Journal of Forensic Medicine and Pathology* 8:350–53.
24. American Board of Pathology. 2008. Information 2008, policies, procedures and requirements. www.abpath.org/index.htm.
25. Tedeschi, L. 1980. The general pathologist's role in forensic medicine: The Massachusetts scene. *Human Pathology* 11:113–21.
26. Hanzlick, R. 2007. The conversion of coroner systems to medical examiner systems in the United States. A lull in the action. *American Journal of Forensic Medicine and Pathology* 28:279–83.
27. Hickman, M., and K. Strom. 2007. *Medical examiners and coroners' offices, 2004.* Bureau of Justice Statistics Special Report. Washington, DC: U.S. Department of Justice, Office of Justice Programs.
28. Hanzlick, R. 2007. *Death investigation systems and procedures.* Boca Raton, FL: CRC Press.
29. Goodin, J., and R. Hanzlick. 1997. Mind your manners. Part II. General results from the National Association of Medical Examiners Manner of Death Questionnaire, 1995. *American Journal of Forensic Medicine and Pathology* 18:224–27.
30. Hanzlick, R., and J. Goodin. 1997. Mind your manners. Part III. Individual scenario results and discussion of the National Association of Medical Examiners Manner of Death Questionnaire, 1995. *American Journal of Forensic Medicine and Pathology* 18:228–45.
31. Hanzlick, R., J. Hunsaker, and G. Davis. 2002. *A guide for manner of death classification.* 1st ed. National Association of Medical Examiners.
32. Molina, D. K., et al. 2008. The sensitivity of computed tomography (CT) scans in detecting trauma: Are CT scans reliable enough for courtroom testimony? *The Journal of Trauma* 63:625–29.
33. Bayer-Garner, I. B., et al. 2002. Pathologists in a teaching institution assess the value of the autopsy. *Archives of Pathology and Laboratory Medicine* 126:442–47.
34. Hunsaker, J. 2006. A word from the president [Editorial]. *American Journal of Forensic Medicine and Pathology* 27:197–99.
35. Peterson, G., and S. Clark. 2006. Forensic autopsy performance standards. *American Journal of Forensic Medicine and Pathology* 27:200–25.
36. Clark S., and G. Peterson. 2006. History of the development of forensic autopsy performance standards. *American Journal of Forensic Medicine and Pathology* 27:226–55.
37. National Association of Medical Examiners. 2003. Accreditation checklist. 1st rev. www.thename.org.

Forensic Medicine and Human Identification

5

D. KIMBERLEY MOLINA

Contents

5.1	Background	61
5.2	Human versus Nonhuman	62
5.3	Establishing Identity	63
5.4	Visual Identification	64
5.5	Circumstantial	64
5.6	External Characteristics	66
5.7	Internal Characteristics	70
5.8	Radiographs	70
5.9	Anthropology	71
5.10	Sources of Comparison	75
5.11	Methodology for an Unidentified Person/Body	76
References		76
Suggested Reading		77

5.1 Background

Establishing the identity of a person may seem like an easy task; the person, or their friends or family, can simply be asked their name. In medicolegal cases, however, there are often reasons why people are either unable to give accurate answers or purposefully give inaccurate ones. In cases of death, a body may also be too disfigured due to trauma to allow for easy identification. This is common in cases of high-velocity crashes (e.g., cars, airplanes), fires, explosions, or decomposed/skeletonized remains. Though sometimes difficult, identification remains a necessary task. Living individuals for whom identification is required may include wanted criminals attempting to elude custody, amnesia victims, comatose victims, victims of disfiguring trauma, or persons who require identity confirmation following identity theft. Deceased individuals requiring identification may include homeless individuals, undocumented immigrants, burned bodies, decomposed or skeletal remains, and individuals who sustained significant facial trauma that precludes visual identification.

Just as identification of a living individual allows for contacting of next of kin if necessary, or processing their medical/legal needs, identification of a deceased person serves many purposes. Family can be contacted, allowing for the grieving process to begin and permitting arrangements for disposition of the body. Correct identification of a decedent also allows for accurate documentation of the death (i.e., filing of the death certificate) and permits the deceased's financial and legal issues to be addressed, including the settling of the estate, filing life insurance claims, and probating a Last Will and Testament. Death investigation is greatly enhanced by knowing the identity of the person. A medical and social history can be obtained from the decedent's family and friends, and medical records can be obtained from local hospitals/clinics. Witnesses may be interviewed to determine when the decedent was last seen. The decedent's residence/property may be examined for further information.

In 2007, the Office of Justice Programs, Bureau of Justice Statistics, determined that there were approximately forty-four hundred unidentified medical examiner/coroner cases in the United States, with approximately one-fourth of those cases remaining unidentified after one year. Thus, establishing the identity of an unknown person is a large concern for forensic scientists. The establishment of identity is a combined duty of law enforcement and the forensic scientist/pathologist. The efforts to establish identity ultimately result in one of the following identification categories: positive, presumptive, or unidentified.

This chapter will address the various methods for establishing the identity of an individual/remains from the perspective of a forensic pathologist (medical examiner); thus, the methods discussed will mainly refer to those used for deceased individuals. Many of the principles discussed, however, can also be applied to living individuals who require identification.

5.2 Human versus Nonhuman

The identification of remains begins with establishing that the remains are human. This may be obvious, as in the case of an unidentified person found by the side of a major highway, but it may also be more complex, as in the case of a single bone recovered from the neighborhood dog park. The most common items that require delineation between human and nonhuman are decomposed bodies, particularly skeletons, and fragmented or dismembered bodies (Figure 5.1).

For skeletonized remains, most major bones can easily be assessed by a forensic pathologist or anthropologist for species identification, though smaller bones may require further analysis. Dismembered body parts can also be difficult to assess, especially if decomposed. Portions of eyes, ears,

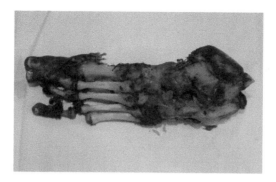

Figure 5.1 A disarticulated foot recovered from neighborhood shrubbery by a dog.

fingers, and toes, as well as complete feet, hands, extremities, and thoraces, may be recovered. Often radiographs can be used to delineate species, the classic example being that of a bear claw resembling a decomposed human hand. Radiographs would reveal multiple sesamoid bones in the paw that are not present in the human hand. Small animals may resemble human fetal remains. Not uncommonly, and especially during hunting season, entrails will be uncovered and the question is again, are they human or nonhuman? Examination of the remains will often yield the answer by the presence of multiple stomachs (e.g., the deer has four), the gastric contents (e.g., grass, leaves), or the anatomy of the organs themselves.

If anthropologic and radiographic examinations fail to differentiate the species, DNA analysis can be performed. Forensic scientists can extract the DNA and look for the hTERT (human telomerase) gene on chromosome 5, which is specific for humans.

5.3 Establishing Identity

Once the remains have been identified as being human, a number of methods can be employed to determine identity.

The most common means for establishing a positive identification are visual, fingerprint, DNA, and dental comparison. The latter three are often referred to as the scientific methods and will be thoroughly discussed in forthcoming chapters, and therefore will not be discussed here (see Chapters 6, 7, and 9). In addition, a *presumptive* identification can be established by numerous methods addressed below, which is often very useful for future utilization of DNA or dental comparison. Although numerous databases exist for the comparison of fingerprint and DNA evidence to establish identity, without a comparison sample, both of these methods are useless. Similarly, dental identification is not possible if no antemortem dental records can be found. The techniques discussed in this chapter can be used either alone or, more often, in combination to establish the presumptive identification of an individual, allowing for either dental, DNA, or fingerprint information to be obtained and compared.

5.4 Visual Identification

Visual identification is perhaps the most commonly used method of identification and is used to establish both positive and presumptive identification. Visual identification is used in cases of automobile accidents or a witnessed collapse and a family member or friend tells the police or emergency medical personnel, "This is John Smith." Law enforcement/medical personnel take the identification as truth, that this *is* who the witness proclaims. However, visual identification is one of the least reliable forms of identification and can be fraught with error. Witnesses/family may be under duress at the time of the incident and make an innocent mistake ("Well, it looked like Uncle Bob"). The family/friends may refuse to examine the remains, may simply glance at the remains rather than truly examining them, or may deny that the remains are those of their loved one as an emotional defense mechanism. Witnesses/family members may purposefully incorrectly identify the decedent for either financial gain or other personal reasons.

Comparison of the deceased to a photograph, whether from a driver's license or personal photo, is another form of visual identification. Visual identification is also employed in public notice campaigns where an individual's photo or a facial reconstruction is broadcast on television or flyers, asking anyone who recognizes the photo/sketch to contact authorities to assist with possible identification.

In many cases, however, the body may not be able to be identified visually. The changes of decomposition (Figures 5.2 to 5.4), including drying of the mucous membranes, bloating of the soft tissues, discoloration of the skin, and skeletonization, may obscure/obliterate features. The face may also have sustained injuries that distort the features, severely limiting visual identification, including trauma or burning (Figures 5.5 and 5.6).

5.5 Circumstantial

The circumstances of death can often assist in identifying a person. The location where a person/body is found in itself can provide helpful clues. If found in a domicile, one can establish who lives there, or if found in a vehicle, investigators can track down the owner. Personal papers, mail, medications,

Forensic Medicine and Human Identification 65

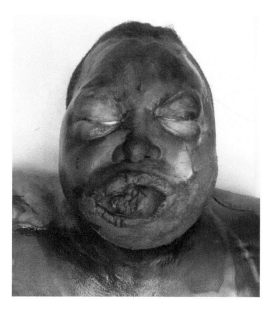

Figure 5.2 Decomposed/skeletal remains that cannot be identified visually.

Figure 5.3 Decomposed remains that cannot be identified visually.

and other items may be present at the location of the remains and may be reviewed for a name or other information (e.g., address, age). The clothing a person is found wearing can be examined for size, brand, or any laundry marks. Clothing can also be compared to accounts from family and friends as to what the decedent was last seen wearing. Jewelry (Figure 5.7) should be examined and can be compared to family/friends' descriptions, or it can be analyzed for personalization or traceable information (e.g., a class ring, an engraved locket). Personal effects with the body can also provide information, such as a cellular telephone, business cards, phone numbers, and keys. Eyeglasses and contact lenses can be examined and compared to the known history of a person.

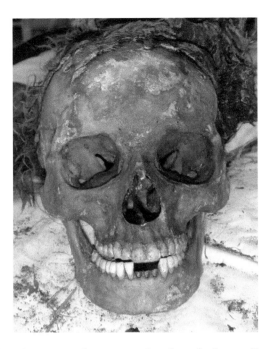

Figure 5.4 Skeletal remains that cannot be identified visually.

Figure 5.5 Burned body that cannot be visually identified.

5.6 External Characteristics

Many people have identifying characteristics on their bodies themselves that are unique enough to establish identity. Things such as body habitus, height, weight, eye color, sex, circumcision, stature, hair type (e.g., curly/straight, long/short), microscopic hair structure (e.g., oval, round, flat), and skin pigmentation can all be helpful attributes. Scars (Figures 5.8 and 5.9) and tattoos

Forensic Medicine and Human Identification

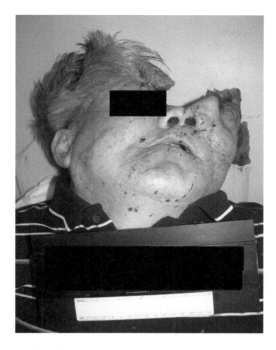

Figure 5.6 Massive head trauma severely limiting visual identification.

Figure 5.7 Jewelry found on a skeleton that was used to assist in identification.

(Figures 5.10 to 5.13) are commonly used as identifying characteristics, especially when they are distinct in either nature or location, for instance, a scar from a burn or an injury, a keloid, or a unique tattoo. Birthmarks or nevi (moles) are often distinct and can be used for identification. The presence, location, and number of piercings may also be helpful.

Occupational stigmata, though more commonly used decades ago, may also be revealed during an external examination. Subungal hematomas involving the great toes are common in dancers and athletes. Onycholysis can

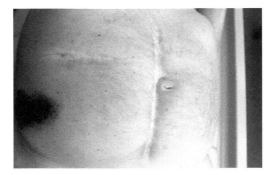

Figure 5.8 Distinct surgical scars.

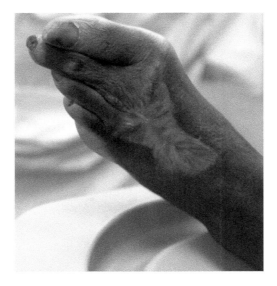

Figure 5.9 Distinct scar from a previous injury.

Figure 5.10 Unique tattoos.

Forensic Medicine and Human Identification

Figure 5.11 Unique tattoos.

Figure 5.12 Unique tattoo.

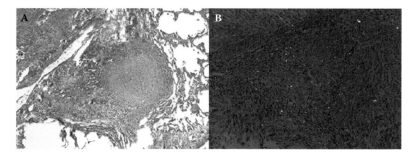

Figure 5.13 Microscopic silica nodule in the lung (A) and polarization of silica particles within the nodule (B).

be seen in occupations that require carrying heavy objects, or in musicians and typists. Inflammation around the fingernails can be seen in hairdressers and gardeners. Contact with chemical agents may cause certain types of skin rashes. Calluses can occur on the hands and feet in certain occupations. Blue scars are often seen in coal miners due to carbon dust entering skin lesions. Small burns on exposed skin may result from steel or foundry work.

5.7 Internal Characteristics

An autopsy examination is commonly performed on unidentified bodies, and in most jurisdictions is required by law to be performed on such bodies. The presence or absence of certain diseases can be helpful in establishing identity, especially when medical records are available. Diseases such as coronary artery disease or cancer may be present. Conditions like cholelithiasis or nephrolithiasis (gallstones and kidney stones) may have been diagnosed prior to death. The absence of organs, due to either surgery or congenital malformation, can be distinctive. While surgeries like appendectomies, hysterectomies, and cholecystectomies are too common to be distinctive, splenectomies, nephrectomies, or other procedures may be more useful. The presence of suture material may also indicate a previous surgical procedure. Implanted devices, such as pacemakers or defribillators, can often be traced through the manufacturer to the recipient. Findings at autopsy may assist in determination of age, including the presence of arcus senilis (opaque ring surrounding the cornea), the presence of osteophyte formation along the vertebral bodies, and the closure of growth plates. Pulmonary anthracosis may indicate the decedent was a smoker, though significant anthracosis may be seen in coal miners who do not smoke. Other inhalational lung diseases may also provide information about the decedent's occupation, such as silicosis (Figure 5.13) (sandblasting, quarrying, stone cutting) and asbestosis (mining, textile workers).

A complete toxicologic evaluation should also be performed, even if not related to cause of death. The presence of certain medications or illicit chemicals may give information regarding lifestyle or possible medical facilities. For example, if methadone is present, treatment facilities could be contacted for helpful information.

5.8 Radiographs

Radiographs are commonly used to establish identity when antemortem radiographs are available for comparison. In such cases, unique structures (e.g., cranial sinuses, sella turcica) can be compared. Identifying characteristics, such as a broken bone resulting in a malunion or varus/valgus deformity of a long bone, can also be used (Figure 5.14). Radiographs may reveal the presence of foreign material, such as old bullets or shrapnel (Figures 5.15 and 5.16), or surgical hardware (Figures 5.17 and 5.18), which can often be traced by a serial number on the device through the manufacturer to the recipient. The presence or absence of growth plates or the extent of osteophyte formation (Figure 5.19) may establish whether the remains are those of a younger or older individual, though forensic anthropology can often be more specific.

Forensic Medicine and Human Identification 71

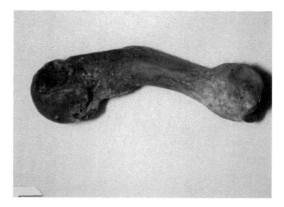

Figure 5.14 Healed phalangeal fracture resulting in deformity.

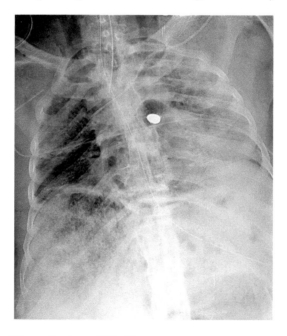

Figure 5.15 Radiograph of an old bullet.

5.9 Anthropology

Forensic anthropology is most commonly used in the cases of decomposed or skeletal remains. The techniques utilized can indicate sex, race, ancestry, and stature.

A brief description of forensic anthropologic discussion follows in Chapter 8. The determination of stature is performed by measuring the long bones and utilizing multiple formulae developed for such a purpose.

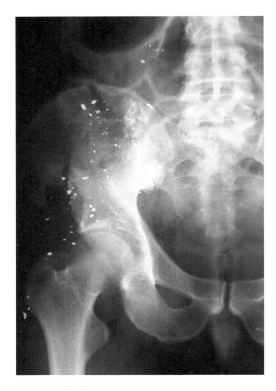

Figure 5.16 Radiograph of old shrapnel.

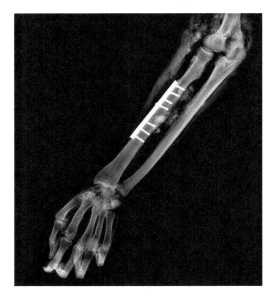

Figure 5.17 Radiograph of surgical hardware within the arm.

Forensic Medicine and Human Identification

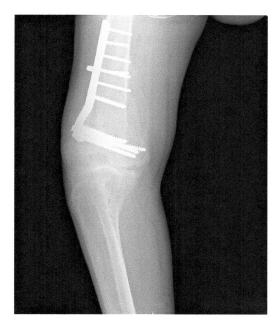

Figure 5.18 Radiograph of surgical hardware within the thigh at the knee.

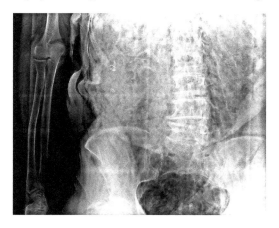

Figure 5.19 Radiograph of osteophyte formation involving the lumbar spinal column.

Pelvic morphology is the best indicator for determining sex, as a woman's pelvis is wider and shallower with an obtuse subpubic angle and an oval inlet. The cranium can also be used to attempt to determine sex, if the pelvis is not available. A male's skull tends to have a receding forehead, prominent brow ridges and occipital protuberance, and a large mastoid process.

If the skeleton is that of a younger person (less than thirty years), growth plates can help delineate age as the plates tend to close (fuse) at certain stages of development. As the person becomes older, age-related changes can be seen, such as osteophyte formation, calcification of the cartilaginous margins of the ribs, and wearing changes of the symphysis pubis. These changes, along with fusion of the cranial sutures, can be used to approximate age.

Ancestry becomes more and more complex as our society becomes more global and less isolated. Traditionally, anthropologists acknowledged three races: Caucasoid, Mongoloid, and Negroid. However, with intermingling of the races, distinct separations are often difficult. For instance, Hispanics often show any combination of the three race categories. Ancestry determinations often combine the classic features of each race, and computer programs are used to determine a likely lineage. Classically, Negroid crania show wide nasal orifices, round/oval orbits, and alveolar prognathism; Caucasoid crania have narrow nasal orifices, parabolic palates, and rectangular orbits; and Mongoloid crania often demonstrate elliptical palates, complex cranial sutures, and square orbits (see Chapter 8).

When a cranium is present, facial reconstruction can be performed in addition to anthropologic analysis. The forensic artist is given the anthropologic data of approximate age, sex, ancestry, and physical attributes (e.g., small, athletic). The morphologic properties of the cranium are combined with the artist's presumptive rendition of the soft tissue features, including eye color and hairstyle, to generate either a sketch or model of how the deceased looked in life (Figure 5.20). A photo of the resulting face can then be distributed to the media or local community in an attempt to find a witness who can identify the individual. Unfortunately, the resulting image does not always accurately approximate the deceased and may end up misdirecting the investigation.

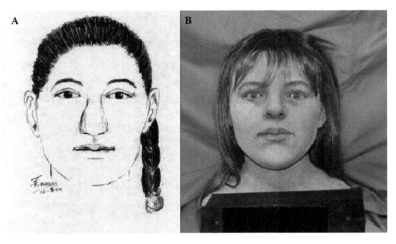

Figure 5.20 Facial reconstruction sketch (A) and model (B) of two different unidentified cases.

Anthropology and facial reconstruction are rarely used alone to establish identity. They are crucial elements in establishing presumptive identifications that can then be combined with DNA or dental analysis for final, positive identification.

5.10 Sources of Comparison

Most commonly, from the circumstances of death, the forensic pathologist or investigating agency has a supposition of who the unidentified person may be and the family is often contacted to provide information (such as doctors, hospitals visited, clothing, tattoos) in order to establish a positive identification. Should no materials be present to allow for such a presumptive identification, several databases exist for comparison of unidentified persons to missing persons. Again, since DNA and fingerprints will be discussed in a forthcoming chapter (Chapters 6 and 7), the Automated Fingerprint Identification System (AFIS) and the Combined DNA Index System (CODIS) will not be discussed here, only mentioned, as they are among the databases used in the attempt to identify persons.

The oldest system is the National Crime Information Center (NCIC) Missing Person File, created in 1975, and the Unidentified Person File, created in 1983. Investigative agencies can submit identifying characteristics to the Criminal Justice Information Services (CJIS) Division's Intelligence Group of the FBI, which oversees the databases. The information is then cross-referenced against known missing persons to find similar, matching cases. If similar cases are found, the investigating agencies are able to contact each other to obtain the information required to establish a positive identification. Unfortunately, not all unidentified persons or missing persons get placed into the system. The FBI has also launched the VICTIMS web-based database in an effort to make the unidentified human remains portion of the NCIC database accessible to the forensic identification community to assist identifying remains. Its purpose is to provide a "…role-based access to enter and search records of the unidentified, while allowing the public access to information that may assist in the identification of these individuals."[1] However, as of 2009, the database is still in its development/data collection stage.

In 2007, the National Institute of Justice began funding the National Missing and Unidentified Persons System (NamUs).[2] This system consists of two databases: (1) unidentified decedents and (2) known missing persons data. Phase III of the NamUs database project is scheduled to occur in 2009, which will link the two databases allowing for comparison of unidentified remains to known missing persons. The database will be searchable by and accessible to medical examiners, forensic scientists, law enforcement, and the general public.

Private organizations may assist in identifying victims of crime, such as the National Center for Missing and Exploited Children, which can be contacted and may provide information helpful in identifying decedents. Various other programs are available for instances of natural or other disasters through agencies such as the Disaster Mortuary Operational Response Team (DMORT), Federal Emergency Management Agency (FEMA), and National Transportation Safety Board (NTSB).

5.11 Methodology for an Unidentified Person/Body

The examination of unidentified persons should be meticulous, well organized, and methodical. Extensive photographs should be taken, including photos of all distinguishing characteristics—eyes, tattoos, scars, etc.—as well as any and all personal property, which should be inventoried and catalogued. In cases of deceased individuals, a full autopsy should be performed, thoroughly documenting all disease processes, presence and absence of organs or tissues, organ weights, and detailed and accurate descriptions of all pertinent positive and negative findings. Full body radiographs should be taken and reviewed. Fingerprints, dental charting, and a DNA sample must be taken on cases, where possible.

One unidentified person can require a great deal of organization and documentation to establish identity; however, in cases of natural or man-made disasters, organization is a necessity. All bodies/body parts, including any and all corresponding property, must be kept separate and uniquely marked. Each body requires the same examination outlined above in an orderly and organized manner, yet it must be conducted as rapidly as possible. In addition, mass disasters may include the need for decontamination of the bodies, a media contact person for the central dispersion of information, as well as a centralized record-keeping system.

It is important to remember that identification is an extremely important task that allows for the grieving process of loved ones to begin as well as a thorough investigation of the death. Identification can only be accomplished through the meticulous examination of the remains. Though there is always the pressure to rush, one must not sacrifice quality for the rapid answer, as the quick answer is often not the correct one.

References

1. FBI VICTIMS database http://www.victimsidproject.org/login.aspx
2. National Missing and Unidentified Persons System (NamUs). http://www.namus.gov.

Suggested Reading

Haglund, W.D. 1993. The National Crime Information Center (NCIC) Missing and Unidentified Persons System revisited. *Journal of Forensic Sciences* 38:365–78.

Knight, B. 1991. The establishment of identity of human remains. In *Forensic pathology*, chap. 3. New York: Oxford University Press.

Spitz, D.J. 2004. Identification of human remains. Part 1. Diverse techniques. In *Spitz and Fisher's medicolegal investigation of death*, ed. W.U. Spitz and D.J. Spitz, chap. 4th ed. Springfield, IL: Charles C. Thomas.

Weedn, V.W. 1998. Postmortem identification of remains. *Clinics in Laboratory Medicine* 18:115–37.

Fingerprints and Human Identification*

6

AARON J. UHLE

Contents

6.1	Introduction	79
6.2	Historical Overview of Fingerprints	80
6.3	Fingerprint Fundamentals	83
6.4	Comparison and Identification of Friction Ridge Impressions	86
6.5	The Postmortem Fingerprint Recovery Process	89
	6.5.1 Inspecting and Cleansing the Friction Skin	90
	6.5.2 Reconditioning Compromised Friction Ridge Skin	90
	6.5.3 Recording Postmortem Impressions	94
6.6	Automated Fingerprint Identification Technology	97
6.7	Conclusion	100
Acknowledgments		100
References		100

6.1 Introduction

Fingerprints have been the gold standard for personal identification within the forensic community for more than one hundred years. The science of fingerprint identification has evolved over time from the early use of fingerprints to mark business transactions in ancient Babylonia to their use today as core technology in biometric security devices and as scientific evidence in courts of law throughout the world. Fingerprints, along with forensic dental and DNA analysis, are also paramount in the identification of unknown deceased individuals and human remains. To this end, recent increases in homicides, mass disaster incidents, and combat casualties from wars in Iraq and Afghanistan highlight the vital role that forensic science plays in human/victim identification. While this responsibility is an emerging challenge for many forensic disciplines, fingerprint analysis

---

* Disclaimer: Names of commercial manufacturers are provided for identification only and inclusion does not imply endorsement by the Federal Bureau of Investigation.

has been the foundation of forensic identification efforts for decades. This chapter discusses the basics of fingerprint identification and how forensic examiners recover friction ridge impressions from remains in order to identify the dead.

6.2 Historical Overview of Fingerprints

> We read of the dead body of Jezebel being devoured by the dogs of Jezreel, so that no man might say, "This is Jezebel," and that the dogs left only her skull, the palms of her hands, and the soles of her feet; but the palms of the hands and the soles of the feet are the very remains by which a corpse might be most surely identified, if impressions of them, made during life, were available.[1]

The patterned ridges present on the bulbs of the fingers have been a source of intrigue for humankind since prehistoric times. The appearance of fingerprints on clay pots and documents throughout early civilization indicates the possible recognition of the individuality and value of fingerprints as a means of personal identification. While this observation is debated by historians, there is no debate that the many instances of fingerprints found in the archaeological record set the stage for the scientific development of modern fingerprint identification in the nineteenth century.

The first practical application of fingerprints as a form of personal identification is credited to Sir William Herschel, a British officer based in India, who used fingerprints and handprints as signatures on native contracts to prevent fraud.[2] Over time, Herschel recognized the value of fingerprint identification and, in a letter written to his superiors in 1877, advocated the implementation of fingerprinting for a variety of civil and criminal endeavors.

At about the same time, Dr. Henry Faulds, a Scottish physician working in Japan, published a letter in the journal *Nature* (1880) discussing his scientific observations on the identification potential of fingerprints. The pioneering research conducted by Faulds was broad in scope, forecasting the forensic use of fingerprints to catch criminals and describing the contemporary method of recording fingerprints using black printer's ink. His theory regarding the evidentiary value of fingerprints was confirmed when he solved a minor crime involving the pilfering of purified alcohol from his hospital laboratory. Faulds was able to compare greasy fingerprints found on a piece of glassware with inked impressions he had collected from his staff, identifying one of his medical students as the offender and making what is considered to be the first fingerprint identification in history.[3]

The work of Herschel and Faulds was further expanded upon by Sir Francis Galton in his landmark book *Finger Prints*, published in 1892. The studies performed by Galton detailed the individuality and persistency of friction ridge skin, providing empirical support to the underlying scientific principles of fingerprint identification. His book also introduced a classification or cataloging system for recorded fingerprint impressions, a necessity for the general acceptance of fingerprints as a means of personal identification. Although the classification system devised by Galton was limited, it became the basis for a number of more functional and contemporary fingerprint classification methods.

One of the most widely accepted and practicable systems of fingerprint classification involved the labors of an Englishman named Sir Edward Henry. By corresponding with Galton, Henry came to recognize the limitations associated with fingerprint classification and determined that, in order for fingerprints to be used as a systematic means of personal identification, a simplified method was required to allow law enforcement the ability to easily file and retrieve numerous fingerprint records. The creation of such a system was accomplished by Henry and two of his subordinates (Haque and Bose) working in India in 1897. Soon afterward, the resulting Henry classification system was adopted by the Indian government, establishing fingerprints as the official means of criminal identification in India.[4] Four years later, in 1901, India's fingerprinting success led England and Wales to implement fingerprints as a means of criminal identification and establish a Fingerprint Bureau at New Scotland Yard.[4]

The first official use of fingerprints in the United States began when the New York Civil Service Commission in 1902 and the New York State Penitentiary System in 1903 adopted the use of fingerprints for civil and criminal identification purposes, respectively.[5] The systematic utilization of fingerprints for personal identification soon became standard operating procedure and spread throughout the United States, as well as the civilized world, culminating in the establishment of a national repository for fingerprints with the newly created Identification Division of the Federal Bureau of Investigation (FBI) in 1924. The repository, which was originally based on a modified version of the Henry classification system, is now an ever-increasing computerized database currently containing over 55 million criminal and 24 million civil fingerprint records.

The identification of human remains through fingerprints is not well defined historically. While Herschel, Faulds, and Galton all suggested the use of fingerprints to identify the dead in the late 1800s, the actual application of fingerprints for this purpose began only after the systematic adoption of fingerprints for personal identification in the early 1900s. One of the first cases involving the use of fingerprints to identify decedents occurred in Birmingham, England, with the identification of a suicide victim in 1906.[3] Another documented example of the early use of fingerprints for human identification is discussed by Henry Faulds in his 1912 book *Dactylography, or the Study of Finger-prints*. Faulds describes the case of a man with no identification documents, apparently killed by an oncoming train, whose badly damaged body was fingerprinted and later identified when the postmortem fingerprints matched an antemortem fingerprint record on file at New Scotland Yard.[6]

In the United States, the U.S. Armed Forces is credited with the early recognition and use of fingerprints as a means to identify the dead. Harris Hawthorne Wilder and Bert Wentworth in the book *Personal Identification*, published in 1918, describe the case of a body floating in the Hudson River near Fort Lee, New Jersey. Visual identification was impossible because of the condition of the remains; however, the victim's clothing led investigators to believe that the man was in the military. As a result, postmortem fingerprints were recorded and forwarded to the War Department, which ultimately matched the recovered prints to an antemortem record on file in Washington, D.C.[7] Wentworth and Wilder also describe the creation of military dog tags containing an etched recording of a soldier's right index impression for the purpose of identifying war dead.[7]

The creation of military identification tags containing fingerprints corresponded with the U.S. entry into World War I (WWI). While no detailed account exists regarding the use of fingerprints to identify U.S. war dead from WWI, there are documented cases involving the use of fingerprints for casualty identification. One such instance concerns the sinking of the SS *Tuscania* in 1918. The *Tuscania* was transporting American troops to Europe when it was torpedoed by a German U-boat, killing approximately two hundred of the twenty-two hundred soldiers onboard. Because of the swirling seas and rocky coastline off the Scottish island of Islay, many of the recovered bodies were damaged beyond recognition, but could be identified through the use of fingerprints.[8] From WWI onward, fingerprints have played a principal role in U.S. military victim identification efforts.

While law enforcement initially adopted fingerprints as a means of criminal identification, their use for victim identification also became important. One of the first FBI identification cases occurred in 1925 when the Portland Police Bureau submitted fingerprints of a decapitated corpse, recovered from the Columbia River in Oregon, to the FBI for identification purposes.[9] When compared to fingerprint records in the national repository, the FBI was able to establish the identity of the body, providing a crucial lead in the open criminal investigation. Eventually, the use of fingerprints for human identification evolved into one of the primary tools used today by law enforcement, medicolegal professionals, and disaster mortuary response teams for personal identification. This expanded civil application of fingerprints was initiated by the FBI, which in 1940 recognized the need for the scientific determination of identity in mass fatality incidents. As a result, the FBI Disaster Squad, a worldwide response team that assists with disaster victim identification efforts through fingerprints, was established. This Disaster

Victim Identification (DVI) team has responded to more than 225 disasters since its inception in 1940, identifying over half of the recovered remains the team has examined through fingerprint analysis.

6.3 Fingerprint Fundamentals

The term *fingerprint* is used to describe a reproduction of the friction ridge arrangement present on the tips of the fingers when an impression is deposited on a touched surface (Figure 6.1). This arrangement of the friction ridge skin is permanent due to the underlying structure of the skin and unique because of complex physiological events, both genetic and environmental, that occur during fetal development. Friction ridge skin is present on the palmar and plantar surfaces of the hands and feet. As such, impressions from the fingers and palms of the hands as well as the toes and soles of the feet can all be used for personal identification purposes.

The friction ridge skin found on the hands and feet differs from the relatively smooth skin that covers the rest of the human body. This corrugated skin, consisting of raised ridges and recessed furrows, assists individuals with grasping objects and gaining traction. Friction skin is composed of two main layers, an outer layer called the *epidermis* and an inner layer called the *dermis*. The epidermis has five different cell layers, whereas the dermis is one large layer consisting mainly of connective tissue and blood vessels. The epidermal ridges are supported by double rows of papillae pegs on the dermis, which can play an instrumental role in the recovery of fingerprints from deteriorating bodies.

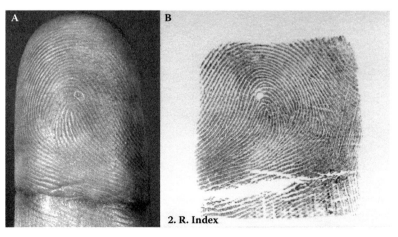

Figure 6.1 Friction ridge skin on the end joint of a finger (A) and its corresponding reproduction, called a fingerprint (B).

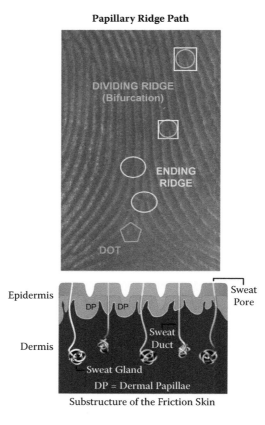

Figure 6.2 Friction ridge detail and structure.

Detailed examination of the friction ridge skin also reveals that ridge path, in most instances, is not continuous across the entire surface of a finger. Some ridges, called *ending ridges*, will flow and abruptly come to an end, while other ridges, called *dividing ridges* or *bifurcations*, will flow and separate into two separate and distinct ridges. Additionally, some ridges are as long as they are wide and are called *dots* (Figure 6.2). These ridge events are commonly referred to as *characteristics* or *minutiae*, and their spatial relationship to one another in a friction ridge impression is the basis for fingerprint comparison and identification. Other features existing in a fingerprint called *formations* are ridge path deviations involving the combination of one or more ridge characteristics. Further examination of the friction ridge skin also reveals irregular ridge contours and sweat pores. Structural and dimensional elements of ridges and pores, when clarity permits, can be used in conjunction with ridge characteristics for comparing and identifying fingerprints.

Friction ridge arrangement at the ends of the fingers generally forms pattern types referred to as *loops*, *arches*, and *whorls* (Figure 6.3). For

Fingerprints and Human Identification

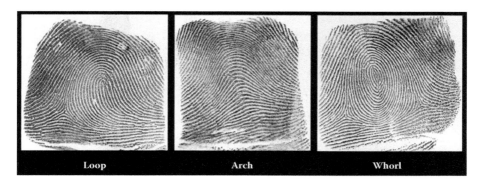

Figure 6.3 Basic fingerprint pattern types.

classification purposes, these basic pattern types can be further divided into eight distinct subgroups based on observed differences within patterns of the same type. Approximately 65% of all fingerprint patterns are loops, 30% are whorls, and 5% are arches. In a loop, the friction ridges enter from one side of the pattern, recurve, and pass out or tend to pass out the same side the ridges entered. An arch has ridges that enter from one side of the pattern, make a wave in the middle, and pass out the opposite side from which they entered. In a whorl, the friction ridges tend to have a circular or spiral ridge flow. It is important for an examiner to note the ridge flow of a print for orientation purposes and the recognition of focal areas that will ultimately assist in the identification process. While pattern configuration alone cannot be used for individualization, it can be used for exclusionary decisions made by an examiner.

Fingerprints fall into three categories: *latent*, *known*, and *plastic impressions*. A latent fingerprint is the two-dimensional reproduction of the friction ridges of the finger on an object by means of perspiration, oils, or other contaminants that coat the surface of the ridges when a finger touches an item. These types of prints generally must be made visible through the use of forensic technology such as alternate light sources, chemical techniques, or fingerprint powders. In some instances, latent prints can be visualized without the use of any fingerprint processing techniques and are called *patent prints*. Latent impressions are deposited by chance and are usually fragmentary in nature with varying degrees of quality. Alternatively, a known fingerprint is the intentional reproduction of the friction ridges of the finger onto a fingerprint card or appropriate contrasting surface. A known exemplar can be recorded using a number of standard techniques, to include black printer's ink, inkless/chemical methods, and LiveScan, which is a computer-based system that creates digital fingerprint images by scanning the fingers. Finally, a plastic print is an impression left in a malleable substrate, such as wax or putty, which retains an image of the friction ridge arrangement.

6.4 Comparison and Identification of Friction Ridge Impressions

While the terminology used to describe the fingerprint identification process has varied over the years, the basic methodology employed by forensic examiners has remained relatively unchanged. One aspect that has changed, however, is an increased awareness of the underlying scientific basis for fingerprint identification. This change has resulted in the standardization of the identification process based on the extensive research of former Royal Canadian Mounted Police Staff Sergeant David R. Ashbaugh, which centers around a quantitative-qualitative philosophy to fingerprint examination called *ridgeology*. Ridgeology is a holistic approach that focuses on the biological uniqueness of friction ridges and involves the sequential examination of the features and spatial relationship of ridges, noting the quality and quantity of the assessed information for identification purposes.[10] This examination is conducted using a methodology that incorporates sound scientific protocols and practices, allowing for accurate and repeatable conclusions that meet rigorous scientific standards. The standard methodology used by fingerprint experts to conduct friction ridge examinations is called ACE-V, for analysis, comparison, evaluation, and verification, which are the four fundamental phases utilized in this process.

Analysis focuses on the examination of the quantity and quality of information present in a print, which can be broken down into three levels of detail. Level 1 detail refers to the overall ridge flow and pattern type of a print. Level 2 detail refers to ridge path, which corresponds to the spatial relationship of ridges and their characteristics in a print. Level 3 detail refers to individual ridge attributes, which involve ridge shapes and pore structure/location in a print. The fingerprint examiner must consider various quality factors, such as distortion, that could alter the reliability of the observed information when determining the suitability of a print for comparison purposes. The information present in the latent or poorest quality print is always examined first, followed by examination of the known or best quality print.

Comparison of friction ridge impressions is a side-by-side assessment of the information analyzed in both prints. The latent or poorest quality print is compared to the known or best quality print to minimize cognitive bias. The examiner first assesses the level 1 information from the analysis of the latent print and compares this with the information gathered from the analysis of the known print. If the information matches, the examiner then assesses the level 2 information from the analysis of the latent print and compares it with the information gathered from the analysis of the known print. Comparison is not a simple "point counting" exercise; in fact, there is no scientific basis for a minimum point threshold or specific number of characteristics that must match in two prints for an identification decision to be reached by an examiner.[11] Level 3 information is usually noted when assessing level 2 detail and, if visible, is compared as well. These comparative measurements begin at a focal point selected by the examiner and progress through the ridges of the entire print in series. For this information to match, the ridge path, of sufficient quantity and clarity, must have the same unit relationship and relative position in both prints. When a fingerprint examiner determines that the information present in a latent and known print is in agreement, with no unexplainable dissimilarities, an identification decision can be reached (Figure 6.4). Due to the pliability of the friction skin, and other environmental factors, friction ridge impressions of the same finger will never look exactly alike.

Evaluation involves rendering a decision based on the results of the analysis and comparison phases of the identification process. There are three possible conclusions that can be reached by an examiner as defined by the Scientific Working Group on Friction Ridge Analysis Study and Technology (SWGFAST).* The first conclusion is *individualization* (identification) and involves the determination that sufficient information present in two impressions matches, meaning that they are from the same source. The second conclusion is *exclusion* and is the determination that the information present in two impressions does not match, meaning that they are not from the same source. The third conclusion is an *inconclusive* decision and is the determination that a conclusive comparison cannot be reached because of a lack of quality or absence of a comparable area in the known exemplar.

Verification is the final step in the ACE-V methodology. Although verification is not technically part of the identification process, it serves as a form of peer review, ensuring reliable and accurate results. All individualizations made by a fingerprint expert are verified, through an independent examination of the identified prints, by a second qualified latent print examiner as a quality assurance mechanism. Verification of exclusion or inconclusive decisions also can be performed but is not required by SWGFAST. Under certain circumstances, the FBI uses a more rigorous form of peer review called *blind verification*, where the verifying examiner is unaware of the evaluation decision reached by the original fingerprint specialist prior to conducting his or her examination.

There are two basic premises that form the foundation of the friction ridge identification process and allow for the use of fingerprints as a means of individualization. These premises concern the individuality and persistency of the friction skin, which have been scientifically validated over time through academic research and the work of experts in the field of fingerprints.

---

* SWGFAST is an organization that sets consensus standards for the fingerprint discipline.

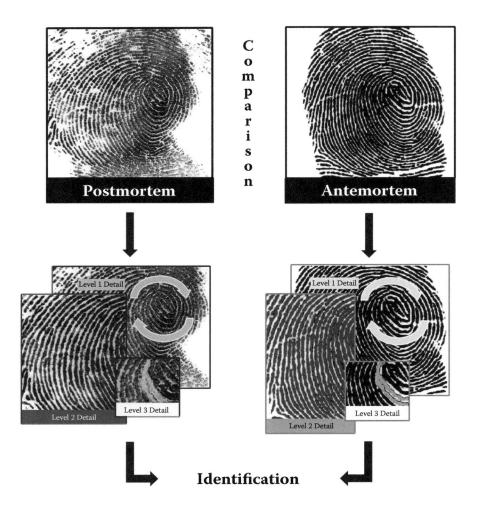

Figure 6.4 The comparison phase of the ACE-V methodology involves the examination of information or detail present in two prints. This information consists of ridge flow (level 1 detail), ridge path (level 2 detail), and if visible, individual ridge attributes (level 3 detail), such as ridge widths. If the information present in the two prints is in agreement with no unexplainable dissimilarities, an identification can be effected.

Individuality refers to the fact that fingerprints are unique; no two areas of friction ridge skin are the same, not even on identical twins. The basis for this statement rests in human embryology and genetics, beginning during fetal development. The physiology of friction ridge skin begins with the development of the volar pads, which are protuberances of tissue that begin to form on the tips of the fingers at about the eighth week of gestation. The degree of complexity of the volar pads (their size, shape, and location on the finger) greatly influences ridge flow or level 1 detail.[12] These volar pads regress or are absorbed back into the finger at about the tenth or eleventh week of gestation, when friction ridges begin to form. Primary ridges develop first, followed by secondary ridge development or the occurrence of furrows between the papillary ridges. Although most of this activity has a genetic component, a nearly infinite number of environmental factors result in the random development (differential growth) of friction ridges and their corresponding level 2 and 3 detail. The end result of these genetic and environmental variances during friction ridge formation is complete biological uniqueness, down to the structure of a single ridge.

Persistency refers to the fact that friction ridges are permanent and remain constant throughout a person's lifetime, until decomposition after death, unless otherwise affected by accidental injury or intentional mutilation. The basis for this statement rests in human anatomy and the histology of the skin. As the body sloughs off dead skin cells, they are replaced by new skin cells generated from the bottom or basal layer of the epidermis.[12] The cells joined together through cell junctions are replaced the same way for an entire lifetime unless scarring occurs. Thus, the basal layer acts as an immutable root system that is the foundation for the permanency of friction ridges and their corresponding level 1, 2, and 3 detail.

6.5   The Postmortem Fingerprint Recovery Process

It has been said by some in the forensic community that there is little difference between obtaining fingerprints from the living and the dead. Those in the fingerprint profession involved with victim identification understand that recovering quality friction ridge impressions from human remains can be one of the most challenging tasks that an examiner can perform. This task differs markedly from printing the living on many levels and requires both mental composure and physical dexterity on the part of the forensic examiner for successful completion.

While fingerprints are obtained from both the living and the dead for identification purposes, the reasoning and mind-set behind the action are different. For example, most examiners can recall the first time they examined human remains, whereas very few can recall with any certainty the first time they fingerprinted the living. The psychological aspects of working with the dead, especially in mass fatality situations, are being addressed by many organizations involved with forensic identification operations throughout the world. Some techniques used to assist examiners in overcoming stresses associated with human identification include mandatory leave, favored by European organizations, and debriefing sessions, favored by U.S. agencies, to include the FBI.

The technical aspects of fingerprinting the living and dead might appear similar on the surface, but in most cases they are considerably different. The majority of identification specialists will record fingerprint impressions from living persons electronically using LiveScan technology or by lightly coating the fingers of an individual with black printer's ink and recording the inked impressions onto a fingerprint card. If the individual has an injury to the friction ridge skin (cut/laceration), an examiner can wait for the skin to heal and record the impressions at a later date. When fingerprinting the deceased, recovered bodies often will exhibit environmental damage to the friction ridge skin, which will contribute to the decomposition/deterioration of the skin and will never heal. In these instances, bodies should be examined promptly and the friction skin reconditioned or returned to a near natural state before quality prints can be recorded.

The examination of human remains is often complex; accordingly, the author has developed a deceased processing methodology to assist forensic examiners in the successful and expedient recovery of postmortem impressions to confirm or establish identity. This three-step process involves:

1. Inspecting and cleansing the friction skin
2. Reconditioning compromised friction ridge skin
3. Recording postmortem impressions[13]

6.5.1 Inspecting and Cleansing the Friction Skin

The first step in processing deceased individuals is visually inspecting the hands to determine if the friction ridge skin has been damaged. In order for the examiner to make this determination, the hands must be cleansed of any contaminant (dirt, blood, etc.) using a sponge and warm, soapy water. A soft toothbrush can be used for removing foreign matter adhering to the fingers, but the examiner must proceed carefully to preserve the integrity of the friction ridge skin. If the friction skin is not compromised, the hands are cleansed and postmortem impressions are recorded. If the friction skin is damaged, the examiner should note the type of damage that has occurred because this will assist in choosing the correct reconditioning technique. The location and nature of the deadly event will offer a good indication of the type of damage observed by an examiner.

6.5.2 Reconditioning Compromised Friction Ridge Skin

Currently, the literature on processing techniques used to successfully recover fingerprints from distressed bodies is often limited and dated. The following discussion includes a modern array of reconditioning techniques designed to assist forensic examiners in the recovery of quality postmortem impressions from damaged friction ridge skin. To simplify this task, the techniques are organized according to the types of damage most often encountered in the examination of human remains. Although medicolegal professionals routinely obtain postmortem fingerprints in death investigations, it is recommended that they seek assistance from a qualified fingerprint examiner when using reconditioning techniques to avoid the possibility of rendering the friction skin unprintable.

One of the most prevalent types of friction skin damage involves maceration and decomposition of human remains related to prolonged water/moisture exposure. This type of damage manifests itself in the creation of creases or wrinkles in the fingers as well as the deterioration of the epidermal layer of friction ridge skin. The primary concern with water-soaked remains is the elimination of wrinkles from the friction skin in order to obtain suitable postmortem impressions. The examiner can pinch/stretch the skin in an attempt to remove the creases or inject tissue builder, a viscous liquid used for embalming purposes, into the fingertips. Tissue builder is injected into the end of the finger by passing the needle through the first joint or medial phalange, resulting in the elevation of depressed areas in the fingertip, thus removing any wrinkles that are present (Figure 6.5).

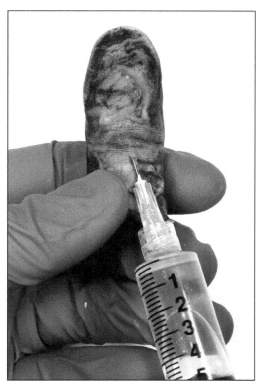

Figure 6.5 The injection of tissue builder into a fingertip to remove wrinkles from friction skin.

The initial stages of decomposition may result in a phenomenon known as gloving, in which the epidermal layer of skin separates from the dermal layer of skin. In this situation, the epidermal skin can be cut from the dermis, dried, and placed over the protected finger of an examiner for recording purposes. If the epidermal layer is completely separated from the dermal layer, recordings of the epidermis and dermis should be taken to ensure that they match and are not from different individuals. This is especially important in disaster situations where a commingling of remains often occurs. In an advanced state of decomposition, the epidermis has usually putrefied, leaving exposed dermal skin. Forensic examiners should not be surprised if they inspect a body and see little or no visible friction ridge detail, as it is often a sign of exposed dermal skin. Recovered dermal prints will appear different than epidermal prints because a dermal impression will have a double row of dermal papillae representing a single epidermal ridge.

Fingerprints can still be recovered from putrefied remains by using the boiling technique, a method that uses boiling water to visualize or elevate ridge detail on the dermis through osmotic rehydration. This process involves bringing water to a boil in a hot pot and then submerging the hand from the body into the water for five to ten seconds (Figure 6.6). The hand is then removed from the water and examined for friction ridge detail, which will be

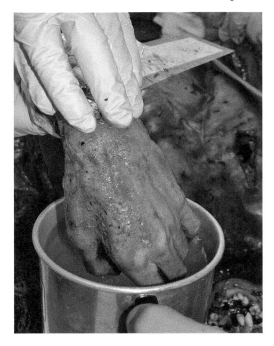

Figure 6.6 Application of the boiling technique to recover friction ridge detail from macerated remains.

visible on the surface of the dermis if it has been successfully reconditioned. If no detail is present, the hand can be placed back into the water for another five to ten seconds. When there is abrasion trauma to the skin, an alternate form of the procedure should be used where the water from the hot pot is indirectly applied to the hand, such as with a sponge, to control development and avoid increasing the size of any cuts that may obscure visible friction ridge detail.[13]

Desiccated or mummified remains are one of the most difficult and time-consuming types of friction skin damage to recondition. Because the body tissue dehydrates and shrinks, often resulting from exposure to arid conditions, the friction skin becomes unusually rigid with severe wrinkling. The body also may display signs of rigor mortis, which must be overcome to successfully examine the remains. In order to break mild rigor, an examiner can forcefully straighten or flatten the fingers of the hand. If this does not work, the examiner can cut the tendon on the inside of the fingers to release the rigor and allow the fingers to straighten. Cases of extreme rigor, such as those involving desiccated remains, require the removal of the fingers from the hand. In order to remove items (digits, clothing, etc.) from a body, permission must first be granted by the medical examiner or coroner. After approval is granted, the examiner may amputate the fingers and place them into sealable jars labeled with the corresponding finger position.

Hands associated with desiccated remains must be rehydrated in order to remove the wrinkles from the friction ridge skin. This is usually accomplished by soaking the fingers in jars containing dishwashing liquid diluted with warm water. Alternatively, a number of different chemical methods can be used to rehydrate the skin, such as soaking the fingers in 1 to 3% sodium hydroxide or in the leather conditioner Lexol® (Summit Industries, Inc., Marietta, Georgia). The rehydration of the fingers may take hours or days, depending on the extent of desiccation. Accordingly, the examiner should regularly check for skin pliability. When the skin has softened, the fingers are removed from the jars and washed clean. The examiner should try to stretch the skin to remove any creases and then use tissue builder to remove any remaining wrinkles, returning the fingers to a near natural appearance. If friction ridge detail is not visible or the fingers have become saturated from soaking, the boiling technique can be used after rehydration to visualize any ridge detail that may be present.

The examination of charred remains can be a delicate task. This type of damage results in brittle friction skin that can be further damaged through excessive handling of the body. When an individual is burned to death, the body will usually exhibit clenched hands. Clenching of the hands is a natural reaction that tends to protect the friction ridge detail on the fingers and possible residual prints left by the victim. Instead of forcing the fingers open, the tendon on the inside of the fingers should be cut and the fingers gently

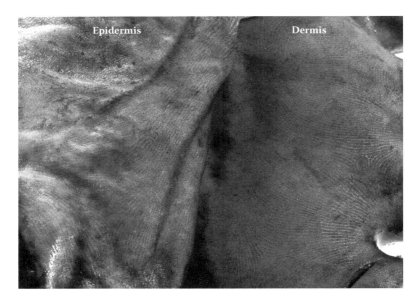

Figure 6.7 Friction ridge detail present on the underside of the epidermis and surface of the dermal skin.

straightened. It may also be necessary to remove the fingers from the hand for examination. At this point, photography is advised to capture any ridge detail that may be present on the fingers. Hardened and loose friction skin may be twisted off the finger, while epidermal skin that has lifted off the dermis but is still attached to the hand should be removed using forceps and curved Metzenbaum scissors for recording purposes.

After the skin has been removed, it should be rinsed with warm water and, if wrinkled, carefully flattened out prior to printing. If the epidermal ridges are unprintable, the underside of the epidermis can, in some instances, be used for recording purposes (Figure 6.7). Recovered prints from the underside of the epidermis, however, will be in reverse color and position. This means that ridge color and ridge flow of the recorded print will be the reverse image of the ridge color and ridge flow of the prints contained on the antemortem standard. In cases of extreme charring with no visible friction ridge detail, the boiling technique can be used as a last resort to clean off the hand and possibly raise ridge detail on the fingers.

6.5.3 Recording Postmortem Impressions

Before attempting to record quality impressions from the friction skin, the examiner must ensure that the skin is dry. This can be a simple procedure that involves blotting the hands dry with paper or cloth towels, or in some instances, where moisture has penetrated deep into the tissue, more intensive

Fingerprints and Human Identification

techniques may be required. One such technique involves the use of isopropyl alcohol to dry water-soaked friction ridge skin. Isopropyl alcohol is applied to the hands, which are then blotted dry with paper or cloth towels. This process should be repeated until the desired results are achieved. Another technique involves using a hair dryer to dry the skin by setting the hair dryer on low heat and blow drying the hands. The final method, called the flame technique, involves the use of a butane grill lighter to dry the skin. The flame is moved back and forth across the friction skin for a few seconds, taking care to dry but not to char the skin. The same results can be accomplished by rolling a finger over a hot light bulb instead of using an open flame.

Since the early days of fingerprinting, the standard method for recording fingerprints has been the application of a thin layer of black printer's ink to the fingers and then recording the friction ridge impressions onto a fingerprint card. Although this technique works well with the living, it is more difficult in its application for printing the dead. The examination of a body usually takes place with the deceased positioned on his or her back for eventual autopsy. This position makes it difficult or nearly impossible for the examiner to apply ink to the fingers using an inking plate, and thus requires that ink be rolled onto a spatula and applied to the fingers. The application of too much ink may result in distorting/smudging of the recorded prints. It is also not feasible to roll the finger impressions onto a fingerprint card, especially when rigor mortis has set in the body. The recording of inked impressions is thus accomplished using a tool called a spoon that can be placed on the end of the finger. Fingerprint blocks are held in the spoon and are used to capture friction ridge detail and create a complete fingerprint record.

The recommended recording strategy for recovering fingerprint impressions from deceased individuals involves the use of black powder and white adhesive lifters. This technique is quick and easy to use, resulting in clear prints compared with those obtained through inking. The first step in the procedure is to lightly coat the fingers with black powder, covering the entire pattern area, using a traditional squirrel hair fingerprint brush or sponge-type paintbrush. Each finger is powdered separately and placed on an adhesive lifter, such as Handiprint (product of CSI Forensic Supply), that is cut to the approximate size of the finger blocks on a fingerprint card. The lifter is placed just below the first joint and then wrapped around the fingertip to record the powder impression. If debris from the finger is being lifted along with the powder and obscures ridge detail, a less adhesive lifter, such as a mail label, should be used. The recorded impression is then affixed to the back of an acetate fingerprint card (Figure 6.8). This type of clear plastic card can be produced by photocopying a standard fingerprint card onto transparency film.

Some alternative printing strategies that are useful in recording quality postmortem impressions from difficult remains involve the use of Mikrosil® (Kjell Carlsson Innovation, Sundbyberg, Sweden) and AccuTrans® (Ultronics,

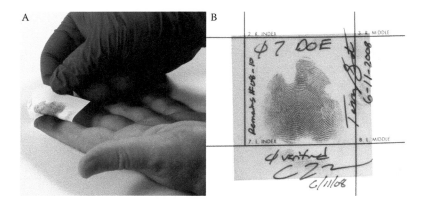

Figure 6.8 The recording of a powder impression using an adhesive lifter (A) and its placement onto a transparent fingerprint card (B).

Inc., Cuyahoga Falls, Ohio). Mikrosil is a casting putty that was originally developed for toolmark examinations before being used in the fingerprint discipline to recover latent impressions from irregular surfaces. AccuTrans is a relatively new polyvinylsiloxaine casting agent specifically designed for the recovery of latent fingerprints and other forensic evidence. Both products have also been used as a way of recording friction ridge impressions from the living and the dead. The casting technique works exceptionally well on desiccated remains containing wrinkles in the friction skin. This technique can be used after the fingers have been rehydrated or at a disaster scene when rehydration is not an option and fingerprints need to be recovered from remains without delay. The first step is to lightly coat the fingers with black powder, followed by the application of Mikrosil or AccuTrans, which is white in color, to the fingers. Mikrosil must first be mixed and then applied to the fingers with a spatula, whereas AccuTrans comes with an automix gun option that allows direct application to the fingers. The casting material must be allowed to dry on the fingers before being peeled off to capture the print (Figure 6.9). Recovered prints will be in correct position and color when compared to an antemortem standard.

When all described recording techniques have failed to produce quality postmortem impressions, images of the friction ridge detail present on the fingers can be captured with digital photography. The proper selection of direct, oblique, reflected, or transmitted lighting schemes will enhance the appearance of ridge detail, often resulting in quality images that can be used for identification purposes. It is also important to capture 1:1 images of the friction skin because the photographs will be compared with antemortem impressions of natural size. If this cannot be accomplished, a scale or object of a known size should be included in the photograph so that image dimensions can be corrected through the use of digital imaging software.

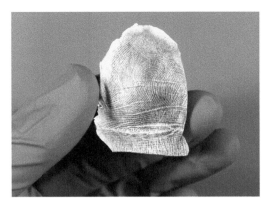

Figure 6.9 Recovery of a powder impression through the use of casting putty.

6.6 Automated Fingerprint Identification Technology

While the recovery of identifiable postmortem impressions from human remains is an integral part of the forensic identification process, it is imperative that these impressions be compared with an antemortem standard in order to have any value in establishing or verifying human identity. The expeditious identification of postmortem remains depends on the most important technological advancement in the history of fingerprinting: the Automated Fingerprint Identification System. This computer system, known as AFIS, has evolved from its early use as a means of searching criminal ten-print records to its use today in identifying suspects of crimes through latent print searches against local, state, and national fingerprint repositories.

In 1999, the FBI released the Integrated Automated Fingerprint Identification System (IAFIS), which consists of a biometric database of millions of fingerprint cards and criminal history records submitted by law enforcement agencies around the country. IAFIS allows the FBI and other criminal justice agencies to electronically access the national fingerprint repository in Clarksburg, West Virginia, for ten-print and latent print searches, meaning that criminals can be tracked by their fingerprints throughout the United States. If individuals have been arrested, it is probable that their fingerprint records are contained in the FBI Criminal Master File (CMF). If individuals have been fingerprinted as part of a background investigation for a job or for military service, it is likely that their fingerprint records are contained in the FBI Civil File (CVL).

Some of the most important criteria in using fingerprints as a means of human identification is the cost-effective and rapid reporting of results, which is directly related to fingerprint computer technology. AFIS, in addition to being a crime-fighting tool, is also instrumental in the identification of the dead. If a dog tag or wallet can be obtained from decedents, the fingerprint record can be located by entering personal identifying information from these items into AFIS and printing off the antemortem record if it exists. The postmortem prints then can be compared manually to the antemortem record to verify identity. In instances of closed-population disaster situations, meaning that the identities of individuals killed in the event are readily known, personal identifying information can be obtained from items such as an airline manifest and entered into AFIS to retrieve fingerprint records. The records can be obtained and manually compared with recovered postmortem impressions, depending on the number of fatalities. Larger disasters often will preclude quick manual comparison of antemortem records, which means that postmortem prints must be searched electronically through AFIS. Postmortem prints are first scanned into AFIS and encoded, meaning that the friction ridge minutiae or characteristics are digitized. Criteria such as pattern type and finger position are then selected followed by the launch of the fingerprint search. Searches of postmortem impressions can take only a few minutes, depending on the submitted criteria, and result in a list of candidates with the closest correlation to the submitted print. Although the I in AFIS represents *identification*, the comparison of the candidates and any identification decision, as it relates to latent print examination, is made by a certified fingerprint examiner and not the computer (Figure 6.10).

The FBI has portable IAFIS terminals that can be deployed to disaster scenes around the world with the capability of searching recovered postmortem impressions through remote access to the national fingerprint repository. In open-population disasters, meaning that the identities of individuals killed in the event are not readily known, recovered postmortem prints should be searched through an automated fingerprint system for identification

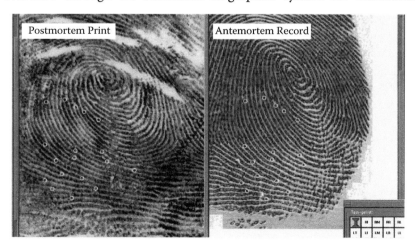

Figure 6.10 The comparison and identification of a postmortem fingerprint searched through an Automated Fingerprint Identification System (AFIS).

Fingerprints and Human Identification

purposes. This is best addressed by a practical look at the deployment of AFIS and the use of fingerprints for mass fatality victim identification in the aftermath of the 2004 South Asian Tsunami in Thailand. Over five thousand people were killed when tsunami waves struck the coast of Thailand on December 26, 2004. Because Thailand is a popular vacation destination, the dead included not only local residents but also many tourists, particularly from Scandinavian countries. The magnitude of the disaster resulted in a worldwide request for antemortem identification records for those believed killed in the catastrophe. As a result, AFIS was established to assist in the massive identification effort because no automated fingerprint system existed in Thailand. Fingerprint cards submitted by various government agencies, as well as latent prints developed on items believed to have been handled by the deceased, were entered into AFIS and used as antemortem standards. Identifiable postmortem fingerprints, recovered from the majority of the bodies using the boiling technique, were then searched against the available antemortem database, resulting in numerous identifications.

An important issue discovered in Thailand when using an automated fingerprint system for victim identification involved dimensional variations associated with recovered postmortem impressions (Figure 6.11). In some instances, the friction skin will expand or shrink to a point that the abnormal size of the recovered prints must be addressed by the examiner in order for a correlation to be made with an antemortem record in AFIS. The lack of antemortem fingerprint records, especially in developing countries, and the

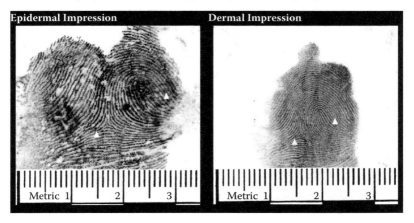

Figure 6.11 An epidermal impression recorded from "gloved skin," which has separated from the dermis because of maceration, may be significantly larger than the corresponding dermal print or antemortem standard. This size difference will greatly affect the accuracy of fingerprint-matching algorithms associated with AFIS, and therefore must be corrected by the examiner through the use of size-scaling algorithms or digital imaging software prior to conducting an automated search.

ability to recover quality postmortem impressions can limit the effectiveness of fingerprints in identifying the dead.

6.7 Conclusion

Fingerprint identification is arguably the oldest forensic discipline known to man. Fingerprints have proved over time to be the most rapid, reliable, and cost-effective means by which to identify unknown deceased individuals, especially in a mass disaster setting. Through the use of various postmortem fingerprint recovery techniques, skilled fingerprint examiners can recover friction ridge impressions even from the most decomposed bodies. The recovered prints can be manually compared with known antemortem records or searched through an automated fingerprint system (AFIS) in order to verify or establish identity. The identification of remains through fingerprints accomplishes the most important and difficult mission of the forensic identification operation: the timely and accurate notification of families regarding the fate of their loved ones.

Acknowledgments

Special thanks goes to Carl Adrian of the FBI Special Projects Unit for his assistance in the preparation of Figure 6.2. I am also thankful for the patience and support of my wife, Lori, and beautiful daughter, Avery, from whom I draw inspiration and without whom I would be at a loss for words.

References

1. Galton, F. 1892. *Finger prints.* London: Macmillan & Co.
2. Herschel, W.J. 1916. *The origin of finger-printing.* London: Oxford University Press.
3. Beavan, C. 2001. *Fingerprints: The origins of crime detection and the murder case that launched forensic science.* New York: Hyperion.
4. Henry, E.R. 1922. *Classification and uses of finger prints.* 5th ed. London: H.M. Stationary Office.
5. Bridges, B.C. 1948. *Practical fingerprinting.* New York: Funk & Wagnalls Co.
6. Faulds, H. 1912. *Dactylography, or the study of finger-prints.* London: Halifax, Milner & Co.
7. Wilder, H.H., and B. Wentworth. 1918. *Personal identification.* Boston: The Gorham Press.
8. Schwartz, S. 2006. SS Tuscania, an American history. http://www.renton.50megs.com/Tuscania/Rememberance/Intro.html.
9. Browne, D.G., and Brock A. 1954. *Fingerprints: Fifty years of scientific crime detection.* New York: E.P. Dutton & Co.
10. Ashbaugh, D.R. 1999. *Quantitative-qualitative friction ridge analysis: An introduction to basic and advanced ridgeology.* Boca Raton, FL: CRC Press LLC.

11. Symposium Report. 1995. Israel National Police: International Symposium on Fingerprint Detection and Identification. *J. For. Ident.* 45:578–84.
12. Wertheim, K., and A. Maceo. 2002. The critical stage of friction ridge and pattern formation. *J. For. Ident.* 52:35–85.
13. Uhle, A.J., and R.L. Leas. 2007. The boiling technique: A method for obtaining quality postmortem impressions from deteriorating friction ridge skin. *J. For. Ident.* 57:358–69.

DNA and DNA Evidence*

BRION C. SMITH
DAVID SWEET O.C.

Contents

7.1	Introduction	104
7.2	Molecular Biology and Inheritance	104
7.3	Laboratory Processing	106
	7.3.1 Extraction (Isolation of DNA)	106
	7.3.2 Amplification (Quantitation and PCR)	108
	7.3.3 Analysis (Electrophoresis, Detection, and Interpretation)	111
	7.3.4 DNA Laboratory Quality Assurance	113
7.4	DNA and the Management of Mass Fatality Incidents	114
	7.4.1 Planning	115
	7.4.2 Establishing the Scope	116
	7.4.3 Communicating with the Laboratory	116
	7.4.4 Evidence Collection Teams	117
	7.4.5 Transportation and Storage	118
	7.4.6 Data Management	119
7.5	Teeth as Sources for Forensic DNA	119
7.6	Saliva and Oral Mucosa as Sources for Forensic DNA	120
	7.6.1 Salivary DNA in Bitemark Evidence	121
	7.6.2 Oral Mucosal DNA and Buccal Swabs	123
7.7	DNA Databases	125
	7.7.1 CODIS	126
	7.7.1.1 Convicted Offender Index	126
	7.7.1.2 Forensic Index	126
Glossary		127
Acknowledgments		131
References		131
Suggested Reading		135

* The opinions and assertions contained herein are solely those of the authors and are not to be construed as official or as views of the U.S. Department of Defense, the U.S. Department of the Army, or the Armed Forces Institute of Pathology.

7.1 Introduction

There are few scientific approaches to human identification that are more effective than a well-trained forensic dentist armed with a set of high-quality dental records and radiographs. Fingerprinting is probably the only other technique used with greater frequency, but as we know, the soft tissue of the extremities does not resist the ravages of time and environment like the enamel and dentin of human teeth. So, in terms of rapidity, degree of certainty, cost-effectiveness, and applicability to a wide range of intact, decomposing, or skeletonized remains, forensic odontology has been the identification method of choice.[1]

For the last two decades, however, a new science has appeared on the forensic stage and, as is the lot for many newcomers, forensic DNA analysis has been both celebrated for its extraordinary achievements and criticized for its complexity and disappointments. Miniscule amounts of biological evidence can be individualized and the results quantified using statistics so staggering that the courts and the public have come to expect the same sort of return on all types of forensic analyses. However, the process of DNA analysis is very slow in comparison to other forensic examinations, extremely expensive, and with few exceptions, must be conducted in highly specialized and fixed facilities. The very power of DNA and the ease with which large population databases can be developed have generated their own set of social problems and ethical concerns. With this new technology comes an increased risk to personal privacy that actually crosses generations, as well as the fear of genetic discrimination in employment and insurance sectors.

The traditional odontology community acknowledges the capabilities of DNA science as applied to human identification, bitemarks, and mass fatality management, and the dividing lines of the relationship are just beginning to become defined. Like the pathologist and the anthropologist, the odontologist's area of expertise includes some of the DNA scientist's favorite targets for analysis. The forensic odontologist will find himself called upon to select and prepare samples for submission to the laboratory and plan for disaster responses that will inevitably include forensic DNA support.

7.2 Molecular Biology and Inheritance

To fully appreciate the power of discrimination offered by the DNA molecule, one must first take a brief moment to grasp its simple elegance. Although earlier investigators initially suspected that proteins were actually the carriers of hereditary information, William Thomas Astbury demonstrated in 1944 that DNA is the sole genetic material of life.[2,3] Although he received only minor recognition for his discovery, his successors became much more famous. James Watson, Francis Crick, and Maurice Wilkins received the Nobel Prize for Physiology and Medicine in 1962 for their elucidation of the structure of DNA. Their revelation that nucleotide base pairing was the functional essence of the DNA molecule, and that those pairings were faithfully consistent throughout biology, was the basis of the award. The nucleotide base adenine (A) always pairs with thymine (T), and guanine (G) always pairs with cytosine (C); the end result is a long molecule composed of two antiparallel strands in a twisted ladder shape that is called a double helix.[4] With the understanding of the molecular structure came a series of techniques in which DNA could be manipulated, including splicing, cloning, sequencing, and even replicating the molecule. These steps led to a new phase of molecular biology called recombinant DNA technology.[3,5-8] Now, scientists could mimic and exploit in the laboratory some of the same changes in the DNA molecule that occur routinely in the natural environment.

The impact of Astbury's discovery on forensic science was the realization that we inherit the code to produce our characteristics but, strictly speaking, not the characteristics themselves. Thus, by uncovering the code that exists within a biological sample, we have a quantifiable and unique basis for individualization. And by focusing on the fundamental code, we remove the subjectivity that arises from analyzing the characteristics, which are the end product of the code and may be significantly impacted by unpredictable environmental forces.

We receive half of our genetic material from our mothers and half from our fathers, but because of the shuffling around of genes (called independent assortment) prior to the creation of the egg and sperm, our parents contribute a different allotment of their genetic material to each offspring. The exception to the rule regarding a unique chromosomal constitution for every person is the occurrence of identical twins. Children that are born from a single fertilized egg that subsequently splits and forms one or more embryos will have the exact same complement of DNA.

The DNA that is passed from one generation to another in this manner is found in twenty-three pairs of chromosomes. Twenty-two pairs of chromosomes are autosomes and the remaining pair consists of two sex-determining chromosomes, which are grouped as X,X (female) or X,Y (male). Both autosomes and sex chromosomes are located inside the cell nucleus and are sometimes referred to as nuclear DNA (nucDNA). Forensic scientists may choose to analyze autosomes for individualization and sex chromosomes for gender determination. Specific analysis of the Y chromosome is an increasingly common practice in sexual assault cases. This allows the analyst to separate the DNA of the male perpetrator from the DNA of the female victim.

The human contains a second, less well-known genome consisting of mitochondrial DNA (mtDNA). As the name implies, this small, circular double helix of DNA is located in organelles called mitochondria distributed throughout the cytoplasm, not the nucleus. Furthermore, this genome is inherited from the mother only, and thus is not unique to the individual. All offspring from the same mother will have the same mitochondrial genome. Maternal relatives, including siblings and maternal cousins, cannot be differentiated on the basis of their mitochondrial DNA. However, mtDNA analysis can be instrumental in resolving identity in certain contexts, including analysis of desiccated skeletal remains, hair, and other severely degraded evidence. Because of the generational constancy of the mtDNA genome, a reference from a living individual today can be used to help identify a maternal relative missing from decades earlier.

No matter whether the DNA is from the nucleus or from mitochondria, or whether the scientist is analyzing autosomes or sex chromosomes, all human DNA contains the same four nucleotide bases paired in the same faithful arrangement described earlier and arrayed in a double helix.[3]

7.3 Laboratory Processing

Upon its arrival at an analysis laboratory, DNA evidence is formally accessioned into the Laboratory Information Management System (LIMS). The evidence will be unpackaged in a biosafety hood, inventoried, reconciled with the accompanying chain-of-custody documents, and assigned a laboratory tracking number. This tracking number is usually an alphanumeric composition that may include a case designator, as well as a sample or specimen designator. Regardless of how the tracking number is derived, it will be the unique identifier used for that sample throughout laboratory processing and reporting. After accessioning, the evidence is documented photographically and repackaged with the submitter's identifying information and the laboratory tracking number. Evidence that does not undergo immediate analysis is sealed with the date and analyst's initials and placed in secure cold storage at −20°C.

7.3.1 Extraction (Isolation of DNA)

The first step in laboratory processing of the evidence is DNA extraction. This is the chemical process by which DNA molecules are removed from the biological source, such as blood, tissue, tooth, hair, etc., and isolated from cellular and protein debris. Bones and teeth are usually pulverized or ground in preparation for this procedure to render the DNA molecules closer to the surface and available for extraction (see Figures 7.1 to 7.3). Hairs may be ground and tissue is usually finely minced. Traditionally, extraction and isolation of DNA is achieved by mixing the prepared sample with stabilizing salts along with detergents that break up cellular membranes and enzymes

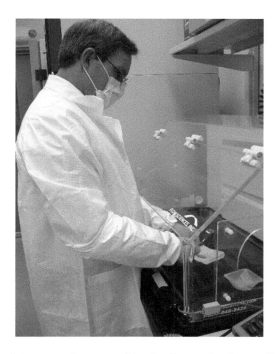

Figure 7.1 Crushing a tooth encased in the fingertip of a sterile nitrile examination glove is a rapid and economic technique to maximize DNA yield once the dentin has been removed from the crown, leaving an enamel shell and any restorations present. This approach has been used successfully at the AFDIL for aircraft accidents when scores of individually recovered teeth required expeditious laboratory processing and some portion of the tooth had to be preserved for return to the family. (Courtesy of the Armed Forces DNA Identification Laboratory.)

Figure 7.2 Removing and retrieving the predentin and other internal aspects of the dentin from a maxillary molar tooth that has been sectioned horizontally using a sterile surgical straight handpiece. (Courtesy of Dr. Kevin Torsky and the U.S. Army Central Identification Laboratory.)

Figure 7.3 Preparing to pulverize a whole tooth in a freezer mill with liquid nitrogen by encasing the tooth in a sterile polycarbonate tube sealed with two metal anvils, including a ferromagnetic plunger. (Courtesy of the BOLD Forensic Laboratory at UBC.)

that degrade the proteins. As a result, the DNA molecules are released from their natural location within the cells and float out into solution. DNA is water soluble, whereas most of the other components in the mixture are soluble in organic solvents. Thus, when phenol and chloroform are added, the aqueous layer containing the DNA separates out and remains on top of the organic layer.[9–11] The aqueous solution is removed and centrifuged through a series of filters to collect the DNA.

Newer extraction techniques employ special beads to isolate the DNA.[14] One particular bead extraction technique is called Chelex. Following an initial boiling step in which the cells are broken apart and DNA is released, the chelating resin beads bind the non-DNA debris. Once the beads have been removed by centrifugation carrying the unwanted material along, a single-stranded DNA layer is left behind for analysis. Another technique uses a lysis buffer to expose the DNA. Silica beads absorb the DNA molecules, thereby taking them out of the sample solution. Several wash steps remove the unwanted cellular debris from the beads. Subsequently, a second buffer elutes the DNA from the beads, which is then available for amplification. This particular approach is amenable to automation and can provide increased efficiency when large numbers of samples require high-throughput robotic processing[9–13] (Figure 7.4).

7.3.2 Amplification (Quantitation and PCR)

Regardless of the outward appearance of evidence that is submitted for analysis, at the molecular level biological materials will exhibit extreme ranges of human DNA content and condition. Depending on the environment in which the evidence was recovered, there may also be nonprimate animal DNA or microbial overgrowth contributing to the total volume of

DNA and DNA Evidence

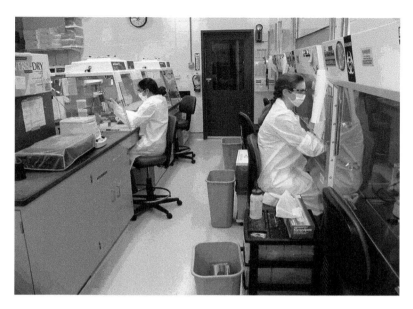

Figure 7.4 Examples of biological fumehoods that are employed during DNA extraction procedures to minimize DNA contamination and protect analysts from organic extraction chemicals. (Courtesy of the Armed Forces DNA Identification Laboratory.)

DNA present in the sample. Prior to moving to the amplification step, it is prudent to determine how much human DNA is actually present so this can be targeted in subsequent procedures using the correct volume and dilution. Excessive quantities of DNA may overwhelm the amplification reaction and interfere with the interpretation of the final data. In the opposite extreme, insufficient template DNA may provide only a partial profile or no profile at all.[9] Several methods have been used to determine the concentration of DNA extracted from the evidence, a step called quantitation. These methods include the older slot blot assay and the more modern real-time quantitative polymerase chain reaction (PCR) assay.

The slot blot assay uses a forty-nucleotide probe that binds to a specific site of interest on human chromosome 17; the amount of bound probe is proportional to the amount of DNA present. A subsequent chemiluminescent (light-producing) reaction is used to expose x-ray film. Then a comparison of the size and intensity of the "blots" on the film are compared with known standards to assist the analyst to determine the quantity of human DNA that is present in the sample. Although specific for primates, this test does not reveal the quality or condition of the DNA that is present in the sample.

Real-time PCR quantitation measures the incremental increase in product after each amplification cycle. During the extension phase of PCR, a fluorescent reporter dye is released and activated. Based on the increasing

Figure 7.5 A bank of thermal cyclers used to perform PCRs on forensic samples in a high-throughput forensic DNA laboratory. (Courtesy of the Armed Forces DNA Identification Laboratory.)

intensity of the fluorescence from this dye, the real-time PCR instrument plots the rate of accumulation of amplified DNA.[9] The point at which this amplification curve crosses a predetermined amplification threshold is used to form a standard curve. The analyst can determine the quantity of human DNA in a given sample by finding where it falls along the standard curve.[9]

Once the analyst measures the quantity of DNA in the sample, he or she can proceed to PCR amplification (Figure 7.5). Not only does this process replicate the target strand of DNA from the evidence, which ultimately provides millions of copies of the original template to facilitate detection by laboratory instruments, but the polymerase chain reaction can also be used to target a specific area of interest for analysis, usually at the exclusion of all other regions in the human genome.

The first step in PCR amplification is to denature or unwind the two strands of the DNA molecule by elevating the temperature, usually to 94 to 96°C. This enables the single strands to be replicated individually. During the second or annealing step the temperature is decreased, usually in the range of 55 to 72°C. This allows short segments of added DNA called primers to bind in a complementary fashion at the ends of the target site of DNA from the sample. In the third step, the temperature is elevated only slightly to 72 to 75°C and extension begins.[10] In the correct chemical environment, the heat-stable enzyme *Taq* polymerase is used to add nucleotides in a chain-like fashion that faithfully follows the code established by the single-stranded template.[11] Thus, wherever there is a thymine (T) in the template strand *Taq* will add an adenine (A), and wherever a guanine (G) is encountered its cytosine (C) counterpart will be added. This step-wise replication of the template strand starts at one primer and continues nucleotide by nucleotide until the ending primer is reached.[11] This completes the first cycle of PCR, and as each subsequent cycle is repeated with denaturation, annealing, and extension, the previous amount of DNA is doubled, which continues with each subsequent cycle.[9–11] The millions of copies of targeted DNA that constitute the product of PCR are called amplicons.

7.3.3 Analysis (Electrophoresis, Detection, and Interpretation)

The principle of electrophoresis as used in the forensic analysis of human DNA is applied whether the sample is undergoing the short tandem repeat (STR) fragment analysis of autosomal or sex chromosomes or the direct sequencing of the mitochondrial genome. In an aqueous environment, the DNA molecule is negatively charged. Thus, in a polarized electrical field and proper medium, the DNA will move toward the positive electrode. As a general rule, the smaller fragments will move faster than the larger ones, allowing the analyst to separate fragments of DNA according to size[9,15] The electrophoresis matrix that is used by most laboratory systems is a viscous polymer. This gel may be poured into an external slab or run inside a single or multiple capillaries (an array). The older slab gel configuration risked bleed-over from one injection lane to the next and took much longer for a technician to prepare. The newer capillary arrays are cleaner and more efficient but also more expensive (Figure 7.6).

Once the nucDNA is sorted or the mtDNA is sequenced by electrophoresis using a polyacrylamide gel matrix, the results must be visually depicted in a way that can be analyzed and compared with other samples. Detection of the separated amplicons varies between the STR analysis of nuclear alleles and sequencing analysis conducted on areas of interest in the mitochondrial genome.

In STR analysis, the pieces of DNA are tagged with fluorescent labels during the amplification process.[11] Later, the sample is injected into the capillary instrument or added to the slab gel and the segments pass through a specific window. At that point a laser detects the fluorescent signal and correlates it to a standard (allelic ladder) of known fragment sizes. The resulting DNA profile is a series of peaks that correlate to these fragments and are sorted by size and dye color. Pairs of peaks usually indicate heterozygosity at that location (locus) on the molecule, whereas a single peak generally indicates the individual is homozygous or has only one variant at that locus.[16]

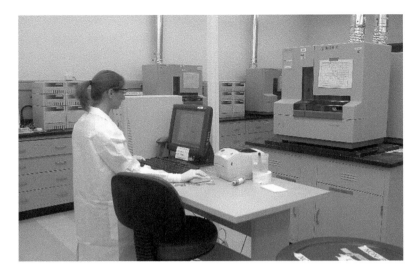

Figure 7.6 An analyst working with the computer-controlled genetic analyzer that differentiates PCR products from thermal cyclers and obtains the final DNA profile. (Courtesy of the Armed Forces DNA Identification Laboratory.)

Direct sequencing of mtDNA is a slightly more complicated procedure. Following PCR the regions of interest on the molecule are subjected to cycle sequencing or dye terminator analysis.[10] The nucleotides are incorporated into a growing strand of replicated DNA from PCR. But unlike traditional PCR, the extension process is randomly halted through the addition of dideoxynucleotides. These nucleotides are unique in that they are missing the 3′-hydroxyl group that would normally permit the extension process to continue. Instead, the incorporation of this dideoxynucleotide terminates the extension process. Although normal deoxynucleotides are also present, only the dideoxy terminators are fluorescently tagged with one of four dye colors for adenine, guanine, cytosine, and thymine. Later, when each piece of DNA passes through the laser window, the instrument detects each fluorescing color and determines whether A, C, G, or T is the terminator nucleotide for that strand. Subsequently, when the strands are sorted according to length, the instrument will compute the actual sequence of the original amplicon.[11,17]

Although many of the laboratory processes can be automated, the one step that still depends on the skill and experience of the DNA analyst is the interpretation of the data. All the peaks that are generated by the instrument actually represent an STR allele in the case of nucDNA or a mitochondrial base in the case of mtDNA. Most laboratories require a minimum of two experienced analysts to review all these data prior to conclusions being reported. It is also the exclusive purview of the analyst to compare data between two samples, draw a conclusion, and to calculate the statistical weight of the opinion.

7.3.4 DNA Laboratory Quality Assurance

The forensic DNA community has achieved an enviable degree of standardization throughout its casework processes. The community adheres to a consistent application of quality assurance measures that include the delineation of roles and responsibilities of laboratory management, minimum education requirements for laboratory staff, established standards for training, annual proficiency testing, guidelines for the validation of new equipment and technologies, and mandatory components for inclusion in the final report.

Although the growth of forensic DNA laboratories began years earlier, the first national effort to ensure the reliability of equipment and casework conclusions was formed by the FBI as the Technical Working Group on DNA Analysis Methods (TWGDAM).[18,19] This group was composed of commercial, academic, and government scientists and in 1989 released the first "Guidelines for a Quality Assurance Program for DNA Analysis." TWGDAM continued to work in conjunction with the National Institute for Science and Technology to develop a set of laboratory reference standards to ensure consistency of DNA equipment performance.[18,20]

In 1994, the DNA Identification Act established the TWGDAM guidelines as the national standards for forensic DNA laboratory operations until the director of the FBI issued his own. To facilitate the development of those federal standards, the FBI established a DNA Advisory Board (DAB) from 1995 to 2000. Subsequently, the DAB released two sets of standards, one for processing crime scene evidence and the other to address the analysis of convicted offender database samples.[21–23] Thus, in October 1998, the Quality Assurance Standards for DNA Testing Laboratories became effective. The Quality Assurance Standards for Convicted Offender DNA Databasing Laboratories followed in April 1999.

TWGDAM was renamed the Scientific Working Group on DNA Analysis Methods (SWGDAM) in 1999. Once the charter for the DAB expired in November 2000, SWGDAM inherited the responsibility for the maintenance and updating of the national standards under the continuing sponsorship of the FBI. Compliance with these standards became a prerequisite for participation in the National DNA Index System program and to be eligible for grants provided by the National Institutes of Justice. Federal, state, local, and even commercial laboratories all began to adapt their operations to accommodate these standards. Between SWGDAM and its predecessor TWGDAM, sequentially updated versions of the Quality Assurance Standards (QAS) were released in 1991, 1995, and 2004.[18] The director of the FBI is expected to approve an updated version in 2008.

All federally funded, federally operated and Combined DNA Index System (CODIS) participating laboratories are required to demonstrate compliance with the QAS in accordance with the DNA Identification Act of 1994. Current national standards require annual audits with a mandatory external assessment in alternating years. A collective effort by the FBI, the American Society of Crime Laboratory Directors/Laboratory Accreditation Board (ASCLD/LAB), and the National Forensic Science Technology Center (NFSTC) produced a DNA audit guide that helped reduce the subjective interpretations of the QAS.[18,19,24,25]

In a step further than demonstration of compliance with the QAS, most U.S.-based laboratories choose to pursue formal accreditation. The most commonly sought accreditations are through ASCLD/LAB or NFSTC. ASCLD/LAB is currently moving to an international certification in partnership with the International Organization for Standardization (ISO), specifically under standards for competence of testing and calibration laboratories (ISO/IEC 17025:2005).[24,26] NFSTC provides similar accreditation through its not-for-profit subsidiary Forensic Quality Services (FQS).[25,27] Other credentialing bodies less common to conventional forensic DNA operations include the College of American Pathologists[28] and the American Association of Blood Banks.[29] However, certifications from these latter bodies are generally found in laboratories that provide medical diagnoses, blood banking, or paternity testing in addition to criminal evidence processing.

If the forensic odontologist is involved in developing a contingency where DNA analysis will impact on his casework, including a mass fatality incident response plan, he or she would do well to research local laboratories' technical capabilities, as well as the status of their QAS compliance and any accreditations that they may hold. Any concerns associated with the qualification of laboratory staff or past audit results should be resolved well before large amounts of critical evidence are submitted to the laboratory.

7.4 DNA and the Management of Mass Fatality Incidents

In many ways the application of forensic DNA testing in a mass fatality incident is a two-edged sword. The technology is very precise and can individualize extremely small fragments of bones and tissue. It seems perfect for sorting out the largest and most complex disaster scenarios. But, the same exquisite capacity for detailed analysis is counterbalanced by a high cost in both time and material resources. Overuse may actually delay the closing of cases. The laboratory facility is almost never collocated with the morgue or incident location, so evidence transfer, communication of the test results, and coordination of DNA data with the other investigative findings are as much a challenge as the laboratory processing requirement itself. And, given these challenges and their inherent potential for delaying final case resolution, repatriation of the victims' bodies, and family notification, where does the case manager draw the line as to how much evidence to test? Importantly, how will reference DNA material for comparisons with the victims be obtained? And lastly, how does the case manager mitigate unrealistic expectations given the exaggerated reputation DNA analysis enjoys regarding turnaround time and the belief that the analyst obtains definitive results in every case?

7.4.1 Planning

The proper response to a mass fatality incident starts well before the event happens. If the odontologist is involved in his jurisdiction's disaster planning process, he should broach the need for planning and coordination for a DNA response capacity with one or more government or private forensic DNA laboratories. Federal, state, and local government laboratories operate on very tight budgets that are tied directly to current-day political and legislative priorities, notwithstanding their busyness with respect to ongoing casework. Although appropriations will rise and fall, no government laboratory is funded to maintain excess capacity in the off chance that a mass fatality incident might occur in the future. In a like manner, commercial laboratories have a profit margin to maintain, and although some are quite good at expanding capacity on short notice, there will generally be a delay and some need for immediate funding to cover the expenses of a productivity surge. Meeting with laboratory representatives to confirm their willingness to be part of a mass fatality contingency is essential. If the laboratory were geographically nearby, it would be prudent to inquire about their continuity of operations plan (COOP). In some circumstances, the very same disaster that they plan to help address could compromise their own facility, and thus the ability to support any relief effort.

One very important topic for discussion with the DNA laboratory during the preevent planning stage includes clarifying the types of samples (soft tissues, bones, teeth, blood, swabs, etc.) that laboratory personnel are trained to handle. For example, the numbers of laboratories that have little or no experience extracting DNA from bones or teeth might surprise planners. Although teeth are not generally the sample of choice because in most cases the human remains are found in a reasonable time after the event, at least one historical over-the-water incident culminated in over 175 teeth samples being submitted for DNA analysis after an extended period of recovery. If a laboratory's throughput capacity and technical procedures do not reconcile with the likely disaster scenarios, the jurisdictional planners should engage additional DNA laboratory resources. If more than one laboratory is included in the disaster plan, authorities should host a meeting between technical representatives so that communication, evidence transfer, data interpretation, anticipated expenses, and turnaround times, as well as compatibility of typing systems and instrumentation, are agreed upon well in advance.[30-33]

7.4.2 Establishing the Scope

A large-scale disaster disrupts the affected jurisdiction in many ways—physically, emotionally, economically, and politically. As soon as possible, however, the disaster response plan must be applied and the journey away from chaos will begin. One of the most significant decisions made by local authorities involves the scope of the medicolegal death investigation. Essentially, a decision must be made regarding whether the identification of all biological material recovered will be sought versus the more direct goal of establishing each victim's identity and a firm cause and manner of death for those involved. This and other decisions, such as balancing speed versus accuracy, will place the DNA laboratory and the case manager on a tightrope between conscientiousness and controversy. Government and elected officials, families of the victims, the media, and even the laboratory staff themselves will ebb and flow between resolve, compassion, and frustration. Establishing realistic expectations in the beginning, even if they seem pessimistic or unpopular, will purchase more patience and credibility as the postevent investigation wears on.[30–32,34]

7.4.3 Communicating with the Laboratory

The single most important fact to remember when dealing with remote DNA support is that the laboratory analyst does not have the opportunity to see, hear, feel, or understand and process the information available to those at the incident location or morgue *unless it is communicated to them clearly*. Some odontologists will empathize with this challenge, knowing that in clinical dental cases dental laboratories must rely almost exclusively on the information submitted on the work request form. Depending on the quality and experience of the dental laboratory, if the clinician submits poor or incomplete information or flawed casts or impressions, then the lack of clarity will certainly be reflected in the final product. It is much the same with forensic DNA laboratories providing services to an operation many miles away. Unfortunately, the surge in samples and the unrelenting public call for immediacy will complicate the communications effort even further. For example, if the morgue submits a sample for DNA processing with the wrong specimen number, the results will come back to the morgue potentially associating that sample with the wrong remains. If the collection team erroneously labels a sample as left tibia when in fact it was from a right tibia, then the DNA lab will unknowingly perpetuate that anatomical error in the report back to the field. The best way to avoid these pitfalls is to be aware that communication is an issue, establish a relationship with the laboratory before an incident occurs, and have trained teams of forensic scientists prepared to select the samples for DNA analysis and complete chain-of-custody documentation accurately.[30–32]

7.4.4 Evidence Collection Teams

Most morgue operations place the DNA collection team near the end of the examination line after the photographs and radiographs have been taken and following the completion of the dental examination and autopsy. The chief of morgue operations or medical examiner/coroner will direct DNA samples to be taken in accordance with the scope of the investigation, as discussed earlier.[30] Usually the postmortem team will take a single sample from relatively intact remains to confirm or augment other identification methods. Team members may also be required to select the best possible material from each of numerous fragmented human remains in order to provide a primary identification or the genetic basis for reassociation of body parts. Most frequently, natural disasters tend to require the former approach to sampling, whereas transportation accidents and terrorist events are more likely to have a greater need for reassociation.

The ideal DNA sample collection team includes two to three persons possessing (1) training in general morgue operations, (2) a rudimentary understanding of DNA science, and (3) thorough familiarity with human anatomy. They must understand the criticality of the anatomical description and the unique numbering of samples, plus be able to handle a Stryker saw, tissue forceps, and scalpel with skill and safety. Odontologists and anthropologists are usually good choices for the collection team, but death investigators and emergency medical personnel are good alternatives. The trauma surrounding the event and subsequent environmental conditions will adversely affect the soft tissue first by fragmentation and later by decomposition. Although skeletal muscle is an easy sample to collect at the morgue and relatively simple to process at the laboratory, the condition of the remains may necessitate the collection of samples of bones and even teeth instead. Clumps of hair, skin flaps, and soft tissue that are predominantly composed of adipose tissue all cause additional steps in laboratory processing and should be avoided when possible.[30] The collection team must use sterile disposable supplies when possible and wipe all surfaces and clean nondisposable gloves, instruments, and working surfaces with a 10% bleach solution between samples.

Tissue, bone, or tooth samples should be placed in a secure container without any preservative. The collection teams at the Armed Forces DNA Identification Laboratory have had greatest success using sterile 50 ml conical tubes with screw caps. Urine specimen cups may leak, glass containers could break, and small plastic bags are subject to puncture. Conical tubes with screw caps easily accommodate 5 to 25 g samples of soft tissue, bone, or tooth, do not leak, and have a smooth exterior surface for handwritten or adhesive labels. Plus, enforcing their use actually limits the amount of sample that an overly enthusiastic anthropologist, odontologist, or pathologist can submit from a single source. This reduces the long-term storage requirements of the laboratory and forces the collection team to focus on the selection of the best quality material while reducing unnecessary cutting of the remains.

The corollary to the morgue operation is the Family Assistance Center and the coordination of the collection of antemortem reference samples that come from the victim's personal effects or family references that provide a DNA profile for comparison purposes. The family members that appear at the Family Assistance Center are not always the best genetic candidates for family references. Furthermore, most out-of-town family members may not linger long after the initial event. For these reasons, as much information regarding the victim's genetic tree and the whereabouts of other relatives must be obtained on the first interview with the next of kin. Given the emotional displacement of family members after a disaster, predesigned forms that include family tree templates will help distraught relatives place themselves and others in the proper genetic relationship. The preferred family reference donors are listed under the section on salivary DNA.[30–32]

7.4.5 Transportation and Storage

Once the collection team has verified the accuracy of the chain of custody, the samples should be transferred to the laboratory as expeditiously as possible. Individually labeled and sealed 50 ml conical tubes are placed in heavy-duty clear zip-lock plastic bags. Depending on the size of the bags, three to five tubes can be placed in a single bag. This second layer of packaging serves two purposes. First, if leakage occurs from a tube, it limits the potentially contaminating exposure to a limited number of other samples, and it also reduces the likelihood that numerous labels will become smudged or illegible. Second, when working with very large numbers of samples, the plastic bag simplifies the moving of evidence to and from the laboratory or in and out of storage. If precoordinated with the laboratory, collection teams can even use the bags to batch samples according to their priority. Bags of samples are best transported in a clean commercial-grade cooler. The 16-quart size or larger allows ample room for bags of ice or reusable ice packs to keep samples cold during transportation. If samples cannot be transferred to the laboratory immediately, they should be kept in a cool, dark, dry environment, preferably at −20°C. Samples may be shipped using a commercial courier, but a courier rotating directly and only between the morgue and laboratory, maintaining wireless communication with both sites, and possessing security clearances at both sites is highly desirable. This arrangement will overcome the business hour restrictions that hamper some delivery services and also will facilitate an unbroken chain of custody.[30–32]

7.4.6 Data Management

The sample collection and laboratory processing portions of the DNA mass fatality response are complicated activities. But neither is more complex than the need to assemble all of the data that are generated, review them, and compare these unknown profiles with the available references, including interpreting the results and assigning a statistical weight to the conclusions. This activity becomes far more challenging depending on the number of different laboratories that are processing samples and the degree of commonality between their procedures. These issues are best addressed during the disaster planning phase. Other factors that will impact data management requirements are the typing technologies applied (autoSTR, Y-STR, mtDNA, etc.), the genealogical proximity of family references to the victims, and decisions regarding the use of partial or incomplete profiles.

Most DNA laboratories are acquiring or developing their own laboratory information management systems that aid the staff in maintaining the chain of custody and facilitate data analysis, statistical computation, and tracking both productivity and turnaround times. The variation in capability from one management system to another is quite extreme, and very few laboratories have systems that are developed specifically to handle mass fatality scenarios.[30–32,35,36]

7.5 Teeth as Sources for Forensic DNA

Teeth are known to survive most postmortem events, including natural phenomena such as decomposition and autolysis, as well as environmental insults, such as water immersion, burial, and fires to as hot as 1,100°C.[37,38] Neurovascular cells of the pulp and, to a lesser extent, odontoblasts embedded in the predentin layer and trapped during mineralization in the dentinal tubules have been shown to be valuable sources of DNA evidence when other bodily tissues are degraded or missing.[39,40] Various methods are advocated to access these various regions of the tooth and effectively recover sufficient amounts of DNA from the different cell types to both minimize the potential loss of dental structures and data and yet maximize DNA recovery.

The spectrum of various methods to access dental cells for DNA extraction are represented by the two most widely used dental DNA recovery techniques. Additionally, current research is focused on other methods to increase further the yield of DNA from degraded and compromised samples. To conservatively access cellular material, Smith et al. advocate horizontal sectioning of the tooth near the cementoenamel junction. This retains the coronal aspect for ulterior forensic examination or historical preservation while exposing the pulp chamber and radicular pulp system, predentin, and dentin for DNA recovery.[41] Once the dentin has been removed from the crown, the enamel shell and any restorations are left intact. Cases that may require this approach are those that involve remains of unique cultural value or museum specimens where the destruction of the material must be minimized. Examples are investigations into the remains of Tzar Nicholas and his family, analysis of dental evidence representing members of George Washington's extended family, and attempts to identify the putative skulls of Mozart and Fredric Schiller (unpublished data). This technique is also applied by the U.S. military in the identification of its missing in action (MIA) soldiers and in mass fatality incidents in which individual teeth are recovered and may represent the sole remains available for return to a family.[42] Attention is being given to other nondestructive methods, especially with respect to ancient invaluable human exhibits. Studies by Krzyżańska use a microfluidic pump to flush cells from the tooth by rinsing the pulp system from the apical orifices through small holes in the occlusal surface.[43] These methods may show promise for forensic exhibits to allow DNA recovery from the pulp while preserving the tooth.

Alternatively, following the application of traditional forensic odontological identification methods, Sweet recommends that crushing or grinding the entire tooth into a fine powder can obtain the maximum quantity of DNA.[44] This must follow adequate documentation of the exhibit with charts, notes, photographs, and radiographs. The method allows recovery of DNA from remnants of pulpal cells as well as the embedded cells present in the hard tissues, especially in cold cases involving skeletal remains.

On a case-by-case basis, then, the laboratory must make informed decisions about the best method to employ to recover DNA. This decision should be made in concert with the forensic odontologist using his knowledge of dental histology and taking into account the presence of any identifiable morphological or restorative traits of the tooth.

7.6 Saliva and Oral Mucosa as Sources for Forensic DNA

In addition to being consulted on the use of teeth and the craniofacial structures as a source for human identification, the forensic odontologist should also be prepared to advise on matters concerning the biological character of oral mucosal and salivary DNA for the same reason. Salivary DNA can be recovered in a variety of scenarios and from a wide range of inanimate objects, including clothing, foods, tobacco products, oral hygiene devices, drink containers, dental prostheses, stamps, and envelopes. Laboratories will generate their own protocols for recovery of salivary DNA from personal effects, but will probably defer to the odontologist in processing a bitemark for DNA. The forensic odontologist may also be approached with questions regarding the buccal swab. As the name suggests, DNA from the oral mucosa of the inner cheek is the target of the buccal swab technique. It is a collection method that is rapidly supplanting the older phlebotomy or finger stick techniques in which whole blood samples are collected for DNA processing.

7.6.1 Salivary DNA in Bitemark Evidence

Saliva derives from three major bilateral glands and hundreds of smaller ones scattered throughout the oropharyngeal region. Over two-thirds of the 1.0 to 1.5 L of the saliva produced daily is from the submandibular gland, whereas the parotid glands account for approximately 25% of the total volume and the sublingual glands only 5%.[9] The biological composition of the fluid varies between the glands, and the volume of salivary secretion at any given time of day will change according to physiological stimulation or medical issues or a pharmacological condition. But generally secretion is highest during eating.

Saliva is composed chiefly of water but also contains electrolytes, buffers, glycoproteins, antibodies, and enzymes.[9,10,45] The detection of one particular group of enzymes, the human α-amylases, which are responsible for initial polysaccharide digestion, is a traditional method for identifying evidentiary stains for the presence of saliva.[9,10,46] Although amylase is found in other bodily fluids, its very high levels in saliva are unique in that they can be as much as fifty times greater than those in other sources. Its stable biochemical activity over time is distinctive as well.[10,47]

Amylase-dependent forensic screening tests are increasingly available in the commercial sector. Some are very specific and are based on monoclonal antibody activity that focuses on human salivary α-amylase, whereas others are more general in relying on the detection of amylase activity to release a colored dye suggesting the presence of saliva.[9] However, all such direct-contact assays risk consumption of portions of the sample that are otherwise destined for DNA analysis. If used, consideration should be given to selecting the most informative product that requires the least volume of sample.[48,49]

Visual screening for evidentiary saliva stains should be augmented by the use of alternative light sources. Although ultraviolet light is helpful in uncovering the presence of biological fluids, short-wave ultraviolet light is a potential health hazard and is also known to disrupt the DNA molecule and compromise the ability to conduct forensic DNA testing. Alternative light sources, such as lasers and high-intensity lights that can be filtered to provide a single wavelength, are probably the best for screening evidence, including skin, for the presence of saliva.[10,50,51] The advantage of an alternative light source in screening for salivary stains on skin is that the potential

DNA sample is physically untouched and that detection can be made even in the absence of a patterned injury.[46]

It is the cellular component of saliva that makes it an attractive target for the DNA analyst. These cells are not secreted by the salivary glands but are incorporated into saliva as part of the shared oral environment. Specifically, oral mucosal cells are sloughed into the salivary mix through normal epithelial turnover and the activity of mastication. Additionally, white blood cells, most commonly the acute inflammatory polymorphonuclear leukocytes, arise from the crevicular fluid secondary to gingivitis. The microbial fauna that perpetually occupy the oropharyngeal spaces carry their own genomes and add to the total sum of DNA, but in the context of forensic analysis, the human-specific DNA primers do not recognize the bacterial DNA sequences.[46]

The ability to obtain forensic DNA data from salivary evidence was well described by 1992, but it was not until Sweet et al. published the double swabbing technique in 1997 that the standard of practice for salivary DNA collection from skin was truly established.[52,53] The double swab procedure takes into consideration the need to moisten and lift the cellular component of an adherent saliva stain as well as addresses the risk that the bitemark victim's skin will be a likely contributor to the DNA in the sample. The technique requires two sterile cotton swabs and 3 ml of sterile, distilled water.[53] A brief description of the procedure follows:

1. Thoroughly moisten the head of a cotton swab in sterile, distilled water.
2. Roll the head of this swab over the area of the saliva stain while using moderate pressure and a continuous circular motion.
3. Allow this first swab to air dry in a contamination-free environment for at least thirty minutes.
4. Within ten seconds of completing the first swab, roll the tip of the second, dry swab across the now moist area of the stain.
5. Use a circular motion and light pressure to absorb the moisture from the skin into the second swab.
6. Allow the second swab to air dry in a contamination-free environment for at least thirty minutes.
7. After drying, both swabs are packaged together, sealed, and marked with unique sample and case numbers.
8. The chain-of-custody document is completed and samples are submitted to the laboratory.

As an addendum to the sample collection from the victim's skin, a DNA elimination sample should also be collected from the victim to allow the laboratory to subtract that profile from any data generated from the skin swabs. This reference sample can be whole blood, a buccal swab (see below), or if the victim is deceased, a tissue sample taken at autopsy.[53] The double swab technique has a confirmed track record of success to include salivary DNA collections from a block of cheddar cheese and a body submerged in water for over five hours.[54,55]

7.6.2 Oral Mucosal DNA and Buccal Swabs

A successful conclusion based on DNA analysis is dependent on the opportunity to compare a questioned profile with a known profile. An investigator has little control regarding the condition and sufficiency of questioned profiles found on the victim or at the crime scene, but a level of quality control can be exercised in obtaining the known reference profile. Whole blood samples drawn through a finger stick or veinapuncture procedure are the traditional methods in obtaining a DNA reference. Well-meaning investigators occasionally have collected head hair believing that they were procuring a simple noninvasive source of DNA. They did not realize that the laboratory processing requirements for hair samples is resource-intensive, compared to the more predictable blood sample. The taking of head hair as a DNA reference should be limited only to those cases where the questioned samples consist of shed hairs and the use of mitochondrial DNA testing is likely.

For routine casework and long-term storage, whole blood applied to a filter paper card provides the greatest amount of reference DNA by volume with the least likelihood of contamination. However, the use of a buccal swab as a rapid, noninvasive alternative to blood collection is increasingly commonplace.[56,57] Because of the anatomical specificity of the forensic odontologist's expertise, he or she may be consulted in the collection, storage, and transportation of the buccal swab. Targets of this collection technique are the stratified squamous epithelial cells that can be rubbed from the inner cheek, although a certain amount of salivary DNA is naturally collected as well and is beneficial.

When buccal swabs are being collected to help identify a missing or unidentified family member, the laboratory will provide a list of the preferred donors according to their relationship with the victim. Exactly which family members are targeted as donors depends on the anticipated DNA typing system to be applied.[30,58] Generally, when autosomal STRs are being used, the preferred reference sources, ranked in order of desirability, include a pristine sample from the:

1. Victim (collected before the event and properly preserved)
2. Victim's biological parents
3. Victim's biological child and this child's other biological parent
4. Victim's biological child
5. Victim's full siblings
6. Victim's half siblings

Combinations of these references may also be extremely helpful, and the laboratory managing the case should be consulted. Recall, too, that earlier in the chapter, paternal lineage markers (Y-STRs) and maternal lineage markers (mtDNA), while not unique to the individual, may still provide inclusionary or exclusionary results. In combination with circumstantial and other forensic evidence, the results can be statistically compelling. The lineage markers also offer the potential for a reference source that spans many generations. For example, a great-grand-nephew may serve as a mtDNA reference for a victim that is missing decades earlier.

There are many commercial buccal swab collection kits on the market today.[59–62] The simplicity and painlessness of the procedure combined with the minimal training requirement for the person performing the collection has launched this method of DNA reference collection to the forefront of use. The other advantages that buccal swabs offer over blood-based collection techniques are the ease of self-collection and greater tolerance of this procedure in children and uncooperative donors. Although the commercial market offers a series of ingenious collection devices, suitable results can be obtained with sterile cotton-tipped applicators found in most medical supply stores.

Most buccal swab collection protocols are essentially the same. Start by documenting the identity of the donor or establish a unique sample number if the identity of the donor is unknown.

Wear gloves and avoid contaminating the swabs by contact with any surface or aerosol other than that of the donor:

1. Have the donor thoroughly rinse his or her mouth with water and expectorate. Repeat.
2. Wipe one side of the buccal mucosa with 2 × 2 sterile gauze.
3. Firmly stroke the dried area of the mucosa ten times with the swab, slowly rotating the cotton tip each time. Avoid the site near the opening of Stenson's duct if possible.
4. Repeat the process with a second swab on the contralateral cheek.
5. Allow both swabs to air dry in a contamination-free environment for at least thirty minutes.
6. Place in a prelabeled paper envelope or wrapper and seal with evidence tape.
7. Verify that the unique labeling and correct contents of the packet are documented; initial and date the seal for continuity purposes.
8. Complete the chain-of-custody form and ship to the laboratory as directed, maintaining a cool, dry, ultraviolet-light-free environment wherever possible.

Rinsing prior to DNA collection reduces the presence of food particles or debris, such as chewing tobacco, and even reduces the possibility of contamination by a nonnative DNA source.[63] Wiping the mucosa also eliminates adherent oral debris, such as materia alba, plaque, and microbial flora, at the site of collection. Importantly, for these same reasons, rinsing or wiping before taking an oral swab should *not* be done if the subject is a suspected rape victim and oral copulation may have occurred. In this particular situation, the goal is not to obtain a reference sample for a donor but rather gain biological evidence of the attacker.

The receiving laboratory will have its own protocol for processing buccal swabs. As the popularity of this collection technique grows, an increasing number of laboratories are adopting a high-throughput platform that accommodates the swab samples. In this way, hundreds or even thousands of samples containing high-quality DNA in a predictable concentration can be processed relatively quickly.

7.7 DNA Databases

Over the past two decades, forensic odontologists have witnessed a series of computer software programs that provide the ability to store highly detailed antemortem and postmortem records. The Computer-Assisted Postmortem Identification System (CAPMI) and Wind-ID are memorable examples, as is the 2004 version of the National Crime Information Center (NCIC) that accommodates more than just dental fields for missing and unidentified remains.[64-66] The International Criminal Police Organization (INTERPOL) contracts with PlassData in Denmark to administer DVI System International, which is a similar database application that stores dental and other identification data, including DNA profiles, for use in disaster victim identification responses worldwide. These programs include elaborate search algorithms that enable the investigator to scan hundreds or thousands of records quickly in search of a match between the questioned and known sets of records. Inevitably, following the generation of these best-possible matches, the associated records are retrieved in original hard copy or high-quality digital form and examined by a qualified forensic odontologist to determine if the threshold for a dental identification has been achieved.

Other forensic specialties have developed similar capabilities. The fingerprint community queries the Integrated Automated Fingerprint Identification System (IAFIS) located at the FBI's Criminal Justice Information System in West Virginia that contains over 55 million subjects.[67,68] Firearms examiners use the Integrated Ballistics Identification System (IBIS) maintained by the Bureau of Alcohol, Tobacco, Firearms, and Explosives at the National Integrated Ballistic Information Network.[67,69] Less familiar forensic databases belie their purposes with names like Ident-A-Drug, ChemFinder, National Automotive Paint File, and Forensic Information

System for Handwriting.[67] Each provides a set of references of class or individual characteristics that hopefully lead the investigator toward individualization of the evidence on his or her own case. Among all the law enforcement databases in the United States, none has had a more dramatic impact on American culture and crime-fighting success than the national DNA database, the Combined DNA Index System (CODIS).[70]

7.7.1 CODIS

CODIS has generated such intense media attention and concerns regarding genetic privacy that some may fail to understand how the program really works. The FBI laboratory began development of CODIS software in 1990 as a pilot project. Initially, the program included only fourteen state and local laboratories. However, when the DNA Identification Act of 1994 was passed, it affirmed the FBI's authority to establish an all-states laboratory network. Now, every CODIS-approved laboratory is required to maintain certain quality control standards in order to contribute to and query the National DNA Index System. The database uses two distinct indexes.

7.7.1.1 *Convicted Offender Index*

One data set is called the Convicted Offender Index that contains the known DNA profiles of individuals submitted by state and federal law enforcement officials. Each sample is obtained in accordance with that state's respective DNA collection statutes. The laws governing sample collection and whether an individual must be convicted of a violent crime or simply arrested before uploading the profile varies from state to state. Matches made between an evidentiary DNA profile and the Convicted Offender Index provide investigators with the real-time identity of a potential perpetrator.

7.7.1.2 *Forensic Index*

The second data set is the Forensic Index. It contains unnamed DNA profiles recovered as crime scene or sexual assault evidence. A match made within the Forensic Index may not lead immediately to the perpetrator's name, but it can link crime scenes together and detect serial offenders whose activities span several jurisdictions. Although local law enforcement agencies are expected to maintain their own freestanding data, the true value of the NDIS program derives from multiple laboratories uploading quality data from thousands of offenders in addition to crime scene evidence, so it can be searched by law enforcement agencies nationwide. In this way, police from all over the country can coordinate their independent investigations and share whatever leads they may have developed in an attempt to defeat criminal activity. Currently, over 170 public law enforcement laboratories at the federal, state, and local

DNA and DNA Evidence

levels perform electronic data comparisons and exchanges through NDIS by using CODIS software. CODIS software has also been loaned to over forty international law enforcement laboratories in over twenty-five countries for their own database programs.[70] The recent passage of the DNA Fingerprint Act of 2005 will expand the reach of CODIS submissions to include non-U.S. citizens detained or arrested by other federal agencies during border protection and homeland security activities.

The successes of the CODIS software and the NDIS program has logically led to an expansion of both in other areas too. In 2000, the FBI laboratory began to plan for another database to be called the National Missing Persons DNA Database (NMPDD) to aid in finding over one hundred thousand persons listed as missing by the Bureau of Justice Statistics and identifying the over forty thousand unnamed remains that are held in various jurisdictions across the United States. The NMPDD will comprise three indices, to include the STR profiles and mtDNA sequences of (1) persons known to be missing whenever a source of DNA is available, (2) close family members of the missing person, and (3) any unidentified human remains.[70,71] To facilitate potential matching of unidentified remains with their relatives' DNA profiles, CODIS software will be upgraded to perform kinship analyses or familial searching. This software will be similar to those previously used by the Armed Forces DNA Identification Laboratory (AFDIL) to identify American war dead and by the New York City Office of the Medical Examiner and AFDIL to identify victims from the terrorist attacks on September 11, 2001.

Glossary

Like most technical specialists in the forensic community, DNA scientists use terms that can make an already complex subject even more confusing. Most odontologists will be familiar with the fundamentals of biology and biochemistry because of their own educational background. However, the words and phrases used in this chapter and explained below will provide the odontologist with the additional vocabulary necessary to understand basic forensic DNA oral and written presentations.[9,10,72]

Allele: The variant in a DNA fragment size or in a DNA sequence at a particular locus is called an allele. The more alleles that naturally occur at a given locus in a given population increase the discriminating power of that locus. When paired (*homologous*) chromosomes each have the same allele at the same locus they are called *homozygous*. When the two alleles are different between the paired chromosomes, they are said to be *heterozygous*.

Amplicon: The amplified segment of DNA that is the product of PCR is called the amplicon. The two boundaries of this target DNA are usual marked by forward and reverse primers before the PCR process begins.

Amplification: The process of increasing the quantity of original DNA template by using the polymerase chain reaction (PCR) is referred to as DNA amplification.

Analysis: The third step in laboratory processing is called analysis. Once a DNA sample has undergone *extraction* and *amplification* and is loaded into a genetic analyzer, data are generated that represent either a genetic profile based on the size of the alleles at each locus tested (autoDNA, yDNA) or the actual sequence of the targeted DNA (mtDNA). The raw data are then reviewed by the analyst, who uses his education, training, and experience to confirm the result and, when appropriate, compare sets of data, draw conclusions, and calculate statistical values. Some laboratories call the instrumentation portion of this process *detection* and reserve the word *analysis* for the final data review step by the analyst only.

Bases: DNA is composed of *nucleotides* strung together in a twisted double strand. Nucleotides are ring-shaped molecules with various combinations of carbon, oxygen, hydrogen, and nitrogen with a phosphate group attached. The nitrogen-containing portion of the nucleotide is called a base. Differences in base design result in four variations of bases in DNA: adenine (A), guanine (G), thymine (T), and cytosine (C). The distinctive sequence in which these four bases occur within specific locations along the DNA strand provides the genetic code for protein production, as well as the power of individualization used by the forensic scientist.

CODIS and NDIS: The Combined DNA Index System (CODIS) is a software application that was developed and distributed by the Federal Bureau of Investigation. The software allows the storage and searching of large amounts of DNA data. Crime laboratories that qualify for CODIS access at the local, state, and national levels can search their own profiles and, more significantly, also search the DNA profiles posted by other laboratories through the shared data in the National DNA Index System (NDIS).

Complementary: In its natural configuration, DNA is double stranded. In the successful pairing of two single strands, the opposing sequences must be complementary. This means that adenine will only align opposite thymine, and guanine will align exclusively with cytosine. If the sequences within the opposing segments of DNA do not meet these exacting requirements, regardless of their length, the single strands are not complementary and will not pair together.

Short segments of complementary DNA are used as oligonucleotide primers to mark the starting point of target DNA during PCR.

Denature: During the laboratory process of amplification (PCR) and again during the analysis step, the double helical strand of DNA must be unwound into separate single strands. The separation of strands, or denaturation, can be accomplished by the addition of certain chemicals or by elevating the temperature to approximately 98°C. The latter approach occurs during PCR. Denaturation is sometimes referred to as *melting*. The opposite of denaturing is annealing, which describes two complementary strands binding together.

DNA: Deoxyribonucleic acid is one of the body's macromolecules and codes for all proteins. Forensic scientists currently focus on three major types of human DNA. All are identical at the molecular level, but each varies in its protein-coding responsibilities and its location within the cell. Y-DNA and X-DNA are the male and female sex chromosomes, respectively. Autosomal DNA (autoDNA) refers to any of the remaining twenty-two pairs of nonsex chromosomes. Sex and autosomal chromosomes are all located within the nucleus of the cell and contain the popular forensic targets called short tandem repeats (STRs) and single nucleotide polymorphisms. Mitochondrial DNA (mtDNA), on the other hand, is found outside the cell nucleus inside organelles called mitochondria, which reside in the cellular cytoplasm. Forensic scientists usually target the regions of the mtDNA genome with known hypervariable sequences or single nucleotide polymorphisms.

Electrophoresis: In the appropriate media, the negatively charged fragments of DNA migrate in an electrical field according to their molecular weight. Generally speaking, larger DNA fragments will move through the medium slower than smaller fragments of DNA. In this way, the analyst uses electrophoresis to separate DNA fragments of different sizes to visualize the results of PCR analysis. The most informative electrophoresis process takes place on a genetic analyzer.

Enzyme: Proteins that facilitate a biochemical reaction are called enzymes. In forensic DNA science, the most commonly used enzyme is DNA polymerase that, in the proper chemical environment, assists with the synthesis of new strands of DNA from an existing template during the polymerase chain reaction. Additionally, an enzyme called proteinase K is used to break down cellular components and expose DNA during the extraction process.

Extraction: The first of three laboratory processing steps in forensic DNA science is called DNA extraction. It involves isolating the DNA by preparing the gross sample, sometimes by grinding or macerating it as in the case of bones, teeth, or tissues. Then proteinases are used to disrupt the cellular barriers and allow the naked DNA to go into solution. Separation of the DNA from the cellular debris and solvents is achieved by alcohol precipitation, filtration, or use of silica or paramagnetic beads.

Gene: Any segment of DNA that is transcribed into ribonucleic acid and eventually into a functional or structural protein is called a gene. Most segments of DNA targeted during forensic analysis are believed to have little or no structural or functional role.

Genome: The full genetic makeup of an organism is called its genome. The human genome is composed of 3 billion base pairs and 20,000 to 25,000 genes.

Genotype versus phenotype: The sum of genetic information in an organism's genome is its genotype. The physical manifestation of that genetic information is called the organism's phenotype.

Locus: The location of a particular gene or segment of DNA is called the locus (plural: loci). A locus can be described by using the chromosome number, designating the short or long arm of the chromosome, and the band or subband on that arm.

Mitochondrial DNA: Mitochondria are organelles that exist outside the cell nucleus in the cytoplasm. Mitochondria possess their own genome that is separate and distinct from the twenty-three pairs of chromosomes inside the cell nucleus. The human mitochondrial genome is 16,569 base pairs in length, carries 13 coding regions, and is inherited along maternal lines. It contains known regions of diversity that help forensic analysts to distinguish individuals of different maternal lineages from one another.

Mutation: An alteration in DNA coding sequence that is usually erroneous and occurs during DNA replication. It may be caused by an environmental influence, viruses, or simply a copying error during cell division.

Nuclease: Enzymes that digest nucleic acids and nucleotides are called nucleases. These proteins occur naturally in many microbial contaminants, and although human-specific primers will not recognize and amplify the bacterial genome, the nucleases within the bacteria can degrade the human DNA sample.

Nucleic acid: A long chain of five-sided sugar rings, nitrogenous bases, and phosphate connectors is a nucleic acid. The structural unit of the molecule is called a nucleotide. Without the phosphate group, the molecule is called a nucleoside.

PCR: The polymerase chain reaction is a laboratory technique that targets a specific segment of original DNA template placed in the appropriate chemical environment along with primers, polymerase, and nucleotides. During a series of carefully orchestrated temperature changes, the template undergoes denaturization, annealing, and extension. At the end of each cycle the quantity of DNA template has been doubled.

Polymerase and *Taq*: Polymerase is the enzyme that adds nucleotides to the new strand of DNA during PCR. The step is call extension. Since the high temperature in the denaturing step of PCR deactivates most polymerases, the isolation of a thermostable polymerase from the thermophilic bacterium *Thermus aquaticus* (*Taq*) allowed multiple cycles of denaturing and annealing to occur without loss of enzyme activity.

Primer: A small segment of complementary DNA that establishes the starting point for DNA synthesis in PCR is called an oligonucleotide primer. It signals the point at which polymerase begins to add nucleotides.

Sequencing: Current forensic mtDNA sequencing is based on modifications of the Sanger chain termination method that allows the scientist to determine the order of the nucleotide bases—adenine (A), guanine (G), cytosine (C), thymine (T)—in a given segment of DNA.

Short tandem repeat (STR): These are defined as two or more nucleotides in a fixed sequence and repeated in a continuous series. For example, if the tetranucleotide (four-nucleotide) sequence [A-G-T-A] is repeated nine times on one chromosome at a specific locus and eleven times at the same locus on the homologous chromosome, the STR profile for that individual can be said to be 9,11 at that locus. Most forensic analytical procedures will target six to seventeen loci on autoDNA and Y-DNA for STR profiling.

Acknowledgments

The authors gratefully acknowledge the support of Dr. Kevin Torsky at the Joint POW-MIA Accounting Command Central Identification Laboratory (JPAC CIL) and LTC Louis N. Finelli of the Armed Forces DNA Identification Laboratory (AFDIL) for their assistance.

References

1. Pretty IA, Sweet D. 2001. A look at forensic dentistry. Part 1. The role of teeth in the determination of human identity. *Br Dent J* 190:359–66.
2. Avery WT et al. 1944. Studies on the chemical nature of the substance inducing transformation of pneumococcal types. 1. Induction of transformation by a desoxyribonucleic acid fraction isolated from pneumococcus type III. *J Exp Med* 79:137–58.
3. Farley MA, Harrington JJ, eds. 1991. *Forensic DNA technology*. Chelsea, MI: Lewis Publishers.

4. Watson JD, Crick FHC. 1953. Molecular structure of nucleic acid—A structure for deoxyribonucleic acid. *Nature* 171:737–38.
5. Lobban PE, Kaiser AD. 1973. Enzymatic end-to-end joining of DNA molecules. *J Mol Biol* 78:453–71.
6. Cohen SN et al. 1973. Construction of biologically functional bacterial plasmids *in vitro*. *Proc Nat Acad Sci USA* 70:3240–44.
7. Sanger FS et al. 1977. DNA sequencing with chain-terminating inhibitors. *Proc Nat Acad Sci USA* 74:5463–67.
8. Mullis KB, Faloona FA. 1987. Specific synthesis of DNA *in vitro* via a polymerase catalyzed chain reaction. *Methods Enzymol* 155:335–50.
9. Li R. 2008. Forensic biology, 135–51. Boca Raton, FL: CRC Press.
10. James SH, Nordby JJ, eds. 2005. *Forensic science: An introduction to scientific and investigative techniques*, 269–78. 2nd ed. Boca Raton, FL: CRC Press.
11. Rudin N, Inman K. 2002. *An introduction to forensic DNA analysis*. 2nd ed. Boca Raton, FL: CRC Press.
12. Sweet D et al. 1996. Increasing DNA extraction yield from saliva stains with a modified Chelex method. *Forensic Sci Int* 83:167.
13. Nagy M et al. 2000. Optimization and validation of a fully automated silica-coated magnetic beads purification technology in forensics. *Forensic Sci Int* 152:13.
14. Baker LE, McCormick WF, Matteson KJ. 2001. A silica-based mitochondrial DNA extraction method applied to forensic hair shafts and teeth. *J Forensic Sci* 46:126.
15. Buel E, Schwartz M, LaFountain MJ. 1998. Capillary electrophoresis STR analysis: Comparison to gel-based systems. *J Forensic Sci* 43:164.
16. Butler JM. 2005. *Forensic DNA typing: Biology, technology, and genetics of STR markers*. 2nd ed. Burlington, MA: Elsevier.
17. Edson SM et al. 2004. Naming the dead—Confronting the realities of rapid identification of degraded skeletal remains. *Forensic Sci Rev* 16:63.
18. Cormier K, Calandro L, Reeder D. 2005. Evolution of the quality assurance documents for DNA laboratories. *Forensic Magazine*, February/March.
19. Bieber FR. 2004. Science and technology of forensic DNA profiling: Current use and future directions. In *DNA and the criminal justice system: The technology of justice*, ed. D. Lazer, chap. 3, 23–62. Cambridge, MA: Massachusetts Institute of Technology.
20. Technical Working Group on DNA Analysis Methods. 1995. Guidelines for a quality assurance program for DNA analysis. *Crime Lab Dig* 22:21–43.
21. Federal Bureau of Investigation DNA Advisory Board. 1998. Quality Assurance Standards for forensic DNA testing laboratories. *Forensic Science Communications* 2000:2.
22. Federal Bureau of Investigation DNA Advisory Board. 1999. Quality Assurance Standards for convicted offender DNA databasing laboratories. *Forensic Science Communications* 2000:2.
23. National Criminal Justice Reference Service. http://www.ncjrs.gov/App/Publications/abstract.aspx?ID=240435.
24. American Society of Crime Laboratory Directors/Laboratory Accreditation Board (ASCLD/LAB). http://www.ascld-lab.org/.
25. National Forensic Science Training Center (NFSTC). http://www.nfstc.org/index.htm.

26. International Organization for Standardization (ISO). http://www.iso.org/iso/home.htm.
27. Forensic Quality Services (FQS). http://www.forquality.org/.
28. College of American Pathologists. http://www.cap.org/apps/cap.portal.
29. American Association of Blood Banks. http://www.aabb.org/Content.
30. Technical Working Group for Mass Fatality Forensic Identification. 2005. *Mass fatality incidents: A guide for human identification*. National Institute of Justice Special Report, NCJ 199758.
31. National Institute of Justice. 2006. *Lessons learned from 9/11: DNA identification in mass fatality incidents*. NCJ 214781.
32. Royal Canadian Mounted Police. 2003. *Disaster procedures: A guide for the forensic identification specialist*.
33. Jensen RA. 1999. *Mass fatality and casualty incidents: A field guide*. Boca Raton, FL: CRC Press.
34. Morgan OW et al. 2006. Mass fatality management following the South Asian tsunami disaster: Case studies in Thailand, Indonesia and Sri Lanka. *PLoS Med* 3:e195.
35. Alonso A et al. 2005. Challenges of DNA profiling in mass disaster investigations. *Croat Med J* 46:540–48.
36. Leclair B et al. 2007. Bioinformatics and human identification in mass fatality incidents: The World Trade Center disaster. *J Forensic Sci* 52:806–19.
37. Duffy JB, Waterfield JD, Skinner MF. 1991. Isolation of tooth pulp cells from sex chromatin studies in experimental dehydrated and cremated remains. *Forensic Sci Int* 49:127–41.
38. Sweet DJ, Sweet CHW. 1995. DNA analysis of dental pulp to link incinerated remains of homicide victim to crime scene. *J Forensic Sci* 40:310–14.
39. Schwartz TR, Schwartz EA, Mieszerski L, McNally L, Kobilinsky L. 1991. Characterization of deoxyribonucleic acid (DNA) obtained from teeth subjected to various environmental conditions. *J Forensic Sci* 36:979–90.
40. Smith BC, Sweet DJ, Holland MM, DiZinno, JA. 1995. DNA and forensic odontology. In *Manual of forensic odontology*, ed. CM Bowers, GL Bell, 283–98. Colorado Springs, CO: American Society of Forensic Odontology.
41. Smith BC, Fisher DI, Weedn VW, Warnock GR, Holland MM. 1993. A systematic approach to the sampling of dental DNA. *J Forensic Sci* 38:1194–209.
42. Shiroma CY et al. 2004. A minimally destructive technique for sampling dentin powder for mitochondrial DNA testing. *J Forensic Sci* 4:791–95.
43. Krzyżańska A, Dobosz T. 2007. Non-destructive method of DNA extraction from teeth and bones. Paper presented at the 5th ISABS Conference in Forensic Genetics and Molecular Anthropology, Split, Croatia, September 7.
44. Sweet D, Hildebrand D. 1998. Recovery of DNA from human teeth by cryogenic grinding. *J Forensic Sci* 43:1199–202.
45. Rice DH. 1982. *Surgery of the salivary glands*, 21–33. Trenton, NJ: BC Decker.
46. Sweet D. 2005. *Bitemarks as biological evidence in bitemark evidence*, ed. RBJ Dorion, 183–201. New York: Marcel Dekker.
47. Barni F, Berti A, Rapone C, Lago G. 2006. α-Amylase kinetic test in bodily single and mixed stains. *J Forensic Sci* 51:1389–96.

48. Quarino L, Dang Q, Hartmann J, Moynihan N. 2005. An ELISA method for the identification of salivary amylase. *J Forensic Sci* 50:873–76.
49. Myers JR, Adkins WK. 2008. Comparison of modern techniques for saliva screening. *J Forensic Sci* 53:862–7.
50. Vandenberg N, Van Oorschot RAH. 2006. The use of Polilight® in the detection of seminal fluid, saliva, and bloodstains and comparison with conventional chemical-based screening tests. *J Forensic Sci* 51:361–70.
51. Auvdel MJ. 1988. Comparison of laser and high intensity quartz arc tubes in the detection of body secretions. *J Forensic Sci* 33:929–45.
52. Walsh DJ, Corey AC, Cotton RW, Foreman L, Herrin GJ, Word CJ, Garner DD. 1992. Isolation of deoxyribonucleic acid (DNA) from saliva and forensic science samples containing saliva. *J Forensic Sci* 37:387–95.
53. Sweet DJ, Lorente JA, Lorente M, Valenzuela A, Villanueva E. 1997. An improved method to recover saliva from human skin: The double swab technique. *J Forensic Sci* 42:320–22.
54. Sweet DJ, Hildebrand DP. 1999. Saliva from cheese bite yields DNA profile of burglar. *Int J Legal Med* 112:201–3.
55. Sweet DJ, Shutler GG. 1999. Analysis of salivary DNA evidence from a bite mark on a submerged body. *J Forensic Sci* 44:1069–72.
56. Walker AH, Najarian D, White DL, Jaffe JM, Kanetsky PA, Rebbeck TR. 1999. Collection of genomic DNA by buccal swabs for polymerase chain reaction-based biomarker assays. *Environ Health Perspect* 107:517–22.
57. Kaidbey KH, Kurban AK. 1971. Mitotic behaviour of the buccal mucosal epithelium in psoriasis. *Br J Dermatol* 85:162–66.
58. National Institutes of Justice. 2006. Guidelines for reference collection kit components and oral swab collection instructions. In *Lessons learned from 9/11: DNA identification in mass fatality incidents*, Appendix E.
59. Lightening Powder Company. http://www.redwop.com/.
60. Bode Technology Group. http://www.bodetech.com/.
61. Whatman. http://www.whatman.com/.
62. Lynn Peavey Company. http://www.lynnpeavey.com/.
63. Martinez-Gonzalez LJ, Lorente JA, Martinez-Espin E, Alvarez JC, Lorente M, Villanueva E, Budowle B. 2007. Intentional mixed buccal cell reference sample in a paternity case. *J Forensic Sci* 52:397–99.
64. Lorton, L, Rethman S, Friedman R. 1988. The Computer-Assisted Postmortem Identification (CAPMI) system: A computer-based identification program. *J Forensic Sci* 33:977–84.
65. Win-ID. *http://www.winid.com/*.
66. National Crime Information Center. *http://www.fbi.gov/hq/cjisd/ncic.htm*.
67. Bowen R, Schneider J. 2007. Forensic databases: Paint, shoe prints, and beyond. *National Institute of Justice Journal 258:34–39*.
68. IAFS. www.fbi.gov/hq/cjisd/iafis.htm.
69. IBIS. www.atf.gov.
70. CODIS. http://www.fbi.gov/hq/lab/html/codis1.htm.
71. Ritter N. 2007. Missing persons and unidentified remains: The nation's silent mass disaster. *National Institute of Justice Journal* 256:2–7.
72. Farkas DH. 1999. *DNA simplified. II. The illustrated hitchhiker's guide to DNA*. Washington, DC: AACC Press.

Suggested Reading

Gaytmenn R, Sweet D. 2003. Quantification of forensic DNA from various regions of human teeth. *J Forensic Sci* 48:622–25.

Loreille OM et al. 2007. High efficiency DNA extraction from bone by total demineralization. *Forensic Sci Int Gen* 1:191–95.

Remualdo VR, Oliveira RN. 2007. Analysis of mitochondrial DNA from the teeth of a cadaver maintained in formaldehyde. *Am J Forensic Med Pathol* 28:145–46.

Tran-Hung L et al. 2007. A new method to extract dental pulp DNA: Application to universal detection of bacteria. *PLoS ONE* 2:e1062.

Woodward ST et al. 1994. Amplification of ancient nuclear DNA from teeth and soft tissues. *PCR Methods Appl* 3:244–47.

Forensic Anthropology

8

HARRELL GILL-KING

Contents

8.1	Introduction	137
8.2	Typical Case Progression	138
	8.2.1 Animal vs. Human and Minimum Number of Individuals	139
	8.2.2 Minimum Number of Individuals	140
	8.2.3 Medicolegal Significance of Human Remains	140
	8.2.4 The Biological Profile	141
	8.2.4.1 Sex	141
	8.2.4.2 Ancestry	142
	8.2.4.3 Skeletal and Dental Age	143
	8.2.4.4 Stature and Physique	146
	8.2.5 Individualization	147
	8.2.6 Postmortem Interval	148
	8.2.7 Trauma	149
	8.2.7.1 Antemortem Trauma and Pathology	150
	8.2.7.2 Perimortem Trauma	150
	8.2.7.3 Postmortem Trauma	154
8.3	Databases	155
8.4	The Future	157
References		158

8.1 Introduction

"Forensic anthropology is the application of the science of physical anthropology to the legal process."[1] Forensic anthropologists provide services to a large community, which includes a variety of law enforcement agencies, from local to federal or even international jurisdictions, medical examiners, coroners, and others charged with the responsibility for the investigation of death. In these endeavors forensic anthropologists cooperate with odontologists, pathologists, radiologists, and other forensic specialists who deal routinely with human remains. In the fourteen years since the eminent William R. Maples scribed the initial version of this chapter,[2] forensic anthropology has experienced a dramatic increase in visibility within the popular culture as a result of media depictions, some fanciful, others accurate and informative. The increasing contributions of forensic anthropology, from unidentified remains cases and homicide investigations to transportation and natural disasters to crimes against humanity, have been best described by its practitioners.[3–8]

As public awareness of forensic anthropology has increased, so has the number of board-certified diplomates, currently about seventy-seven, and the number of institutions offering advanced degrees in forensic anthropology, or physical anthropology with a forensic emphasis.[1] In the 1970s most practitioners of forensic anthropology held academic positions and offered only occasional assistance to investigative agencies. Once rare, forensic anthropology service laboratories affiliated with universities are no longer unusual. Some organizations employ a number of full-time forensic anthropologists (e.g., Joint POW/MIA Accounting Command/Central Identification Laboratory [JPAC/CILHI], National Transportation Safety Board [NTSB], and private disaster management corporations), and an increasing number of large medical examiner establishments employ full-time forensic anthropologists. Consequently, the presence of forensic anthropologists providing case reports, depositions, and expert testimony in civil and criminal courts and in tribunals around the world has increased dramatically in the past two decades.[9]

8.2 Typical Case Progression

Cases requiring the services of forensic anthropologists arise in a variety of ways. Excluding mass fatality scenarios, the appearance of unknown human remains may involve skeletal components and scavenged fragments scattered about the landscape, clandestine burials, submerged remains, or the occasional skull upon a mantel kept as a *memento mori* discovered incidentally during execution of a warrant for an unrelated cause. Anthropologists are increasingly summoned by arson investigators for *in situ* examination and recovery of fragile remains prior to transport.

When remains come to light, law enforcement may have a theory about the identity of the decedent, or perhaps about the manner in which the decedent came to an end. In such cases, someone may be missing from the community, and circumstances lead investigators to believe that the remains might be that individual. Additional information about the putative cause or manner of death may also have been developed. In such instances, experienced forensic anthropologists will follow something akin to the null hypothesis approach. As the examination progresses, the anthropologist attempts to defeat or disprove the *a priori* theories offered. In this way, the careful examiner avoids

Forensic Anthropology 139

any inclination to notice only the data that support the favored theory while ignoring observations that might not fit the official mindset.

When remains are presented to the anthropologist with no background information whatsoever, the task is to perform the most thorough examination possible with the materials available. In some instances an anthropologist may be asked to examine a skull, a set of postcranial remains, or some skeletal component when additional remains are actually available. Experience dictates that the better course of action is to insist upon examining all of the materials available. In this way, the most complete and consistent report may be rendered. This approach is particularly important when the remains may be reexamined by subsequent investigators. If additional case-related remains emerge during the course of investigation, these should be immediately made available to the original examiner.

8.2.1 Animal vs. Human and Minimum Number of Individuals

The first steps in examination of skeletal remains will usually be determining whether the specimen is animal or human, and if so, how many individuals are represented. Differentiating animal remains from human remains usually amounts to an examination of the epiphyses of long bones, or simply the recognition of a particular species as itself based upon the examiner's skill as a comparative osteologist. In practice, this task usually devolves to the usual suspects, i.e., bear, pig, turtle, primate, or some species of bird, for the latter are often confused with fetal bones by unskilled observers. When fragments reveal little or no distinguishing anatomy, the examiner may resort to histological/microscopic means or other distinguishing physical or chemical properties (Stewart 1979, 45-58). If it is necessary to go beyond simply stating that a specimen is nonhuman, comparative skeletal atlases, some region specific, are readily available for those willing to do the necessary taxonomic keying.[10-13] On occasion, the anthropologist is presented with a specimen such as a small amulet or other artifact that is allegedly made of human bone. The author recalls a scrub stone said to have been made from the "compressed sweepings from the ovens at Treblinka." On another occasion, the artifact was a small crucifix supposedly "carved from a human femur" in one of the Nazi death camps. In the first instance, x-ray fluorescence and mass spectroscopy revealed a combination of artist's plaster and charcoal. Under microscopic examination, the crucifix proved to be of walrus ivory, and probably produced from a die. Based upon conversations with several colleagues, there is apparently a significant prevalence of *pseudobone* objects driven by underground marketing of "holocaust" artifacts. Anthropologists in university settings will find an array of analytical equipment and techniques applicable to these problems no more distant than a phone call to a colleague in the chemistry or physics department.

8.2.2 Minimum Number of Individuals

Determining the *minimum number of individuals* represented in a collection of bones usually requires looking for *duplication of a component* (e.g., two left humeri, two right upper third molars, etc.) or *excessive asymmetry* between paired components that cannot be explained by pathology, e.g., developmental or traumatic stunting. Here it is important to recognize that attribution of some bones may be challenging, e.g., digits, sesamoids, etc., and that the human skeleton presents an impressive number of normal variations—more or fewer cervical and lumbar vertebrae, presence of cervical ribs, etc.[14,15] On more than one occasion a forensic anthropologist has received skeletal remains that may have become commingled on a shelf after many years of storage. In some cases, evidence custodians clearing out old specimens have unintentionally associated components from different individuals in a box or other container that is then presented to the anthropologist as the remains from a single case. The author was once cautioned by a forensic pathologist who worked in Hawaii that "an extra patella" in a skeletal submission might not be surprising since many traditional Hawaiians carried one as a good luck object![16] Statistical statements may be needed to support a conclusion about the minimum number of individuals. This is particularly likely when there is a reasonable probability that a set of recovered remains may be commingled.[17] Given a total number of bones that can be precisely identified as to their exact location in the skeleton, and the number of bones in that category actually found in the sample, the probability of commingling can be calculated by hand.[18]

8.2.3 Medicolegal Significance of Human Remains

Not all human skeletal material that comes to light is of forensic significance. Law enforcement personnel, road construction crews, and others have occasionally encountered buried human remains from archaeological contexts. These have ordinarily undergone sufficient taphonomic modification, e.g., loss of collagen, diagenesis, exfoliation, etc., that their antiquity is evident. Techniques for establishing the postmortem interval for long-dead remains, ranging from gross inspection to physical and chemical methods, have been described ably elsewhere.[19] Remains of historic or contemporary age may be accidentally unearthed when the locations of private cemeteries are unknown or have not been properly recorded, or when ground markers have been removed or have fallen into ruin. Such interments can usually be easily distinguished from coffin parts, embalming artifacts, etc.,[20] and are of no forensic significance, although such events may occasionally give rise to civil proceedings.[21]

8.2.4 The Biological Profile

Having certified that a known number of sets of human remains are at hand, the forensic anthropologist establishes a *biological profile*. This is a qualitative and biometric description of the remains that, ideally, includes, in order, a diagnosis of *sex, ancestry* (population membership), *skeletal and dental age*, and a description of *stature and physique*. The biological profile may be complete or partial, tentative or robust, depending upon the developmental status (i.e., child, adult, etc.), quality, and quantity of the remains and the skill of the investigator. As the biological profile is constructed, the anthropologist will typically enumerate any additional features that might be used as *unique identifiers*, e.g., old injuries, embedded projectiles, orthopedic appliances, congenital or developmental anomalies, genetically determined variations, etc. Unique identifiers associated with the dentition are best noted and referred to the forensic odontologist, who will perform the case-related charting and comparisons with antemortem records of possible matches that may become available. In the author's laboratory, standard dental charts and digital bitewing radiographs are made a permanent part of each case file. In this way, information can be transmitted electronically to odontologists around the world for rapid comparison with suspected matches for unidentified remains.

8.2.4.1 Sex

Typically, sex will be determined first. The most reliable diagnostic features are the innominate bones (*os coxae*) of an adolescent or adult. Depending upon the completeness of the specimen, sex may also be determined from the cranium, long bone dimensions, discrete features, general size criteria, and several discriminant function tests that compare bone dimensions to their means within databases populated by individuals of known sex. It is important to note that a significant number, approximately 5%, of individuals in most populations will be androgynous, i.e., will possess an equal number of male and female skeletal traits[63] (Angel 1985). Natural selection has exaggerated differences in those aspects of skeletal anatomy most closely related to reproduction. While male pelvic structure is selected to withstand compression, the female pelvis must not only tolerate the compressive loading of locomotion, but also provide the expansibility and protective architecture required by late gestation and the birthing process. Hence, female pelves display flared ilia, a large pelvic outlet, a wide subpubic angle (i.e., the arch formed by the two ischial bones), and sacrum that extends dorsally, increasing the x-sectional area of the birth canal.[22] Not infrequently, the skull might appear to be of one sex while the pelvic bones indicate the opposite conclusion. In this case, the pelvis is the more reliable predictor of sex. When the sex is judged to be female, the anthropologist will look for *evidence of parity*. Passage of the term infant through the canal stretches ligaments transecting the pelvic outlet, resulting in pitting on the dorsal surface(s) of the pubic bones, modification (*lipping*) of the sacroiliac joint, and deepening of the preauricular sulcus, producing a *triad of parity*.[23-26] Establishment of parturition gains added importance in the era of DNA analyses. Offspring will bear the maternal and paternal nuclear haplotypes as well as the mtDNA signature of the mother, the latter being of added significance when the bones are badly degraded. All determinations of sex should be accompanied by a statement of *statistical confidence* of the diagnosis based upon the technique(s) used. The determination of sex in skeletonized fetuses, neonates, and children prior to adrenarche is difficult at best. In these instances, evaluation of the *amelogenin locus* is the most reliable method.

8.2.4.2 Ancestry

The question here is "How would others have classified the decedent?" as to group, type, race, or some other folk typology during life. From the anthropologist's perspective, the task is assignment of the decedent to a *population* or *biotype* in the biological/genetic sense. In practical terms, this amounts to describing a set of phenotypic characteristics that falls within a folk taxonomy regardless of its biological reality. Complicating the task is the fact that investigative agencies operate within a different vernacular and simply want to know whether the decedent was Black, Hispanic, Asian, etc.—categories that lack any real biological meaning in the genetic sense, but which have found their way into official reporting formats around the world. In current practice, most anthropologists have abandoned the term *race* in favor of *biotype*, *population*, or *ancestry*, terms that denote as closely as possible the genetic relationship of an individual to a group that shares genes within itself. Close gene sharing (i.e., breeding by distance) produces *characteristic average features* that might place a living individual within a more broadly, if unscientifically, recognized group. The most difficult cases to assess are those involving admixture, e.g., Negro plus Mexican Indian, or Negroindio; Amerindian plus French/European plus Negro, or Creole; etc. As in the case of sex, population characteristics are shaped by natural selection. Most of the consistently observable skeletal differences between human populations, e.g., stature, limb proportions, facial characteristics, and the like, are the result of climatic adaptations to the environments in which these populations originally evolved. Thus, bodies with high surface area (e.g., long legs and arms, tall stature, long heads, etc.) should typify original equatorial populations, while those adapted for heat retention (think of Northeast Asians and other arctic indigenes) should have rounded bodies on short skeletal frames with attenuated extremities, rounded skulls, and flattened facial profiles. Numerous other *nonmetric variations* can also be associated with populations as allele frequencies for those traits increase as a result of gene sharing within a circumscribed geographical area. Physical anthropologists have described these diagnostic skeletal variations and their incidence within many populations, subgroups, and admixed groups elsewhere.[27–31] *Metric or statistical assignment* of ancestry involves comparisons of measurements from the unknown remains with various means and ranges for the same measurement within collections of known populations. Anthropologists sometimes employ discriminant *function tests* in which many measurements from an unknown are used collectively to assign a *biological distance* from the mean values of the same measures in a collection of control individuals of known ancestry. Such tests usually provide a statement about the likelihood of membership in the reference group. In many instances, various *limb proportion indices* may strengthen the statistical analysis, and these should be considered when the inventory allows. Finally, no assignment of group membership is complete without reference to *nonmetric traits* whose incidence within a particular population approaches diagnostic threshold (e.g., *os japonicum*, shoveled and winged incisors, tertiary third molar anatomy, rocker mandible, etc.). Just as the pelvis is the complex of choice in assigning sex, the skull provides the single most useful set of structures for attribution of ancestry. The dentition, when present, provides an additional rich source of variation to support or refine the assignment of population. Ultimately, the most difficult aspect of the ancestry issue is the translation of detailed and often complex anatomical and statistical findings into common "folk" or other vernacular typologies that usually do not reflect biological reality, or into overly narrow database categories that do not allow for findings of admixture or other useful information. Forensic anthropologists' reports should include the assignment of biotype, within the limitations of the data, along with any additional information about suspected admixture. Only when the *biological* population has been described should the findings be translated into more widely used, if less accurate, descriptive categories.

8.2.4.3 *Skeletal and Dental Age*

Tissues, organs, and systems mature at different rates. Some undergo renewal throughout life while others decline or disappear altogether under the varying effects of wear, disease, nutrition, and trauma. Attempts to determine the chronological age of a decedent at the time of death by any combination of methods involving the hard tissues will result, at best, in an estimated range. Under ideal circumstances, sufficient materials would be present to allow both osteological and dental approaches to age determination. Dental techniques, reported elsewhere in this volume, should correlate well with skeletal assessments of age in individuals up to around fifteen or sixteen years of age, and should provide reasonably comparable ranges up to about twenty. Because full maturation of the skeleton requires half again as long as the dentition, the former becomes increasingly more reliable as a basis for estimating age. The dentition is the only part of the skeleton that articulates directly with the outside environment. Therefore, the variable effects of diet, disease, traumatic insult, and accessory use are more apt to reduce the value of teeth in determining age in individuals beyond the mid-third decade, and in groups with chronically poor oral hygiene (who tend to appear older than their actual chronological age). The *sex and population membership* of a decedent must be determined before applying any aging technique because these parameters significantly influence rates of development, necessitating recalibration of the result. The details of osteological aging techniques are beyond the scope of this chapter and should be left to experienced practitioners. A general approach to determination of age follows:

Fetal period: Estimation of fetal developmental age assumes forensic importance in most jurisdictions because it is usually an indicator of viability. In instances of criminal death of a pregnant individual courts may decide whether to prosecute more than one homicide depending upon the age (i.e., viability) of the fetus. Knowing the age of a discovered fetus may also assist in matters of identification. Usually, *diaphyseal lengths* may be used in various algorithms to estimate crown–rump length, which may then be translated into lunar age. The timing of appearance of primary and some secondary ossification centers is also of use. Several sources give good accounts of the statistical reliability of various bones and measurements for both gross and radiographic fetal age determination.[32,33]

Birth to sixteen years: Dental eruption timing and sequence for deciduous and adult dentition are reported elsewhere in this text. As noted, dental and osteological age should correlate well within this developmental interval. In recent years anthropologists and odontologists have become increasingly aware of differences in rates of skeletal and dental maturation among various populations,[34] and have begun to apply adjustments to their age estimates accordingly.[35] Also see the dental age estimation link at the University of Texas Health Science Center at San Antonio's Center for Education and Research in Forensics.[36]

In this interval, anthropologists will make use of diaphyseal lengths, appearance and attachment (fusion) of secondary ossification centers, and the obliteration of some synchondroses.[37] Radiological age standards have proved useful, especially for the hand and wrist, from early childhood to late adolescence; however, these skeletal components are among the first to be removed by scavengers and are often unavailable.

Sixteen to thirty years: As attachment of primary and secondary ossification centers occurs throughout the skeleton, attention turns to the completion of fusion of these centers. Numerous investigators have established rating scales that describe the degree to which growth cartilages (metaphyses) have been replaced by bone, signaling the completion of that skeletal element or an articulation.[19,38] Likewise, staging techniques have been developed for the ossification of various symphyses, synchondroses, and sutures (e.g., sternal rib cartilages, pubic symphysis, cranial sutures, basilar synchondrosis, etc.). These techniques usually take the form of a semantic differential, which describes changes in the appearance characteristics of a particular structure at known points in time based upon controls.[39] It is worth noting that the value of a technique is strictly limited by the population sample upon which it is based. Thus, aging techniques must be recalibrated as secular trends (e.g., improved public health measures, available nutrition, antibiotics, etc.) modify maturational rates and longevity.

Thirty years and beyond: As the last epiphyses (usually the medial clavicles) complete development, skeletal age may still be estimated, albeit with increasing error. Unlike the dentition, the skeleton undergoes remodeling throughout life. As a result of endocrine-driven cellular interactions that constantly remove bone and replace it, the skeleton continues to "turn over" approximately every seven to ten years, remodeling itself to accommodate gravity and the mechanical habits of its owner. Alongside this process, the skeletal cartilages that separate and cushion bones undergo increased hardening with resulting grossly observable wear at the articulations, i.e., generalized *osteoarthritic changes*. Age-related changes in the weight-bearing joints (ankle, knee, hip, sacroiliac, spine, etc.), as well as mineralization of nonarticular connective tissues (e.g., thyroid, cricoid, arytenoid cartilages, etc.), have been documented along with appropriate caveats having to do with differential effects of lifestyle, disease, and diet.[40] Other time-dependent structural changes may be appreciated *radiographically*. For example, as the marrow of long bones assumes a larger part of the hemopoetic burden with age, one can observe an advance of the apex of the marrow cavities in femora and humeri. This is seen as increasing radiolucency toward the proximal epiphyses. This progression has been documented and timed.[41] Radiographs may also complement and extend developmental staging techniques even after epiphyseal completion by revealing remnants of growth plates that have not yet reached uniform density.[42] Finally, as skeletal remodeling continues over the course of a lifetime, *histological changes* in cortical bone may be correlated with age. As old cortical bone is scavenged by osteoclasts to maintain mineral homeostasis, new vascular pathways import osteoblasts that replace it. As the skeleton moves through time, the amount of unremodeled lamellar bone seen in microscopic cross sections of cortex will diminish, and the number of partly replaced structural units of old bone, *osteon fragments*, will increase. These changes have been documented and calibrated by various authors for a number of sites in the skeleton,[43–45] and are of use in the aging skeleton because the process of turnover on which it is based extends throughout life. As is the case with other techniques, error in the calculated age range by histomorphological methods increases with time.

8.2.4.4 Stature and Physique
The calculation of stature from the skeleton involves determining the length(s) of long bones, which are then used individually or in combinations in regression equations to determine living stature.[46,47] The anthropologist will be able to select the correct stature algorithm(s) only after determining sex and ancestry since the long bones have proportionally different relationships to overall stature in different sexes and populations. Since the estimated stature derives from long bone lengths that do not change significantly after maturity, this approach gives a range that does not take into consideration loss of stature from compression of the spinal fibrocartilages. A correction is usually applied for individuals whose age is estimated to be over thirty years.[48] Statures for children whose long bones have not completed longitudinal development are based upon the diaphyseal length. When long bones are incomplete because of trauma or taphonomic effects, it is sometimes possible to estimate the vital length of some bones by proportionality techniques.[49] The estimated lengths are then used in calculation of stature with the caveat that this approach introduces additional error in final calculated stature, thus widening the range estimate. In general, the best estimates of stature are based upon multiple bones, which are used in algorithms derived from population data reflecting current secular trends.

Anthropologists are sometimes asked to render estimates of the living weight of a decedent who has been reduced to bones. Putting aside ancillary information such as belts, shoes, and other clothing that may accompany remains, the answer will require strong qualification. Since an individual may lose as much as 50% of his or her body mass over a relatively short time (e.g., cachexia, starvation, etc.), such estimates are always questionable, and particularly so when the decedent appears to have been an indigent. Accordingly, statements about frame size, proportionality, and the distribution of muscularity are preferable. Cross-sectional thickness or simply the weight of various bones in combination with proportionality ratios can provide information about how much soft tissue weight an individual might be expected to carry. This picture may be refined by a careful examination of entheses, the points of tendon insertion, which are modified by muscular activity over a period of time. Thus, one may arrive at an estimate of how well developed an individual may have been for a particular frame size and stature at *some point in life*. Examination of the *pattern* in which the skeleton has reinforced itself in response to habitual or repetitive biomechanical action has sometimes proved useful in the inclusion or exclusion of certain occupations, sports, or other activities performed over a period of several years, which may alter a list of suspected matches.[50]

8.2.5 Individualization

Establishing the intersection set between sex, age, ancestry, and stature into which the decedent falls will eliminate a substantial number of suspected matches and false leads, and may help in redirecting the investigation into the identity of an unknown. An estimate of the postmortem interval (see below) will reduce the list further. When these data, combined with a list of *unique identifiers*, are compared to a database of missing persons, the list of possible matches usually reduces to a manageable few. At the discovery of unknown human remains, the authorities will either have a theory about the identity of the decedent or not. If there is a suspected match, all pertinent antemortem data will be assembled. This will include dental charts, bitewings and panographic images if available, old x-rays, or other medical images (e.g., CT scans, MRI, etc.). When images are unavailable, medical records describing prostheses, pacemakers, shunt devices, cosmetic implants, orthopedic devices, and the like may be sufficiently detailed for comparison to the postmortem evidence. Antemortem records of diseases that would be expected to leave evidence in the hard tissues are also useful, particularly when the incidence of a disorder is known. Detailed descriptions of conditions (e.g., fractures, lesions, etc.), procedures, and appliances (including serial numbers) are useful. These antemortem data should confirm or exclude a potential match. In some cases of suspected identity, when none of the foregoing is available, it is sometimes useful to perform a skull-to-photograph superimposition. Although traditionally used to *exclude* matches, some have successfully employed video superimposition to achieve positive identifications when a complete skull and good quality photographs from several angles are available.[4] If open-mouth photographs are of sufficient resolution and detail, a direct comparison between the antemortem and postmortem anterior dentition may be possible. This approach rises to the standard of positive identification when combinations of features such as treatments (e.g., crowns, cosmetic modifications, extractions, etc.) and anatomy (e.g., diastemas, rotations, embrasures, etc.) provide multiple points of comparison. This technique is best deployed jointly by the anthropologist and the odontologist.

The identification process follows a Bayesian statistical model. The likelihood of an individual being a particular sex, age, ancestry, and stature is roughly the product of the individual probabilities of being any one of those things. When individual identifiers are available, those with a known incidence can be entered into the calculation, reducing the set of possible matches toward unity. Identifiers that can be *traced directly* to a decedent provide the basis for a *positive identification*, e.g., an intramedullary rod or a pacemaker with a serial number that matches a surgeon's record for a particular patient. In such instances it is imperative that *direct association* between the decedent and the device can be established. For example, an orthodontic or orthopedic device affixed to the remains is preferable to one that has become detached. Experience in mass death incidents involving scattered and commingled remains bear this out. In some instances an implant, orthopedic device, or prosthesis may be found in a decedent without a suspected match. If the medical artifact bears a serial number and can be attributed to a particular manufacturer, it is sometimes possible to trace the device to a particular treatment facility, and thence through surgical records to a recipient. Whatever the means of identification, in the *post-Daubert* era, all conclusions and the techniques from which they are derived will require robust statistical support. As an example, though used as a basis for positive identification for years, comparison of ante- and postmortem frontal sinus x-rays has only recently been validated.[51] When unknown remains come to light for which there is no suspected match, the only remaining course of action is submission of the osteological and odontological findings to the appropriate database (see below).

8.2.6 Postmortem Interval

Forensic anthropologists are sometimes asked to estimate the *postmortem interval* (time since death) for a set of remains. The main reasons for estimating the postmortem interval are (1) the inclusion or exclusion of suspects, (2) reduction of the number of possible matches in a database, and (3) determination of the forensic significance of a set of remains, i.e., is the decedent an archaeological or historical specimen, or a contemporary case that requires investigation? On most occasions, when an anthropologist is asked to determine the postmortem interval, the decedent will have been dead for weeks to years. Ideally, the remains will be pristine, and it is for this reason that many examiners prefer to attend the recovery, whether it may be an exhumation, collection of scattered bones, or even submerged remains. The anthropologist may supervise and document the process, collecting relevant samples, e.g., insects, botanical specimens, clothing, coins, or other "time givers," which will be transmitted to appropriate specialists for further analysis. But, most importantly, he or she will want to assess the remains *in context* before any processing occurs. General observations will include *corporal* (from the body) as well as *environmental* information: What is the quantity and quality of the remains? Do the remains express any odor? To what extent are the remains scavenged and weathered? What are the characteristics of the local weather, terrain, water sources, and fauna, all of which will influence the rate of decomposition or disassembly of remains? In addition to these two major sources of information, there are two general approaches to timing a death: *rate methods* and *concurrence methods*. The degree to which bone has lost mineral and organic content, the change in sound or electrical conduction properties of bone, changes in specific gravity, and the amount of total lipid lost are examples of features that change with documentable rates. The details of these and other rate techniques are beyond the scope of this discussion, but may be found elsewhere. Concurrence estimates of the postmortem interval depend upon establishing an association between the remains and an object or event for which time can be fixed. An individual will not have died before the most recently minted coins in his pockets; there may be a scattering of leaves upon the body from nearby trees, which places its death before leaf fall, a natural event whose timing will be known to local botanists. The state or type of clothing may reveal season of death as well as time of day or night, etc. When an elderly decomposing, mummified, or even skeletonized individual is discovered indoors, one often need look no further than the oldest letter in the mailbox. Good summaries of concurrence methods are available.[52]

Whatever the approach taken, the time interval estimates must become broader as the actual postmortem interval lengthens. Because the estimate may be used to establish or exclude possible matches, or entered into a database along with other information, it is better to err on the side of more inclusive estimates than to exclude a true match through overconfidence.[19,53] Above all, it is important to avoid a mindset about what to expect. In 1999, while relocating some prison burials from ca. 1900, the author encountered an individual with nearly complete integument, copious adipocere, and a substantial amount of pink acellular skeletal muscle. Most of the other decedents were, as expected, represented by little more than dental fragments and coffin splinters. Just short of proclaiming the burial a much more recent one, he was reminded of an almost identical experience described by William Bass, who in the 1970s encountered similar findings in a Civil War era burial.[54] Important timing information will almost always be lost in the process of recovery, transport, processing, and storage of remains. When the anthropologist is asked to examine remains at the end of this process with little or no reliable information about context or procedure, it is prudent to refrain from any except the most general estimate of postmortem interval.

8.2.7 Trauma

An important part of the anthropological analysis will be an assessment of traumatic injuries and other diseases or disorders sustained by the decedent. While recognizing that there are far more causes of death that will *not* be reported by the hard tissues, those that do affect the skeleton or dentition represent the most enduring kind of evidence. Hard tissue injuries are designated as *antemortem*, *perimortem*, or *postmortem* according to time of occurrence.

8.2.7.1 Antemortem Trauma and Pathology

Antemortem injuries and diseases will often be of use as identifiers (see above). The classic examples are oral or orthopedic pathologies and their respective treatments, prostheses, etc. Certain chronic disorders, e.g., rickets, DISH (diffuse idiopathic skeletal hyperostosis; Belanger 2001, 258–267),[64] advanced rheumatoid disease, etc., which exhibit skeletal facies, are also valuable in individualization when these have been noted in the medical history of a suspected match. Some chronic antemortem conditions may extend to the end of life, and on a few occasions, may even contribute to death. Obviously, such findings assume added importance when a clear cause of death cannot be shown. A skeleton with a pacemaker beneath the disarticulated bones of the thorax was recently encountered by the author. Subsequent tracking of the serial number identified an elderly decedent with a long history of cardiovascular disease. Though not as diagnostic as an atheromatous set of coronary arteries in the hands of a pathologist the day after death, the finding suggests, at least, a contributing cause. Although one occasionally finds old projectiles (bullets, shotgun pellets, etc.) embedded in the skeleton, most traumatic antemortem injuries will have been due to blunt force since one is more apt to survive these than blade or firearms assaults.

8.2.7.2 Perimortem Trauma

Perimortem injuries are those that occur at or near the time of death and are most likely to be associated with the true cause of death. Accordingly, such injuries become a critical focus. The most frequently encountered fatal perimortem defects are induced by gunshot, blade, or a blunt object forcibly applied. Each of these produces more or less characteristic defects. As a two-phase material (calcium hydroxyapatite and collagen), bone withstands compression and stretch. Under slow loading of force, the struck surface compresses while the opposite side stretches. Because bone is weaker under tensile forces, the stretched side fails first, often producing *concentric cracking* (as in the flat bones of the skull) or *concoidal* (wedge-shaped) fracture lines emanating from the point of failure. Under rapid loading (as in a bullet strike), the bone responds as a brittle material. In the latter instance one may see *radiating* cracks across the bone surface, or none at all.[55]

Gunshot injury: The rules for interpretation of gunshot wounds (GSWs) in hard tissue differ somewhat from those in fleshed cadavers. In most instances, given an adequate sample of remains, one should be able to determine (1) entry and exit sites, (2) the approximate angle of entry of a projectile, (3) the order of entry defects if in the same surface, and (4) an approximation of caliber, or at least the elimination of certain calibers. Because the soft tissue has disappeared, and because garments may not be available for inspection, determining range of fire is often not possible. Except when a projectile has struck an intermediate target, the entry defect should provide, at least in one dimension, the approximate diameter (caliber) of the round. Variations in the shape of an entry from circular to elliptical report the approximate angle of entry. The exit is generally distinguished by the presence of an outward bevel. Usually, the exit defect will be irregular and somewhat larger than the entry because of deformation of the round during its transit through the target. Both entry and exit bevels will have edges that slope approximately 45° from the incident angle. This feature owes to the manner in which fracture lines propagate through the hydroxyapatite crystal. Notable exceptions to this rule include the *keyhole defect* produced by a low-angle strike tangent to the skull. In this case, a furrow resembling a keyhole is produced. Although the round may not enter the skull, a bevel is produced on both the outer and inner surfaces.[56] Double beveling may also be observed when the projectile strikes an intermediate target such as a glass pane, screen door, etc., causing the round to lose stability, thus imparting its energy into the target in an unpredictable manner. Detailed descriptions of the interaction of projectiles and bone may be found in several sources[65] (DiMaio 2003, 175–83). If garments accompany the remains, they should be examined for defects overlaying any ballistic injuries for possible indications of range of fire, such as soot or scorching. Ballistic metal usually transfers some of its substance to the bone through which it passes. Rounds entering the body and skull are often fragmented as they strike bone tissue. For this reason, remains believed to contain ballistic materials should be radiographed before an examination begins. This is especially important when GSW defects are present in the skull. Some small-caliber rounds often fail to exit the skull. When this is the case, following x-ray, the skull should be opened and examined to determine the path of the round and to retrieve it for ballistic examination.

Blade injury: When death appears to be the result of sharp force injury, a close examination of all bone surfaces is imperative. Imagine the torso from chin to the pubic bones (the vital area), then picture the subtending bones (vertebrae, sternum, ribs, clavicles, scapulae) painted upon this surface. When the torso is morphed into a round target, and the underlying bones into a bull's-eye, the latter comprises about 65 to 75% of the target. The forensic implication is clear. In theory, in a fatal blade injury one would expect bone to be marked in the majority of such cases. The extremities, especially the hands and forearms, and occasionally the legs, may bear defensive injuries as well. Therefore, all bones should be examined, to the extent possible, before cleaning and again afterward. Care must be taken when macerating remains not to remove or damage the periosteum. Blade marks are often seen in this fibrous tissue sheath overlaying undamaged bone cortex. These injuries are best viewed in oblique light under low-power magnification, or under oblique fluorescence. Defects may be captured for comparison before further cleaning by making a cast with polysiloxane or similar material. Unlike ballistic metal, blades rarely transfer any of their substance to bone. However, when a sharp edge is applied to a bone obliquely, microscopic examination will usually reveal a "bar code" effect, resulting from defects in the cutting edge as the blade is applied. The same is true when a serrated blade strikes bone at a low angle. These markings may be matched to a suspect blade using a comparison microscope. As is the case with soft tissue blade injuries, it is unlikely that the defects will yield information about the dimensions of the offending blade. When the tip of a blade is forced into a bone, it is sometimes possible to determine whether it is backed or sharp on two edges (as in a dagger). Cases involving postmortem dismemberment will usually present several kinds of blade injuries, from knives, manual and power saws, cleavers, and even axes.[57] These must be distinguished from perimortem blade assault. Careful microscopic examination will usually differentiate knives from saw blades, and manual saws from circular and reciprocating saws. Further differentiation between various types of saw blades may also be made microscopically.[58] On a few occasions, the author has observed *vital reaction* in the periosteum adjacent to the site of dismemberment of an extremity. This finding, far more common in soft tissue, indicates active circulation, though hopefully not consciousness, during the removal of the limb. When multiple blade injuries are present, some investigators will create a *cut map*, i.e., a three-dimensional representation of the skeleton and all of the injuries. This reporting format is useful in the context of possible witness accounts, and may help in differentiating fatal, nonfatal, and defensive injuries. Finally, in the case of remains outdoors, it is important to distinguish between blade injuries and *pseudotrauma* caused by animals with scissoid mouth parts, e.g., turtles and carrion birds, gnawing by scavengers with carnacil dentition, and trampling by animals.[59] Animal chewing will usually produce markings on opposite sides of a bone as the jaws occlude, whereas true blade marks will appear on only one surface. As in the case of all penetrating injuries, blade or ballistic, it is important to examine any garments associated with the remains for defects that may correspond to the injuries.

Blunt force injury: Blunt force injuries are the most common form of mechanical trauma. Caused by relatively slow loading rates, they allow bone to deform before failure, producing characteristic damage patterns. Because the energy (E) transferred to the bone is half the product of the mass (m) of the object striking it and the square of the velocity with which it is delivered (v^2), *velocity* will make the greatest contribution to the damage observed. Keeping this relationship in mind, one can reduce the number of possible scenarios leading to a particular injury. A second important consideration is the *area through which the energy is delivered*. A blow of 75 ft.lbs.s^{-1} delivered by the flat side of a 12 × 2 inch plank will do less damage than the same amount of energy delivered by the 7 square inch face of a sledge hammer. The same is true of comparable energies delivered to *curved vs. flat surfaces*, e.g., when the hypothetical plank is slammed against the body wall vs. the curve of the parietal bone. Likewise, equal force applied to a healthy vs. diseased bone (tumor, osteoporosis, etc.) will often result in different degrees of damage. When examining unidentified skeletal remains, it is important to remember that some mechanical injuries may be *incidental*. These will usually appear as perimortem injuries, although they do not contribute to death. For example, severe skull or cervical fractures may have been caused by falling down a staircase after a fatal coronary, or from a utility pole after a lethal electrical shock. The author once had the opportunity to examine skeletal material recovered from a collapsed area of a long abandoned historic mine. Though many of the bones were broken, it was impossible to know whether the two victims had expired from asphyxia, dehydration, or the crushing effects of the collapsing shaft.

It is important to consider that force applied to one part of the skeleton may be transferred, causing damage elsewhere. Shock from a hard landing may be transferred through the legs, damaging the bones of the pelvis or spine, and vertical loading of the spine from below has sometimes resulted in ring fractures of the skull base. A blow to the left gonial angle may cause a hinge fracture of the right mandibular ramus when the head is arrested against an unyielding surface, and the same principle applies in the classic *contrecoup* skull fracture. Reconstruction of a shattered skull, though time-consuming, may provide information about the number and order of strikes, or reveal a pattern that suggests the nature or class of weapon used. In a recent case, the decedent's skull was crushed by the right rear wheel of the vehicle from which she "fell." Arrested later, other occupants of the car alleged that she "opened the door in an intoxicated state and fell beneath the wheel." An alert pathologist observed that there was too little blood at the site where the tire rolled over the skull, and called for an anthropological examination. Reassembly of some eighty-five fragments revealed three suspicious patterned injuries that later proved to have been caused when the victim was struck repeatedly with a socket wrench. She had exsanguinated elsewhere before being dumped on the road. It is essential that all fragments be examined carefully for transferred evidence or for a more detailed toolmark analysis. Experienced examiners will recall instances of wood splinters, glass fragments, bits of paint, etc., embedded in bone later to be associated with a bat or a broken bottle with complementary bone chips, hair, or dried blood. Occasionally, one encounters remains that bear blunt injury defects that appear to have been made by more than one kind of object. Such findings may indeed represent the work of more than one assailant, but most often will have been caused by repeated application of the same object at different striking angles, the classic example being crescent and round depressions on a skull from application of the edge and flat face, respectively, of the same hammer. *Compound implements* are of particular interest in this regard. The author once examined remains bearing several crescent depressions and one elongated full-thickness fracture on the skull, and a small rectangular punch-out on the sternum. The implement, later associated with the assault, was a tire tool. The lug wrench end, applied at an angle, produced the crescent fractures while the handle had created the elongated depressed parietal break as well as a defensive fracture of the ulna. The nib, rectangular in cross section, was a perfect fit for the defect in the sternum. Other examples are provided by roofing hammers, ball peen hammers, single-bladed hatchets, etc., all of which produce at least two kinds of patterned injuries, depending on which side is applied to the target. Where blunt force injuries are concerned, a three-dimensional imagination and an occasional stroll through the local hardware store are the examiner's best analytical tools.[60]

8.2.7.3 *Postmortem Trauma*

Postmortem trauma is an important category of damage in skeletal remains that must be distinguished from insults occurring near the time of death. Although, strictly speaking, the "fall following a coronary" cited above qualifies as postmortem trauma, the phrase is most often used to describe modifications of remains that occur some time after death. Forensic anthropologists will recognize several categories of effects stemming from natural and anthropogenic causes. (1) In cases where remains are exposed, skeletal components may be damaged by *movement* due to natural forces. As bones disarticulate, they may be scattered by water or wind, depending on the slope of the terrain and the amount of water running across it. Fluvial transport often results in damage to ribs and the delicate structures of the skull base, depending on water velocity and distance traveled. As bones dry, some of the flat elements of the skeleton may warp and crack, producing damage that might be confused with injury. Similarly, buried remains subject to many cycles of wetting and drying may display breakage of ribs, spinous processes, and other effects. The weight of soil above a collapsed coffin may produce damage to the rib cage or pseudotrauma in the anterior dentition or delicate bones of the maxillofacial area. (2) The most common source of scattering and postmortem damage in exposed remains is *animal activity*. Large and small mammalian scavengers leave characteristic dental markings, usually perpendicular to the long axis of a bone. In some cases, bones will be crushed by powerful jaws, e.g., bears, alligators, feral hogs. When recovering scattered remains, it is wise to ask what kinds of animals inhabit the area. Some familiarity with the dentition and the characteristic patterns of scavenging of animals within the area of search is useful.[61,62] It is important to remember that most *scavengers will be attracted to wound sites*, and that evidence of injury in bone, fatal or otherwise, may thus be altered, obscured, or eliminated altogether. (3) *Anthropogenic damage* is often the result of "discovery by bulldozer backhoe." Operators of heavy equipment in rural areas often mangle remains in the act of discovering them, causing additional difficulty in distinguishing actual perimortem injury. One colleague wryly noted that "if one wants to find remains in a large field, one has only to instruct someone to 'brush hog' or till the area." Colleagues in coastal regions often describe postmortem "propeller" damage inflicted on floating remains. The most problematic instances of anthropogenic damage are those that produce *recovery and processing artifacts*. A ground probe may produce what appears to be a bullet hole in shallow burial. Shovels and trowels in the hands of inexperienced investigators may induce what appear to be blade or chopping defects. Cases involving remains that have been intentionally disarticulated by knife, saw, etc., are often seen by a pathologist before the anthropologist is consulted. On these occasions the initial examiner must carefully note and describe *any additional cuts* that have been made with the autopsy saw for sampling or other purposes, lest these be confused with original marks made by the assailant. Although most anthropogenic artifacts are easily distinguished from perimortem damage, they often provide a skillful cross-examiner with opportunities to confuse a jury, and at the very least, may call into question the skills of those responsible for the recovery and analysis of the victim.

8.3 Databases

When the anthropological and dental analyses do not produce an identification, the biological profile, unique identifiers, and postmortem interval data are usually entered into a database. The most widely known U.S. database is the National Crime Information Center (NCIC), which includes descriptive information for analyzed unknown remains as well as missing individuals. Many states operate databases and missing persons clearinghouses for their own jurisdictions. Still other databases specialize in a particular demographic segment of the national population, e.g., the National Center for Missing and Exploited Children (NCMEC).

In recent time, databases utilizing DNA technologies and powerful search engines have been created for a number of populations. These include the Armed Forces DNA Identification Laboratory (AFDIL) and the Combined DNA Index System (CODIS). The latter contains reference samples consisting of nuclear and mitochondrial markers from relatives of missing persons as well as mitochondrial and, usually, genomic markers from unidentified human remains. The CODIS system utilizes biological profile and dental information submitted with skeletal samples from unknown remains as *metadata*. When the system detects a possible hit, i.e., a match between a reference sample and a set of remains, it may be weak (i.e., achieve a less than desirable level of statistical certainty), owing to the badly degraded condition of the remains. Or, the system may find *several* possible matches either for the same reason or because the original reference samples were taken from individuals who were not first-degree relatives of the decedent. Given several possible matches, or one weak one, the anthropological and dental profiles are used to parse the list or to strengthen the weak match.

Presently, a number of problems reduce the effectiveness of databases. Most do not interact with others because of incompatible formats, proprietary issues, or matters of confidentiality and access between jurisdictions and entities operating the various databases. The most important limitation on the use of any database in identifying unknowns, live or dead, is its *inclusiveness*. The best chance a missing individual or set of remains has of being identified resides in whether these have been submitted to a database with as much accompanying information as possible. Obviously, unidentified remains must have an accurate analysis. A significant problem arises because of the differing skill levels of those who initially develop the profile. If the unknown remains are sufficiently complete and "fresh" to allow accurate determination of sex, age, ancestry, and stature visually, then a report from a pathologist may be sufficient for use as critical metadata. However, when remains are incomplete, fragmentary, degraded, etc., the biological profile should be completed by an experienced forensic anthropologist. Errors in the assignment of ancestry or age, improper dental charting, or other misinformation entered into a database will likely result in false elimination of a correct identity match, i.e., "garbage in, garbage out." Likewise, DNA reference samples should be from the closest possible relative of the missing individual. When bone or dental samples from unknown human remains are submitted to the CODIS database, these should be accompanied by biological profile information. Although it is not always possible to accurately determine all of the features of the profile, an effort must be made.

The "submission" issue is a large one. Reasons for the U.S. population's rejection of national forms of identity databasing have deep roots in a variety of historical, cultural, political, psychological, and religious factors. Many countries require that some form of personal identifier be entered into a national database; e.g., in Korea persons are fingerprinted at the age of eighteen. Though many large organizations and governmental agencies require and store various forms of personal identification, whether biometric, fingerprints, DNA, or others, these have very limited access and do not interact. Thus, their value in large-scale searches for the missing and unknown remains is negligible. In this connection, an increasing number of states have enacted statutes requiring that remains not identified within a particular interval must be submitted to the National Missing Persons (CODIS) Database. The promise of databases will not be realized until the problems of accurate data entry and interconnectivity as well as broad public acceptance and participation are resolved.

8.4 The Future

Changes in training and technology in the United States and elsewhere have already produced a generation of forensic anthropologists who have moved beyond osteoarchaeology applied to identification, and increasingly toward an amalgam of bone biology and chemistry, molecular analysis, and ever more sophisticated software and instrumentation. Contemporary forensic anthropologists will be as comfortable interpreting x-ray fluorescence data, analyzing stable isotope ratios, and reading MRI scans as their predecessors were with calipers and flat plate radiographs. Several predictions seem worthwhile:

1. Increased use of anthropological findings as *metadata* within a molecular and biometric database identification framework will mandate more *comprehensive validation studies* to strengthen elements of the biological profile in a *post-Daubert* environment.
2. Growing realization that "one size does *not* fit all" will result in the *dissemination of taphonomic research facilities* into an increasing number of biotic provinces, including montane, marine, and lacustrine environments, to address a variety of problems, from determination

of postmortem interval to improved evidence location and recovery methods. These will translate into greater admissibility in an era of rising evidentiary standards.
3. There will be a need for *expansion of contemporary osteological study collections* to support ongoing validation studies as well as research and teaching. These collections will need to be *more diverse*, and the numbers of individuals in each population represented will have to increase to statistically useful levels. As improvements in public health and nutrition occur in third world populations, rapid secular changes in growth and life span will have to be reflected in such collections. Particularly critical is the *almost total absence of fetuses, neonates, and children* in U.S. skeletal reference collections. This will be remedied only through increased public awareness of need combined with improved and expanded remains solicitation programs.
4. One can anticipate an expansion of the already prominent role of forensic anthropology in *mass fatality incidents*, whether natural or manmade, e.g., coastal storms, transportation disasters, terrorism, etc., and in the investigation of crimes against humanity (e.g., Argentina, Bosnia, Rwanda, Chile, etc.).

Perhaps most importantly, one can predict that agencies responsible for death investigation and identification will develop cadres of specialists, including anthropologists, odontologists, pathologists, molecular biologists, and others whose contributions form a seamless *team approach* to these problems on any scale.

References

1. American Board of Forensic Anthropology. http://www.theabfa.org.
2. Maples, W.R. 1997. Forensic anthropology. In *Forensic dentistry*, ed. P. Stimson and C. Mertz, 65–80. Boca Raton, FL: CRC Press.
3. Joyce, C., and E. Stover. 1991. *Witnesses from the grave: The stories bones tell.* 1st ed. Boston: Little, Brown.
4. Maples, W.R., and M. Browning. 1994. *Dead men do tell tales.* 1st ed. New York: Doubleday.
5. Rhine, S. 1998. *Bone voyage/a journey in forensic anthropology.* 1st ed. Albuquerque: University of New Mexico Press.
6. Steadman, D.W., and W.D. Haglund. 2005. The scope of anthropological contributions to human rights investigations. *J Forensic Sci* 50:23–30.
7. Koff, C. 2004. *The bone woman: A forensic anthropologist's search for truth in the mass graves of Rwanda, Bosnia, Croatia, and Kosovo.* 1st ed. New York: Random House.
8. Stewart, T.D. 1970. *Personal identification in mass disasters.* Washington, DC: National Museum of Natural History, Smithsonian Institution.

9. Bernardi, A. 2008. *2008 Annual report: Argentine forensic anthropology team.* New York: EEAF.
10. Gilbert, B.M., P. Wapnish, and H. Savage. 1996. *Avian osteology.* Columbia: Missouri Archeological Society.
11. Hesse, B., and P. Wapnish. 1985. *Animal bone archeology: From objectives to analysis.* Manuals on archeology 5. Washington, DC: Taraxacum.
12. Gilbert, B.M. 1990. *Mammalian osteology.* Columbia: Missouri Archaeological Society.
13. France, D.L. 2008. *Human and nonhuman bone identification: A color atlas.* Boca Raton, FL: Taylor & Francis.
14. Bergman, R.A., S.A. Thompson, and A.K. Afifi. 1984. *Catalog of human variation.* Baltimore: Urban & Schwarzenberg.
15. Buikstra, J., C. Gordon, and L. St. Hoyme. 1984. Individualization in forensic anthropology. In *Human identification: Case studies in forensic anthropology*, ed. T. Rathbun and J. Buikstra, 121–35. Springfield, IL: C.C. Thomas.
16. Odom, C. Personal communication to the author, August 20, 1991.
17. Snow, C. 1984. The Oklahoma City child disappearances. In *Human identification: Case studies in forensic anthropology*, ed. T. Rathbun and J. Buikstra, 253–77. Springfield, IL: C.C. Thomas.
18. Snow, C.C., and J.L. Luke. 1970. The Oklahoma City child disappearances of 1967: Forensic anthropology in the identification of skeletal remains. *J Forensic Sci* 15:125–53.
19. Krogman, W.M., and M.Y. Iscan. 1986. *The human skeleton in forensic medicine.* 2nd ed. Springfield, IL: C.C. Thomas.
20. Berryman, H., S. Symes, and O. Smith. 1997. Recognition of cemetery remains in the forensic setting. In *Forensic taphonomy: The postmortem fate of human remains*, ed. W. Haglund and M. Sorg, 165–76. Boca Raton, FL: CRC Press.
21. Rogers, T.L. 2005. Recognition of cemetery remains in a forensic context. *J Forensic Sci* 50:5–11.
22. St. Hoyme, L., and M. Iscan. 1989. Determination of sex and race: Accuracy and assumptions. In *Reconstruction of life from the skeleton*, ed. M. Iscan and K. Kennedy, 5394. New York: Wiley-Liss.
23. Kelley, M.A. 1979. Parturition and pelvic changes. *Am J Phys Anthropol* 51:541–46.
24. Putschar, W.G. 1976. The structure of the human symphysis pubis with special consideration of parturition and its sequelae. *Am J Phys Anthropol* 45:589–94.
25. Houghton, P. 1975. The bony imprint of pregnancy. *Bull NY Acad Med* 51:655–61.
26. Houghton, P. 1974. The relationship of the pre-auricular groove of the ilium to pregnancy. *Am J Phys Anthropol* 41:381–89.
27. Coon, C.S. 1962. *The origin of races.* 1st ed. New York: Knopf.
28. Coon, C.S., and E.E. Hunt. 1965. *The living races of man.* 1st ed. New York: Knopf.
29. Molnar, S. 2001. *Human variation: Races, types, and ethnic groups. Rethinking childhood.* 5th ed. Upper Saddle River, NJ: Prentice Hall.
30. Gill, G.W., and J.S. Rhine, ed. 1990. *The skeletal attribution of race.* Maxwell Museum of Anthropology Papers 4. Albuquerque: University of New Mexico Press.
31. Saunders, S.R. 1989. Nonmetric skeletal variation. In *Reconstruction of life from the skeleton*, ed. M. Iscan and K. Kennedy, 95–108. New York: Wiley-Liss.

32. Fazekas, I.G., and F. Kósa. 1978. *Forensic fetal osteology*. Budapest: Akadémiai Kiadó.
33. Scheuer, L., and S.M. Black. 2000. *Developmental juvenile osteology*. San Diego: Academic Press.
34. Mincer, H., E. Harris, and H. Berryman. 1995. Molar development as an estimator of chronological age. In *Manual of forensic odontology*, ed. C. Bowers and G. Bell. 3rd ed. Saratoga Springs, FL: American Society of Forensic Odontology.
35. Demirjian, A., H. Goldstein, and J.M. Tanner. 1973. A new system of dental age assessment. *Hum Biol* 45:211–27.
36. Lewis, J.M., and D.R. Senn. 2008. UT-Age, age estimation database. Version 2.0.16. http://utforensic.org/ageestimate.asp.
37. Ubelaker, D.H. 1989. Estimation of age at death from immature human bone. In *Age markers in the human skeleton*, ed. M. Iscan, 55–71. Springfield, IL: C.C. Thomas.
38. Schwartz, J.H. 2007. *Skeleton keys: An introduction to human skeletal morphology, development, and analysis*. 2nd ed., New York: Oxford University Press.
39. Johnston, F., and L. Zimmer. 1989. Assessment of growth and age in the immature skeleton. In *Reconstruction of life from the skeleton*, ed. M. Iscan and K. Kennedy, 11–23. New York: Wiley-Liss.
40. Stewart, T.D. 1958. The rate of development of vertebral osteoarthritis in American blacks and whites. *Leech* 28:144–51.
41. Acsádi, G., and J. Nemeskéri. 1970. *History of human life span and mortality*. Budapest: Akadémiai Kiadó.
42. Sorg, M., R. Andrews, and M. Iscan. 1989. Radiographic ageing of the adult. In *Age markers in the human skeleton*, ed. M. Iscan, 169–94. New York: C.C. Thomas.
43. Stout, S. 1989. Histomorphometric analysis of human skeletal remains. In *Reconstruction of life from the skeleton*, ed. M. Iscan and K. Kennedy, 41–52. New York: Wiley-Liss.
44. Ingraham, M. 2004. Histomorphological age estimation from the midshaft of the clavicle. Thesis, University of North Texas, Denton.
45. Singh, I.J., and D.L. Gunberg. 1970. Estimation of age at death in human males from quantitative histology of bone fragments. *Am J Phys Anthropol* 33:373–81.
46. Trotter, M. 1970. Estimation of stature from intact long limb bones. In *Personal identification in mass disasters*, ed. T.D. Stewart. Washington, DC: National Museum of Natural History, Smithsonian Institution.
47. Jantz, R., and S. Ousley. 2005. Fordisc III™ in Department of Anthropology Forensic Database Algorithms. Knoxville: University of Tennessee.
48. Bass, W.M. 1987. *Human osteology: A laboratory and field manual*. 3rd ed. Special Publication 2. Columbia, MO: Missouri Archaeological Society.
49. Steele, D.G., and T.W. McKern. 1969. A method for assessment of maximum long bone length and living stature from fragmentary long bones. *Am J Phys Anthropol* 31:215–27.
50. Capasso, L., K. Kennedy, and C. Wilczak. 1999. Atlas of occupational markers on human remains. In *Journal of Paleontology Monographic Publication 3*. Teramo, Italy: Edigrafital SPA.
51. Christensen, A.M. 2005. Assessing the variation in individual frontal sinus outlines. *Am J Phys Anthropol* 127:291–95.

52. Nashelsky, M., and P. McFeeley. 2003. Time of death. In *Handbook of forensic pathology*, ed. R.C. Froede, 69–78. 2nd ed. Washington, DC: College of American Pathologists.
53. Haglund, W.D., and M.H. Sorg. 1997. *Forensic taphonomy: The postmortem fate of human remains.* Boca Raton, FL: CRC Press.
54. Bass, W. 1984. A civil war burial. In *Human identification: Case studies in forensic anthropology*, ed. T. Rathbun and J. Buikstra, 136–47. Springfield, IL: C.C. Thomas.
55. Henry, T.E. 2003. Blunt force injuries. In *Handbook of forensic pathology*, ed. R.C. Froede, 139–47. 2nd ed. Washington, DC: College of American Pathologists.
56. Quatrehomme, G., and M.Y. Iscan. 1999. Characteristics of gunshot wounds in the skull. *J Forensic Sci* 44:568–76.
57. Stahl, C.J. 1977. Cutting and stabbing wounds. In *Forensic pathology: A handbook for forensic pathologists*, ed. R. Fisher and C. Petty, 151–59. Washington, DC: National Institute of Law Enforcement and Criminal Justice.
58. Bromage, T.G., and A. Boyde. 1984. Microscopic criteria for the determination of directionality of cutmarks on bone. *Am J Phys Anthropol* 65:359–66.
59. Behrensmeyer, A.K., K.D. Gordon, and G.T. Yanagi. 1986. Trampling as a cause of bone surface damage and pseudo-cutmarks. *Nature* 319:768–71.
60. Zugibe, F.T., J. Costello, and M. Breithaupt. 1996. Identification of a killer by a definitive sneaker pattern and his beating instruments by their distinctive patterns. *J Forensic Sci* 41:310–13.
61. Sorg, M.H. 1985. Scavenger modification of human remains. *Curr Res Pleistocene* 2:37–38.
62. Rossi, M.L., et al. 1994. Postmortem injuries by indoor pets. *Am J Forensic Med Pathol* 15:105–9.
63. Angel, L. 1985. Postdoctoral seminar in Forensic Anthropology. Smithsonian Institute. Washington, DC. personal communication.
64. Belanger, T.A., and D.E. Rowe. Diffuse idiopathic skeletal hyperostosis. *J Amer Acad Orthopedic Surg* 9:258–267.
65. DiMaio, V.J.M. 1999. *Gunshot wounds: practical aspects of firearms, ballistics, and forensic techniques*, 2nd ed., 71–102.

Forensic Dental Identification

9

MICHAEL P. TABOR
BRUCE A. SCHRADER

Contents

9.1	Introduction	163
	9.1.1 Fingerprints	164
	9.1.2 Personal Items	165
	9.1.3 Tattoos and Scars	165
	9.1.4 DNA and DNA Evidence	165
9.2	History of Dental Identification	166
9.3	Philosophy and Legal Basis for Dental Identification	166
9.4	Steps in Dental Identification	167
	9.4.1 Postmortem Examination	167
	9.4.2 Antemortem Examination	168
	9.4.3 Comparison	170
9.5	Statistical and Mathematical Models	172
9.6	ABFO Guidelines and Standards	177
9.7	Technological and Scientific Advances	178
9.8	Ten Tips and Cautions for Dental Identification	181
9.9	Summary	183
References		184

9.1 Introduction

Identification of an individual can be confirmed by several different methods. These include visual identification, personal effects, tattoos, scars, anatomical structures, medical devices, and implants, as well as fingerprint, DNA, and dental comparisons. Molina subdivides the methods of identification as visual, circumstantial, external characteristics, internal characteristics, radiographs, and anthropology, and notes that DNA, fingerprint, and dental comparisons are considered the scientific methods of identification (see Chapter 5).

Although visual methods of identification are commonly used, they should be used with caution. "Visual identification is one of the least reliable forms of identification and can be fraught with error" (Molina in Chapter 5). Facial and other characteristics can change due to trauma, swelling, fragmentation, and decomposition. Certainly hair color, skin color, and other physical descriptors can be useful, but should never be used alone to confirm identification when disfiguring has occurred. In such cases, most medical examiners will not attempt a visual ID since it may create significant emotional trauma to family members. This can also lead to misidentification of the individual.

Case Report: Misidentification/Delayed Identification

"On April 26, 2006 Whitney Cerak, Laura Van Ryn and seven other people were involved in a car accident in Indiana. Five people perished in the accident."[1] Cerak's parents were informed by authorities that their daughter Whitney had died in the accident. The Van Ryn family anxiously waited at what they thought was Laura's bedside. After five weeks of intensive care it became obvious that the girl in the hospital was not Laura Van Ryn. Whitney Cerak was alive and the Cerak family had unknowingly buried Laura Van Ryn after a closed-casket funeral that drew several thousand mourners.

According to the local coroner, the mistake occurred at the scene. Personal belongings were strewn throughout the crash site and both girls had similar facial features, blonde hair, and similar body weight. At the accident scene, Laura Van Ryn's ID had been associated with the individual transported to the hospital.

The hospitalized girl had considerable facial trauma and swelling. Van Ryn family members were unable to recognize the person they thought was Laura and did not realize the mix-up until weeks after the crash. They were so emotionally involved that they had no reason to doubt the identity. No scientific identification techniques were utilized. DNA, fingerprint, or dental record comparison was not employed to confirm the identity of any of those involved in the crash. After the error became obvious, dental record comparisons confirmed the identities of both girls. The Van Ryn and Cerak families, including Whitney Cerak, corroborated on a 2008 book chronicling the events and the effects those events had on the families.[2]

9.1.1 Fingerprints

Fingerprint identification is a dependable and efficient forensic identification technique, but it is sometimes impossible to record postmortem fingerprints from decomposed or burned bodies. A body that remains immersed in fresh water can decompose rapidly, depending on the water temperature, and may preclude the recording of postmortem fingerprints. Of course, it is not possible to record fingerprints from skeletonized decedents.

For fingerprints to be a useful identification tool in a specific case, the person in question must have antemortem fingerprints on file. There is still a significant segment of the U.S. population for whom there are no fingerprint records. Sources estimate that about one out of every six people in this country has a fingerprint record on file in the Integrated Automated Fingerprint Identification System (IAFIS). Conversely, five of every six have no fingerprints on file.[3] For those individuals fingerprint analysis and comparison will be nonproductive. Postmortem fingerprints are collected whenever possible, but comparison depends on the existence of prior fingerprint records. For additional information on fingerprint analysis see Chapter 6.

9.1.2 Personal Items

Personal effects such as driver's licenses, photographs, car keys, or monogrammed items are often useful clues in researching a decedent's identity. These should be used with caution, however, and should never be the sole determinant in the identification process. For example, after a fiery motor vehicle fatality, determining that the vehicle is registered to the same person named on a driver's license found at the scene can be a very valuable clue. There are, however, documented cases where such deaths have been staged for various fraudulent purposes. Names on items of clothing are clues but not identification. After the World Trade Center attacks in 2001, firefighters who perished were found to be wearing other firefighters' turnout coats. Identification errors could have been made from attempting to make positive identifications using personal effects alone.

9.1.3 Tattoos and Scars

Tattoos and scars provide clues for forensic identification. Scars may be from previous trauma or surgical intervention and can be further investigated by the pathologist. Tattoos, if sufficiently unique, can be used as an identifier of an individual or may indicate that individual belonged to a particular group or gang. See Chapters 5 and 11 for information concerning tattoos and scars and for methods for better imaging this evidence.

9.1.4 DNA and DNA Evidence

Analysis of DNA evidence for identifications has become a widely used forensic technique and is considered by many to be the gold standard. The ability of forensic investigators to obtain antemortem DNA samples even after the death of an individual is a distinct advantage for DNA analysis. A swab from a close relative, stored blood, or material from the decedent's hairbrush or toothbrush may provide adequate comparison material. DNA profiles of decedents can be compared to various databases (see Chapter 7). There are limitations for all forms of identification, and for DNA the major limitations are the time required and the costs involved, which may rapidly exceed the limits of already challenged medical examiner and coroner budgets. DNA analysis has also become a valuable tool used in conjunction with bitemark identification analysis (see Chapter 14). By providing efficient, accurate, and cost-effective human identification, forensic dentists play important roles in death investigation.

9.2 History of Dental Identification

The oldest known example of the identification of an individual confirmed by teeth was reported by the Roman historian Cassius Dio (c. 165–c. 235 A.D.). Aggripina the Younger, wife of Emperor Cassius and mother of future Emperor Nero, contracted for the killing of a perceived rival, Lollia Paulina. In his account Dio reported, "She did not recognize the woman's head when it was brought to her; she opened the mouth with her own hand and inspected the teeth, which had certain peculiarities."[4] Additional historical information on the development of forensic dental identification can be found in Chapter 2.

9.3 Philosophy and Legal Basis for Dental Identification

The confirmation of the decedent's identity serves several important purposes.

Bringing closure to a tragic or unexpected event will often give some peace and closure to the immediate family members in their time of grief. The anguish of not knowing is difficult for families. Although confirmation of death may be terrible, it ultimately leads to the possibility of resolution of a difficult time for family members.

For the legal settlement of estates, a death certificate is usually required. A death certificate cannot be issued without confirmation of identity. Payment of life insurance policy benefits also requires verification of death. The cause and manner of death may be very important items of information for life insurance companies and to the decedent's family. Increased benefits for accidental death or clauses precluding payment for deaths from suicide, acts of war, or engaging in dangerous activities often mean that these cases are decided in courts of law.

In multiple fatality incidents, identification of decedents is often difficult and commingling of remains may occur. In these situations great care must be taken to correlate all body fragments to the appropriate decedent. Dental identification may provide an identified fragment to which other unknown fragments can be compared. Even if no antemortem DNA profile is available for that individual, fragments with the same DNA profile can be associated. The combined use of DNA and dental comparison can help to ensure that all

Forensic Dental Identification

possible fragments are associated and returned to families for proper burial. See Chapter 16 for additional information concerning the legal aspects of forensic dental identification.

9.4 Steps in Dental Identification

9.4.1 Postmortem Examination

The durability of the human dentition, including the ability of teeth to survive decomposition and withstand drastic temperature changes, makes dental evidence comparison one of the most dependable and reliable methods of identification. The mechanism of this process involves comparing features of an unknown specimen to those of a known individual. The durability and longevity of human teeth make this process possible.

An accurate and detailed evaluation of the postmortem specimen will afford the best possibility of successfully comparing that information to antemortem information. Attention to detail at the postmortem examination precludes errors that can lead to nonidentification and the need to repeat steps to get an accurate postmortem record. "Document, document, document" should be the mantra of examiners of forensic materials. By following a step-wise examination checklist that includes photography, dental radiography, and dental charting, a forensic odontologist or forensic dental team can create consistently accurate postmortem dental records.

Photography of a specimen can provide the ability to view specific features without having to revisit the morgue. These photographs should be taken to allow orientation as well as closeup photographs of the dental structures. This photo documentation can prove extremely valuable in cases where the handling of the specimen could lead to further degradation of fragile remains. This is often the case with dental structures that have become desiccated or carbonized from extreme heat. Using macro or closeup photographs of the dentition can also provide additional information that may lead to identification and is discussed in more detail in the comparison section of this chapter.

The goal of the postmortem dental examination is to locate, identify, and document anatomical structures, dental restorations, and dental appliances that will aid the comparison process. The more information documented in this examination, the greater the likelihood of successful comparison to an antemortem record. Depending on the condition of the remains, this task can be simple or complex. In a fully intact body with no injuries to the facial structures, the ability to locate specific dental structures will be simple in comparison to cases in which individuals have been subjected to explosions, rapid deceleration injuries, extreme heat, or crushing forces.

In situations involving specimens that are not readily identifiable as human facial structure, dental remains can be located with the assistance of large-format radiographs. This radiograph facilitates locating radiopaque structures that can assist in locating dental structures within the specimen or body bag. Once these items are identified on this large radiograph, the examiner can use this image to assist in locating structures of interest, including teeth, dental prosthetic items, and bony structures of the mandible or maxilla.

Once all available dental material is found, forensic odontologists should take dental radiographs in an effort to reproduce similar exposures and angulations anticipated in the antemortem dental record. Since the source and type of radiographs may be unknown at this point in the process, it is recommended that a full series of radiographs be obtained. This series of films should include posterior and anterior periapical radiographs and bitewing radiographs. If the specimen is fragmented, the radiographer should consider the necessary film placement and tube head angulations to replicate those normally obtained in a clinical setting. For convenience in image capturing the examiner may also find it helpful to radiograph the maxillary and mandibular teeth separately in bitewing radiograph projections. If the examiner is taking images of fragmented remains, care should be taken to ensure that consistent projection geometry is maintained by placing the film or digital sensor on the lingual aspect of the specimen. Again, carefully following protocol is important, as the examiner may not have a second chance to complete the radiographic examination. Attention to detail is necessary, and any images that are not adequate should be retaken so complete radiographic information is available for comparison to antemortem data. If digital radiography is available, the examiner will find the retake of images to be straightforward since the need to process films is eliminated and the image is instantly viewable. Additional details concerning forensic dental radiography techniques can be found in Chapter 10.

The postmortem record, whether digital or paper, should be recorded in a form that will assist in the comparison process. It should be a format that records and displays the relevant features of the dentition to demonstrate missing teeth and the restored surfaces (Table 9.1).

9.4.2 Antemortem Examination

When an investigating agency determines that a dental identification may be required, the agency attempts to locate and obtain the dental record. This action, securing antemortem dental records, is a crucial step in dental identification, and the quality of these records is totally dependent on practicing dentists keeping accurate records of the dental status of their patients. Most

Table 9.1

Postmortem Dental Record

ID#:_____ Agency _____

Date: _____ Sex: _____ Race: _____ Estimated Age: _____

R 1/32 2/31 3/30 4/29 5/28 6/27 7/26 8/25 | 9/24 10/23 11/22 12/21 13/20 14/19 15/18 16/17 L

Univ #	Code	Description	FDI #	
1			18	
2			17	
3			16	
4	A		15	55
5	B		14	54
6	C		13	53
7	D		12	52
8	E		11	51
9	F		21	61
10	G		22	62
11	H		23	63
12	I		24	64
13	J		25	65
14			26	
15			27	
16			28	
17			38	
18			37	
19			36	
20	K		35	75
21	L		34	74
22	M		33	73
23	N		32	72
24	O		31	71
25	P		41	81
26	Q		42	82
27	R		43	83
28	S		44	84
29	T		45	85
30			46	
31			47	
32			48	

WinID Codes	
Primary Codes	**Secondary Codes**
M - Mesial	A - Annotation
O - Occlusal	B - Deciduous
D - Distal	C - Crown
F - Facial	E - Resin
L - Lingual	G - Gold
I - Incisal	H - Porcelain
U – Unerupted	N - Non-Precious
V - Virgin	P - Pontic
X – Missing	R - Root Canal
J – Missing Cr MPM	S - Silver Amalgam
/ - No Data	T - Denture Tooth
	Z - Temporary

Comments: _____

forensic dentists consider this reconstruction of the antemortem record the most challenging and time-consuming step in dental identification.

Dentists are required by law, in most jurisdictions, to maintain their patients' original records. This places a dental practitioner in an uncomfortable position when asked to release an original dental record for comparison and possible dental identification. Dentists are almost always anxious to help but concerned about the proper management of their patients' health records.

In addition to state regulations regarding record keeping, since 2003 the Health Insurance Portability and Accountability Act (HIPAA) and the accompanying and ironically named Administration Simplification (AS) provisions address the security and privacy of health data, particularly protected health information (PHI). HIPAA and most state dental regulations provide exemptions for the release of original records for the purpose of identifying the dead.[5] Investigators must be educated to explain the rules and politely demand that dentists or the dental office staff provide the original dental record, including financial ledger, written records, health and dental history forms, and all original radiographs. With the current ability to digitize a dental chart, the duplication of an original record can be relatively easily performed. Duplication of the record in this fashion provides the dental office and the forensic investigator with a digital copy of the record and the amount of time that the treating office is without the original record is minimized. Also, after the record is duplicated and the original record is returned to the dental office, a phone consultation can be performed with the treating dentist to allow for clarification of any of the notes or charting peculiarities. During the process of antemortem material collection, the practitioner should also be asked if there are dental models or appliances that may be useful in the identification process.

Meticulous evaluation of the original materials facilitates the creation of an accurate record of the status of the patient's mouth at the time of the last dental visit. It is important to review all written records and radiographs and to give special attention to the most recent procedural notes, patient ledgers, and radiographs. Dental treatment is regularly performed after the latest radiographic and clinical examination.

The antemortem forensic record should be recorded in a format that accurately portrays the latest known status of that patient's dental status (Table 9.2).

9.4.3 Comparison

After the postmortem and antemortem records are completed, the comparison process can proceed. Although this is commonly done manually for individual identifications, in multiple fatality incidents it is likely that a computer program will be used for the search and comparison of the antemortem and postmortem records. WinID3© is a computer program commonly used in North America that will assist the forensic dentist or forensic investigator to establish and maintain antemortem and postmortem databases.[6] WinID was developed by Dr. Jim McGivney as an expanded and enhanced Windows version of the earlier DOS-based CAPMI program developed by

Table 9.2

Antemortem Dental Record

Last: _____ First: _____ MI: ___ SSN/ID#: _____

Date (this record completed)_____ Sex: _____ Age/DOB: _____

R 1/32 2/31 3/30 4/29 5/28 6/27 7/26 8/25 | 9/24 10/23 11/22 12/21 13/20 14/19 15/18 16/17 L

WinID Codes	
Primary Codes	**Secondary Codes**
M – Mesial	A - Annotation
O - Occlusal	B - Deciduous
D – Distal	C - Crown
F – Facial	E - Resin
L – Lingual	G - Gold
I – Incisal	H - Porcelain
U – Unerupted	N - Non-Precious
V – Virgin	P - Pontic
X – Missing	R - Root Canal
J – Missing Cr	S - Silver Amalgam
/ - No Data	T - Denture Tooth
	Z - Temporary

Univ #	Code	Description	FDI #		Univ #
1			18		1
2			17		2
3			16		3
4	A		15	55	4
5	B		14	54	5
6	C		13	53	6
7	D		12	52	7
8	E		11	51	8
9	F		21	61	9
10	G		22	62	10
11	H		23	63	11
12	I		24	64	12
13	J		25	65	13
14			26		14
15			27		15
16			28		16
17			38		17
18			37		18
19			36		19
20	K		35	75	20
21	L		34	74	21
22	M		33	73	22
23	N		32	72	23
24	O		31	71	24
25	P		41	81	25
26	Q		42	82	26
27	R		43	83	27
28	S		44	84	28
29	T		45	85	29
30			46		30
31			47		31
32			48		32

Comments_____

Lorton et al.[7] WinID3 employs several modes of searching the database for matching records. This allows the comparison of a record (antemortem or postmortem) against all opposing (postmortem or antemortem) records in the database, with the resulting possible matches displayed and linked for further visual comparison. The search results are displayed in five separate tables as *most dental hits, least dental mismatches, most restoration hits, most identifier matches*, and *fuzzy dental logic*. After selecting a record for comparison, the examiner is then able to view the specifics of each record and view the case identifiers, the odontogram, and an associated graphic/radiograph of the cases in a side-by-side fashion. WinID3 has been paired with digital intraoral radiographic software to produce an efficient and effective system for multiple fatality incidents that allows the use of the more sophisticated image management software features (see Chapter 12).

Whether the initial comparisons are made manually or with the aid of a computer, a visual comparison of the records should be made by the forensic odontologist. The terminology for conclusions resulting from the comparison and correlation process should follow the guidelines of the American Board of Forensic Odontology (ABFO).[8] Information regarding guidelines for body identification and missing/unidentified persons can be found on the ABFO website at www.abfo.org and in the American Board of Forensic Odontology *Diplomates Reference Manual*.

Forensic dentists may assist medical examiners and coroners by comparing the dentition of decedents with antemortem photographs showing the teeth. The technique was suggested by Dr. R. Souviron, who has long used what he calls smiley-face photographs to compare to unidentified bodies. The grin line method (a smile may not show teeth) using Adobe® Photoshop® has been developed, evaluated, and utilized. The method and its uses have been presented in the Odontology Section of the American Academy of Forensic Sciences.[9,10] This method is not a stand-alone method of identification but is to be used in conjunction with other information to assist medical examiners and coroners to establish identification.

9.5 Statistical and Mathematical Models

Most dentists accept that one person's teeth are different from those of another. Even with identical twins, the sharing of a common DNA profile does not equate with identical phenotypic representation. This is true of all anatomic features, including fingerprints and the teeth (see Chapters 6 and 14).

Forensic examiners have considered the mathematical probability of an individual dentition having a unique combination of missing or present teeth, restored and unrestored teeth, or restored or unrestored surfaces of those teeth. The number of different combinations possible in this type of mathematical sampling is very large. One of the most well known of these studies was completed by Dr. Soren Keiser-Nielsen, a Danish forensic odontologist. In the abstract for this paper Keiser-Nielsen cautioned that a dental expert "cannot base his identification of an unknown body on the relative frequency of occurrence of any singular dental feature, its particular discrimination potential.

Instead, he must make a quantitative and qualitative evaluation of the combination of features involved."[11]

Dr. Keiser-Nielsen's interest was initiated by the acceptance of fingerprint analysis in courts all over the world based on their accuracy and reliability. Although standards varied in different jurisdictions at that time, in most, when twelve concordant fingerprint characteristics could be demonstrated in an antemortem and postmortem comparison, it was maintained that the material must have originated from the same person. Keiser-Nielsen sought to develop a parallel application for the field of forensic dental identification. Specifically, he sought to quantify the probability that any two individuals would have the same combination of teeth missing and present and teeth restored and unrestored.

For the most fundamental problem, that of the presence or absence of teeth, the mathematical expression is $K_{M,X} = \frac{M}{1} x \frac{M-1}{2} x \frac{M-2}{3} x \to \frac{M-(X-1)}{X}$, where M is the number of teeth considered, usually 32, and X is the number of teeth missing. The same calculation could be done for other features, for instance, the number of teeth with restorations.

Applied to the human dentition, Keiser-Nielsen proposed that even greater discrimination could be established using combinations of features. If only teeth missing/present and teeth restored/unrestored were considered with no regard to which surfaces were restored, the formula involved calculating the possible combinations for missing teeth, then multiplying that number by the number determined by calculating the number of possible combinations of restored teeth in the remaining teeth. Using Keiser-Nielsen's example of an individual missing four teeth and having four of the twenty-eight remaining teeth restored, the formulae are:

For the four missing teeth, $K_{32,4} = \frac{32 \times 31 \times 30 \times 29}{1 \times 2 \times 3 \times 4} = 35,960$, representing the number of different possible combinations for four missing teeth.

For the four restored teeth in the remaining twenty-eight, $K_{28,4} = \frac{28 \times 27 \times 26 \times 25}{1 \times 2 \times 3 \times 4} = 20,475$, representing the number of different combinations of four restored teeth among twenty-eight teeth.

Then by multiplying the two, $35,960 \times 20,475 = 736,281,000$, Keiser-Neilsen calculated that the number of possible combinations of a person with four missing teeth and four restored teeth is 736,281,000.

In other words, the likelihood of any two people having the same four teeth missing *and* the same four teeth filled (surfaces not considered) is 1 in over 736 million, more than twice the number of people alive in the United States (July 2008).

Table 9.3 below is based on Keiser-Nielsen's mathematical results for the possible combinations of teeth showing or not showing a feature, for instance, missing vs. present or restored vs. unrestored, etc.

In other words, if there are thirty-two teeth present, and x represents the number of teeth that have been extracted or filled, then the table gives the

Table 9.3 Possible Combinations: Teeth Exhibiting or Not Exhibiting a Feature

Number of Teeth Showing Feature	Possible Combinations	Number of Teeth Not Showing Feature
	1	32
1	32	31
2	496	30
3	4,960	29
4	35,960	28
5	201,376	27
6	906,192	26
7	3,365,856	25
8	10,518,300	24
9	28,048,800	23
10	64,512,240	22
11	129,024,480	21
12	225,792,840	20
13	347,373,600	19
14	471,435,600	18
15	565,722,720	17
16	601,080,390	16
17	565,722,720	15
18	471,435,600	14
19	347,373,600	13
20	225,792,840	12
21	129,024,480	11
22	64,512,240	10
23	28,048,800	9
24	10,518,300	8
25	3,365,856	7
26	906,192	6
27	201,376	5
28	35,960	4
29	4,960	3
30	496	2
31	32	1
32	1	

odds of the occurrence and allowance that a duplicate exists. *This calculation assumes that each tooth has the same likelihood to be missing or filled.*

If an examined body showed that seven teeth were missing (antemortem, not peri- or postmortem) and sixteen teeth were filled, the odds of two people having the exact same configuration of those features would be

$$K_{32,7} = \frac{32}{1} \times \frac{32}{2} \times \rightarrow \frac{32-(7-1)}{7} = 3,365,856$$

multiplied by

$$K_{32,16} = \frac{32}{1} \times \frac{32}{2} \times \rightarrow \frac{32-(16-1)}{16} = 601,080,390$$

$$3,365,856 \times 601,080,390 = 2,023,150,037,163,840$$

The likelihood then is 1 in 2×10^5, a figure that is more than the all-time world population estimated by Haub to be approximately 106 billion.[12]

Keiser-Nielsen suggested taking the analysis even further, by considering the different combinations of restorations in each of the five restorable surfaces of each tooth. If other features such as endodontic treatment and the types of restorative material used were also considered, the numbers of possible combinations of features become enormous.

Keiser-Nielsen also suggested that forensic odontological identification data should be quantifiable, like in fingerprint analysis, and that forensic dentists should follow comparable standards. He suggested that twelve points of comparison be recorded before establishing certainty in identification.

Drs. Soren Keiser-Nielsen and Riedar Sognnaes were dedicated and visionary practitioners. Their contributions to forensic odontology were considerable and perhaps unsurpassed by other forensic dentists.[11,13–15] In a sincere effort to establish and promote scientific and mathematical bases for forensic dental identification, they made assumptions that failed to carefully consider the nature of the dental features they analyzed for their statistical conclusions. They assumed that these characteristics occurred independently and that their resulting values could be multiplied to produce expected frequencies. In a 2003 paper discussing Keiser-Neilsen's and Sognnaes' statistical conclusions, Adams stated, "This type of statistical assessment suggests that all of the various combinations of missing and filled teeth occur randomly and that they are equally probable in the population, an assumption that is not valid." He further stated, "Statistical arguments of this type regarding the possible number of combinations of missing and filled teeth have been especially stressed by Keiser-Nielsen and Sognnaes for forensic purposes;

unfortunately, these types of statistics are improperly applied, misleading, and should be avoided."[16]

Steadman et al. in 2006 stated, "Since individual dental characteristics are not independent (e.g., the chance of having a restored molar is not the same as the chance of having a restored canine), the resulting dental patterns formed by missing, filled, and unrestored teeth are not completely random events. If this were the case, all conceivable dental patterns would be equiprobable in the population, a trend that is certainly not valid."[17]

The assumption that dental treatment and loss of teeth occur randomly throughout the mouth is incorrect. If that assumption were true, the maxillary central incisors would be missing with the same frequency as first molars or third molars and dental restorations would be seen in the facial surfaces of incisors as frequently as in the occlusal surfaces of molars. Also, interproximal restorations would occur with frequencies independent of occlusal restorations in the same tooth. These assumptions are obviously improbable in clinical dentistry.

While it is undoubtedly possible to develop a statistical analysis that considers all of the variables involved in predicting the frequency of occurrence of dental features, the method would be very complex. Perhaps a better method already exists.

The Joint POW/MIA Accounting Command/Central Identification Laboratory, Hawaii (JAPAC/CILHI) was developed to account for Americans lost during past U.S. conflicts. CILHI has developed a program to assist in the assessment of patterns of dental conditions and treatment. The program, called OdontoSearch, is described in a paper published by Adams in 2003.[18] Use of OdontoSearch allows investigators to input dental patterns, then the program provides a frequency value for that pattern in the databases. The program is available online at no cost.[19]

In an OdontoSearch analysis of the probability of dental feature occurrences a formula is used that is conceptually similar to the Keiser-Neilsen formula: $^nC_r = \frac{n!}{r!(n-r)!}$. In this equation n is the overall sample size or number of teeth being considered and r is the number of occurrences of the feature being considered. In the case of twenty-eight teeth being considered and considering five missing teeth the equation would be $^{28}C_5 = \frac{28!}{5!(28-5)!} = 98,280$. Unlike the pure mathematical approach offered by earlier methods, the data in OdontoSearch analyze the frequency of occurrence for dental and dental treatment patterns in populations. According to CILHI, "The program works by comparing an individual's pattern of missing, filled, and unrestored teeth to a large, representative sample of the U.S. population. The methodology and rationale behind the OdontoSearch 2.0 program is very similar to the procedures that have been established for mitochondrial DNA comparisons."[19] Two databases have been created: one reflects a U.S. military population for which the data were collected from 1994 to 2000 and the other a U.S. civilian population, data collected from 1998 to 2004. When searching this database for a particular combination of dental characteristics, the user can select one or both of the databases for comparison. The OdontoSearch online program uses the universal numbering system and does not consider the third molars. These databases and search protocols consider twenty-eight adult teeth only and do not consider any teeth distal to the second molars. Searches can be performed by selection of one or both of the databases and entering a combination of desired dental information for the twenty-eight teeth. In a detailed search this information can include any combination of restored surfaces of M, D, F, and L for anterior restorations; M, O, D, F, and L for posterior restored teeth; MDFL and MODFL, respectively, for anterior and posterior crowns; and X for missing and V for unrestored or virgin teeth. If a generic search is desired, R is selected for a tooth that has any type of restoration instead of using the specific surfaces. The trial entry of data in these fields produces for the user a probability of the specific combination of coding existing in the given database. To apply the technique, the strength of a match is considered a function of the frequency in which the given dental pattern is observed in the reference database(s): $\frac{X+1}{N+1} \cdot 100$, where X is the total number of pattern matches and N is the sample size of the data set(s). If no matches are found, $X = 0$, and the observed pattern is unique within the database. Since the pattern in question is known to exist in at least one person (the case individual), the number of matches (X) and the sample size (N) are increased by one. Inverting this frequency gives the likelihood ratio of identification. By utilizing this empirically derived probability value, matches can be quantified in a manner that is easily defensible in a court of law.[17]

9.6 ABFO Guidelines and Standards

In 1994, the American Board of Forensic Odontology established guidelines and standards for dental identification. These guidelines have been published in the *Journal of the American Dental Association* and are available online at the ABFO website.[20,21] The guidelines offer the specific and detailed information outlined below:

    I. Collection and preservation of postmortem dental evidence
       A. The remains—examination procedures
       B. Photography

C. Jaw resection
   D. Techniques for dissection/resection
   E. The postmortem dental record
      1. Dental examination
      2. Narrative description and nomenclature
      3. Dental impressions
      4. Dental radiology
II. Sources for antemortem data
   A. Local agencies
   B. State agencies
   C. Federal agencies
   D. International resources
   E. Insurance carriers
   F. Other sources
III. Comparison of antemortem and postmortem evidence
   A. Dental features useful in identification
IV. Categories and terminology for body identification
   A. Positive identification
   B. Possible identification
   C. Insufficient evidence
   D. Exclusion

9.7 Technological and Scientific Advances

The advances in dental technology and materials used in clinical practice have led to many changes in the delivery of patient care. These include computer hardware and software, intraoral photography and videography, digital radiography, intraoral laser technology, dental implants, high-strength porcelains, advances in bonding agents, and continued changes in dental esthetic resins, to name a few. These same advances in dental technology can also be useful in dental identification.

Flint et al. conducted a pilot study to determine the ability of a computer program to compare dental radiographs as a method for dental identification. This program enables the comparison of radiographs taken from different projection geometries. In their study, longitudinal radiographs of individuals were used to test the ability of the program to correctly identify images of the same individuals. Following this aspect of the study, clinical radiographs of patients were used to further test the ability to correctly identify radiographs from the same individual. Through their study, the examiners found different threshold levels of the program to properly identify radiographs of the same patient when evaluating different anatomical areas of the mouth. The authors show this method of intraoral image comparison to be both objective and reliable.[22]

As stated earlier in this chapter, there is great demand for esthetic restorations in today's dental environment. The frequent placement of esthetic resin restorations in anterior and posterior teeth can further complicate the ability to make a positive dental identification due to the possibility of the forensic examiner being unable to identify a resin restoration clinically or radiographically. Pretty et al. used quantitative light-induced fluorescence (QLF) in the identification of dental composite resin in extracted teeth. They further explored some of the difficulties encountered in the postmortem examinations that include tooth-colored restorations. In Pretty's study, extracted and previously unrestored teeth were utilized. Resin restorations were placed in half of the tooth specimens while the other teeth remained unrestored. Digital photographs and QLF images were taken of the specimens under both wet and dry conditions. Dental examiners were asked to evaluate the images to determine if teeth were restored or unrestored. The results of the study showed a significant improvement in the ability of the examiners to identify restored surfaces when QLF was employed.[23]

Hermanson et al. have demonstrated the usefulness of UV light in the visualization of dental resins. The evaluation using visible light from the UV spectrum is a qualitative method that can be used to locate esthetic resin fillings or pit and fissure sealants that might otherwise go undetected on clinical examination.[24]

Bush et al. investigated the ability to identify specific dental restorative resins by their formulation. Manufacturers' formulations of dental resins differ with respect to filler particle size and elemental composition. Through the use of scanning electron microscopy and energy-dispersive x-ray spectroscopy, this study demonstrated that the makeup of dental resin materials is potentially useful as an aid in dental identification cases. Through the evaluation of several dental resins, they were able to determine that even under conditions of extreme heat, the element composition of these products was relatively unaltered. The data from this study were used to initiate the creation of a database for use in identification of resin restorations in teeth through the elemental analysis of their composition, and further reinforces the need for the dental practitioner to maintain thorough written records regarding specific materials used in tooth restoration.[25]

The same team of researchers conducted an additional study to demonstrate the utility of dental resin identification and the creation of a dental resin database using portable x-ray fluorescence (XRF). This study again used the evaluation of the elemental composition of dental restorative resins and their ability to withstand high-heat conditions. Using XRF, the examiners examined resins placed in cadavers both prior to and after the placement of the cadavers in a crematorium. In both situations the study demonstrated the ability to identify the different resins by brand. This study further reinforces the usefulness of forensic dental science in its ability to aid in the identification of remains even after exposure to extreme conditions.[26]

These types of examinations can be expected to become more necessary and more useful as the placement of esthetic resins continues to expand. If esthetic dentists do their jobs well, visual and radiographic detection of these materials may present a challenge to forensic dentists.

Another dental material that is not generally considered beyond its radiographic appearance in a postmortem dental examination is root canal filling materials.

Bonavilla et al. examined the usefulness of these type of materials in identification using scanning electron microscopy and energy-dispersive x-ray spectroscopy (SEM/EDS), as listed in the initial resin study above. This study reports that 40 million endodontic fillings are completed in the United States annually. This provides a large possibility that postmortem examinations will result in the discovery of this type of previous dental treatment. In the same fashion as the previous study, these materials were evaluated both pre- and postcremation. It may appear that this group of researchers has pyromaniacal tendencies, but this research has practical and useful applications. The most difficult dental identifications may be those that involve severely burned remains and, in the most extreme cases, cremains. The researchers noted that endodontic obturation materials, gutta percha, silver points, root canal sealers, and endodontic files were identifiable following exposure to high temperatures through the analysis of their elemental composition.[27]

These initial studies allowed the development of a database of dental resins. The next step was to develop a means by which the elemental analysis and brand determination could be determined in the field with a portable unit. This study combined the laboratory use of scanning electron microscopy and energy-dispersive x-ray spectroscopy (SEM/EDS) with the Spectral Library Identification and Classification Explorer software (SLICE). SLICE is an application that archives, examines, and compares x-ray spectra of the EDS. Blind tests were performed in the assessment of dental resins in an effort to determine the resin brand. The application of this technology in a dental application was novel but proved to be successful following the creation of the dental resin database for this type of dental material. Research of this kind continues to expand the possibilities of contributions that can be made in investigations involving dental remains.[28]

Three-dimensional (3D) radiographic imaging in dentistry has greatly increased since the publication of the first edition of *Forensic Dentistry*. Radiographic 3D imaging is more fully discussed in Chapter 10. However, other 3D imaging advances are worthy of mention because of their potential application to forensic odontology. Improvements in computer technologies have

Forensic Dental Identification 181

led to the development of hardware and software that can create 3D models following the scanning of objects. The objects can be a tooth, several teeth, or a complete dental model. A scanner developed by 3M, the Lava™ Chairside Oral Scanner, and other intraoral image capture systems of this type are used to create accurate 3D models to be used for the indirect fabrication of fixed dental prosthetics. It is not difficult to envision that these same components or their derivatives could be utilized to make 3D scans of postmortem cases to compare to archived 3D antemortem scans, leading to positive dental identifications. This same technology may become useful in bitemark analysis (see Chapter 14).

These and other advances in dental and forensic technologies coupled with optimal clinical practices that stress meticulous record keeping, including recording the specific dental materials used in patient care, will become important tools for forensic dentists, facilitating identifications and the return of loved ones to their families.

9.8 Ten Tips and Cautions for Dental Identification

1. *You cannot unfill a tooth.* As obvious as this may seem, if, in a postmortem case, a tooth is unrestored, all antemortem records showing a restoration in that same tooth can be eliminated. A restoration may have been placed *after* the most recent antemortem record, but not the opposite—teeth do not heal. Consideration must be given to errors in antemortem charting.
2. *Extracted teeth cannot grow back.* If an antemortem radiograph demonstrates that a tooth is missing and that tooth is present in a postmortem radiograph, the forensic odontologist can exclude that postmortem record.
3. *A body without an antemortem dental record cannot be identified by dental means.* As in all methods of identification, postmortem information must be compared to a known individual's antemortem information. Not all unidentified bodies will ultimately be identified.
4. *Antemortem records without a body = no ID.* Hundreds of dental records exist in missing persons databases, including the NCIC database in Washington, D.C. Until the body of a specific individual is found and the data compared, no identification is possible.
5. *The quality of the antemortem dental record is critical to dental identification.* Charting errors happen! As long as records are kept by humans there will be human error. Dr. Jim McGivney, creator of WinID, characterizes the most common charting errors as Flips, Flops, and Slides. Flips occur when dentists or assistants erroneously

record treatments to the contralateral tooth, the tooth on the other side of the arch, i.e., 19 to 30 (36 to 46 FDI [Fédération Dentaire Internationale]) or 13 to 4 (25 to 15 FDI). Flops occur when restored surface notations are transposed, i.e., MO instead of DO or OF in place of OL. Slides occur, most commonly, when teeth have been lost and other teeth drift into their spaces. For instance, tooth 15 (27 FDI) has drifted mesially and is erroneously recorded as tooth 14 (26 FDI). With the increased use of digital radiography, the incidence of lost radiographs should decrease. Digital dental software programs internally record the date an image is entered into the file. That information is permanently attached to the image. Dentists are encouraged to keep complete and accurate dental records to facilitate dental identification and to protect the dentist in case of legal actions.

6. *The older the antemortem record, the higher the potential for inconsistencies.* When examining antemortem radiographs forensic dentists should remember that additional treatment may have been completed on that patient in the interim period. Ignoring that possibility may lead to recording points of discrepancy that are, in fact, explainable. Every practical attempt should be made to procure the latest available treatment record.

7. *Advances in dental material science have changed dental identification methodology.* The technological advances in dental resin materials coupled with the increased demand for esthetic restorations have further complicated some forensic comparisons. The use of amalgam filling material is declining and is being replaced with the use of increasingly varied composite resin materials. Dental techniques, including microdentistry and the use of flowable resins, have made the postmortem examination process more complex. The radiopacity of dental resins varies, and some are almost radiolucent. These radiolucent restorations can resemble tooth decay on radiographs. Forensic dentists must inspect restorations during postmortem examinations with a great deal more scrutiny than in the past, when "tooth colored" restorations were usually only seen in anterior teeth. Technological aids for identifying composite resins are discussed above.

8. *Things are not always as they appear.* The restored surfaces of a tooth may appear more extensive in the postmortem exam than is recorded in an antemortem record for a specific tooth. The forensic dentist that assumes this to be a discrepancy may make the error of forgetting that additional dental treatment to that tooth may have occurred after the latest antemortem record entry. Multisurface amalgam restorations may have been replaced with full crowns by a different dentist. This is a common occurrence in forensic casework. These occurrences can

be labeled as "explainable discrepancies" or "logical progressions" and are not necessarily reasons to exclude a record in the comparison process *if* other overwhelming information indicates that the two records are records of the same individual. When examining the antemortem radiographs, care must be taken to ensure that the x-rays are oriented correctly. If the antemortem x-rays are duplicates and not labeled as L or R and the film dimple location is indeterminable, then the forensic dentist must request additional information from the submitting dentist, most often the original films. Original films should always be acquired and examined since duplicate films are very often incorrectly oriented (see Chapter 10).

9. *One unexplainable discrepancy is more important than many consistencies.* Dental evidence is useful for both identification and exclusion. Even if five restorations are consistent among the antemortem and postmortem records, significant doubt must be raised if one unexplained exclusionary item is noted. For example, if the antemortem records show a full crown on a certain tooth and the postmortem record shows an occlusal amalgam on the same tooth, the comparison results in exclusion. One unexplainable discrepancy leads to exclusion.

10. *You may only get one chance to get accurate postmortem dental information.* Attention to detail in the postmortem examination is mandatory. Once a body is released, it may be buried or cremated before discovery that a record is inadequate or an image is substandard. Exhumation to recover information that should have been obtained is embarrassing. Attempting to recover the same information from cremains may be impossible.

9.9 Summary

Postmortem dental examinations are sometimes unpleasant but always necessary exercises. The accurate reconstruction of the antemortem record is an equally important phase of the identification process. With careful attention to detail, dental identifications can be completed in a relatively short time period and at a reasonable cost when compared to other means of identification. In some instances, the forensic dentist may find it useful to consider the new technologies available to assist in the comparison process. With advances in this and other forensic identification sciences, new methods will become more commonplace. Through the cooperative efforts of medical examiners, coroners, law enforcement officials, and forensic odontologists, dental comparisons can be efficiently and accurately completed to identify or exclude.

References

1. Cerak, W. Mistaken identity. http://www.mahalo.com/Whitney_Cerak.
2. Van Ryn, D. 2008. *Mistaken identity: Two families, one survivor, unwavering hope.* Waterville, ME: Thorndike Press.
3. Fingerprinting. http://science.howstuffworks.com/fingerprinting.htm/printable.
4. Cassius Dio, C., Earnest, F., Baldwin, H. 1914. *Dio's Roman history.* The Loeb classical library. London: W. Heinemann.
5. 45 CFR 164.512(g)(1). *Standards for privacy of individually identifiable health information.* Final rule. Uses and disclosures for which consent, an authorization, or opportunity to agree or object is not required, uses and disclosures about decedents. http://ecfr.gpoaccess.gov/cgi/t/text/text-idx?c=ecfr&tpl=/ecfrbrowse/Title45/45cfr164_main_02.tpl. 2003.
6. McGivney, J. 1997–2008. WinID3. St. Louis. http://www.winid.com.
7. Lorton, L., Rethman, M., and Friedman, R. 1988. The Computer-Assisted Postmortem Identification (CAPMI) System: A computer-based identification program. *J Forensic Sci* 33:977–84.
8. ABFO. 2009. *Diplomates reference manual.* www.abfo.org.
9. Bollinger, S.A., P.C. Brumit, B.A. Schrader, and D.R. Senn. 2005. Grin line identification using digital imaging and Adobe Photoshop. Paper presented at American Academy of Forensic Sciences, Annual Meeting, F7, New Orleans.
10. Friday, M.F., P.C. Brumit, B.A. Schrader, and D.R. Senn. 2006. Practical application of the grin line identification method. Paper presented at American Academy of Forensic Sciences, Annual Meeting, F6, Seattle.
11. Keiser-Nielsen, S. 1977. Dental identification: Certainty V probability. *Forensic Sci* 9:87–97.
12. Haub, C. 1998. World population: A major issue for the millennium. *Glob Issues* 3:17–19.
13. Keiser-Nielsen, S. 1969. [Forensic dentistry]. *Zahnarztl Mitt* 59:447.
14. Keiser-Nielsen, S. 1974. Dental evidence in the reconstruction of identity. *Int J Forensic Dent* 2:78–81.
15. Sognnaes, R.F. 1975. Oral biology and forensic science. *Annu Meet Am Inst Oral Biol* 126–39.
16. Adams, B.J. 2003. The diversity of adult dental patterns in the United States and the implications for personal identification. *J Forensic Sci* 48:497–503.
17. Steadman, D.W., B.J. Adams, and L.W. Konigsberg. 2006. Statistical basis for positive identification in forensic anthropology. *Am J Phys Anthropol* 131:15–26.
18. Adams, B.J. 2003. Establishing personal identification based on specific patterns of missing, filled, and unrestored teeth. *J Forensic Sci* 48:487–96.
19. Joint POW-MIA Accounting Command, Central Identification Laboratory. Odontosearch. http://www.jpac.pacom.mil.
20. American Dental Association. http://www.ada.org.
21. American Board of Forensic Odontology. www.abfo.org.
22. Flint, D.J., et al. 2009. Computer-aided dental identification: An objective method for assessment of radiographic image similarity. *J Forensic Sci* 54:177–84.
23. Pretty, I.A., et al. 2002. The use of quantitative light-induced fluorescence (QLF) to identify composite restorations in forensic examinations. *J Forensic Sci* 47:831–36.

24. Hermanson, A.S., et al. 2008. Ultraviolet illumination as an adjunctive aid in dental inspection. *J Forensic Sci* 53:408–11.
25. Bush, M.A., P.J. Bush, and R.G. Miller. 2006. Detection and classification of composite resins in incinerated teeth for forensic purposes. *J Forensic Sci* 51:636–42.
26. Bush, M.A., et al. 2007. Identification through x-ray fluorescence analysis of dental restorative resin materials: A comprehensive study of noncremated, cremated, and processed-cremated individuals. *J Forensic Sci* 52:157–65.
27. Bonavilla, J.D., et al. 2008. Identification of incinerated root canal filling materials after exposure to high heat incineration. *J Forensic Sci* 53:412–18.
28. Bush, M.A., et al. 2008. Analytical survey of restorative resins by SEM/EDS and XRF: Databases for forensic purposes. *J Forensic Sci* 53:419–25.

Forensic Dental Radiography

10

RICHARD A. WEEMS

Contents

10.1 Introduction	187
10.2 Dental X-ray Generators	188
10.3 The Radiographic Image and Image Receptors	189
10.4 Film Image Processing	191
10.5 Intraoral Radiographic Techniques	192
10.6 Biological Effects of X-rays	192
10.7 Radiation Protection	194
10.7.1 Maximum Permissible Dose	194
10.7.2 Personnel Monitoring	195
10.7.3 Handheld X-ray Devices	196
10.8 Radiographic Equipment in the Forefront of Forensic Dentistry (CBCT)	196
10.9 Radiographic Pitfalls and Tips	199
10.9.1 Skeletal and Carbonized Remains	199
10.9.2 Issues with Duplicated Radiographs	199
10.9.3 Jaw Fragments	200
10.9.4 Panoramic	200
References	200

10.1 Introduction

Wilhelm Conrad Röentgen is deservedly recognized as the discoverer of x-rays, but he was certainly not the first to produce these invisible rays of energy. In fact, research and development spanning from 1831 until 1895 incrementally led to his discovery. This included work by Faraday, Geissler, and Hittoff in creating and developing the first high-tension electrical evacuated tubes, which produced what were named cathode rays within the device. The cathode rays produced a spark caused by a stream of high-speed electrons traversing a small gap and striking a metal target. This work was followed by Sir William Crookes and Professor Heinrich Hertz, who demonstrated that cathode rays produced florescence and heat within and without the tube. Without doubt, Crookes and Hertz were unknowingly producing x-rays at the time. However, Röentgen did, in fact, discover that other invisible rays emanating from the device possessed the ability to penetrate solid objects and produce photographic shadows of flesh and bones. "This discovery, marking as it did a distinct epoch in the Science of Medicine, was received by the world with incredulity and amazement, for its reported possibilities savored almost of the occult."[1]

Radiographs of nondiagnostic quality were made of teeth within days of Röentgen's discovery, but the first practical and routine use of dental radiography in a dental practice is attributed to C.E. Kells of New Orleans.[2] The first published use of dental radiography as a means of achieving a human identification was in 1943.[3] Today it remains as the leading and most reliable tool for human identification through dental findings. When there is a conflict between the written dental record and antemortem radiographs of a subject, deference is given to the radiographs as the gold standard having less potential for human error than charted dental information. This chapter on the basic theory of dental radiography is presented at a level such that the principles pertinent to the topics and themes most important to forensic dental investigations are emphasized.

10.2 Dental X-ray Generators

X-ray photons are generated when electrons in an evacuated tube are subjected to an extremely high voltage (kV) potential and bombard a tungsten target at a high rate of speed. Thus, electrical energy is converted to kinetic energy, which is then converted to electromagnetic energy. Only 1% or less of the bombarding electron energy is converted into x-radiation, with the remaining energy resulting in a very large gain of heat, which is the greatest cause of tube failure. This fact accounts for the absolute need to follow the manufacturer's recommended duty cycle by waiting the specified amount of time between exposures. The setting of a mass disaster morgue is more likely to destroy a tube head than working on typical dental patients, where the time for placing films after each exposure helps protect the duty cycle.

The resulting x-ray beam is comprised of millions of photons of varying energy (wavelengths) and is referred to as having a continuous or polychromatic spectrum. Older x-ray units produce even more variation in the uniformity of the beam as the alternating current rises and falls. Photons in these units are also produced only when the alternating current (AC) flows in one of two directions. Newer x-ray generators remedy this by utilizing sophisticated electronics to rectify the AC so that the device functions at a constant and continuous kilovolt (kV) potential and are often described as

Forensic Dental Radiography 189

direct current (DC) or constant potential units. These units are more efficient and provide more high-energy, diagnostically useful photons and cut exposure times roughly in half. Older units also have difficulty in producing the extremely short exposure times (usually tenths of a second) required by digital x-ray sensors, which require significantly less radiation than film. One very simple but effective method to accomplish this is to cover the opening of the tube head collimator in an old unit with round sections of rare earth screen material until the beam is weakened sufficiently to allow longer exposure settings comparable to the unit's timer capabilities.

X-rays emanate from the target of the x-ray source in a diverging pattern. Therefore, there is always a varying amount of magnification of the object in any plane film image. The degree of magnification is determined by the ratio of the x-ray source-to-object distance and source-to-film distance. The larger the distance from the source to the image receptor, the less magnification occurs. Likewise, the closer the object to the receptor, the less the magnification and the sharper the image will be. The diverging pattern also affects the radiation safety involved for the operator. That is because the energy of the quickly diverging beam will weaken mathematically as a square of its distance. Therefore, changing the distance of an individual to the x-ray source from 1 foot to 4 feet reduces the dose or intensity of the radiation to 1/16th of the original dose.[4] This should be considered when attempting to configure operator safety in the mass disaster morgue.

New technology in the form of handheld generators that are truly powered by direct current from rechargeable batteries is now in great use in forensic dentistry (Aribex™ Nomad™) but will be discussed further later in this chapter and in other chapters in this textbook.

10.3   The Radiographic Image and Image Receptors

Unlike photography, the radiographic image is not created by energy reflected back to a film but by the transmissive energy of the beam after passing through an object. Therefore, the total of the external and internal structures of the object is represented in the image and not simply the surface area. This is significant in that a radiographic image reveals objects that cannot be perceived with the naked eye. This also means, however, that dental radiographic images require interpretation by the observer because the image is presented as a two-dimensional representation of a three-dimensional object.

Radiographic images of the teeth and maxillofacial structures can only be created due to the fact that the beam of electromagnetic energy is attenuated in varying degrees, depending on the absorption characteristics of different structures through which it passes, and that recording media will react differently depending on the energy received. Thus, an amalgam restoration absorbs much more energy than its surrounding enamel and dentin, allowing less energy to reach the receptor at that location and creating an invisible latent image. Through some mode of processing of the receptor, the amalgam will later be displayed in an image that can be detected visually. With dental film the processing involves chemicals; with digital sensors it may involve modern electronics and lasers.

Dental x-ray "receptors" have undergone numerous changes over the history and advances in dental radiography. Included have been chemically coated glass photographic plates, intraoral films covered with silver halide emulsions of various grain sizes and image speeds, extraoral silver halide films requiring light-emitting screens, direct solid-state digital charge coupled device (CCD) and complementary metal-oxide semiconductor (CMOS) detectors, and photostimuable phosphor plates.

Each receptor system has its advantages and disadvantages. Film is inexpensive, easy to place, and produces an exceptional image if proper techniques are used, but requires several minutes of chemical processing, ideally in a darkroom setting. Also, film can be fogged by heat, light, pressure, and overactive chemicals. Unused film eventually must be discarded if it becomes out of date. Direct digital sensors (CCD and CMOS) provide an immediate image with no chemical processing or darkroom needs. The ability to enhance the image density and contrast often reduces retakes. It provides an image with spatial resolution ranging from 12 to 20 mm/lp.[6] Yet a comparative study showed that digital images are of equal quality to film when evaluating interproximal caries.[5] However, digital sensors are delicate, expensive, in some cases difficult to place, and require enormous amounts of digital storage for the resultant images. Most agree that direct digital is the best image receptor in a mass fatality incident due to it providing an immediate image and dealing effectively with the exaggerated flow of victims in the morgue setting (Figure 10.1).

Phosphor technology provides an excellent image with placement ease similar to that of film but requires several minutes of processing time as the plate must be scanned by a laser drum to produce the image. The phosphor plates also scratch easily and must be replaced fairly frequently, depending on care of handling.

The most useful feature of the three digital systems is the ability to marry the resultant images with third-party software such as Photoshop™ and WinID™, and the abilities gained through an increasing technology, teleradiography, where radiographic images are interpreted at a distance via the Internet. Digital imaging also produces 256 shades of gray. While humans and some computer monitors cannot discern that many gray levels, interpretive software can. This may eventually lead to software with the ability to "match" dental images in determining a dental identification at a rate that would far exceed human ability. To ensure interoperability with third-party

Forensic Dental Radiography

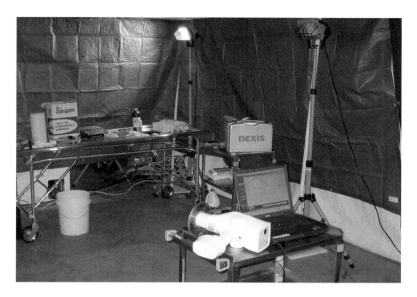

Figure 10.1 Direct digital x-ray system integrated with WinID in the DMORT (Disaster Mortuary Operational Response Team) Katrina East Morgue. No x-ray film or paper records were used in collecting postmortem data.

software, only devices that are DICOM3 compatible (Digital Imaging and Communications in Medicine) should be used.

Finally, if direct digital radiography is not available, scanning traditional radiographs into digital format with a digital flatbed scanner is an alternate technique. The scanner must have transparency/direct positive capabilities. Once digitized, these images can be integrated with any software system used in forensic dentistry with abilities equal to direct digital.

10.4 Film Image Processing

Since traditional film technology does not allow digital image improvement, it is imperative that processing be conducted with strict adherence to manufacturer's specifications. In general, manual processing of radiographs at 70°F requires a five-minute development cycle followed by a thirty-second rinse and a ten-minute fixation cycle. A twenty-minute wash cycle is necessary to produce films with archival quality. If not washed thoroughly, the fixer solution will continue to act on the film after processing and will eventually tint or discolor the image and can destroy its diagnostic content. Automatic processors most commonly produce a dry, processed film in about five minutes. Manual processing chemicals should never be used in automatic processors and vice versa. Inexpensive film processors with small chemical tanks and a passive water wash tank should not be used in mass fatality incidences where the film volume will rapidly weaken fixer solutions and contaminate the wash tank. Likewise, "endo" processor settings should never be used due to underprocessing and an eventual loss of archival film quality.

Darkroom safelights must also be used properly, including matching the appropriate safelight filter with the type of films and extraoral screen being used. The Kodak GBX-II is safe for all dental films currently available. However, it is important to note that films are not totally insensitive to the light emanating from appropriate safelights. Regardless of the filter used, the safelight must be positioned at least 4 feet away from the work surface area, and bulbs within the safelight should be no stronger than 15 watts. Working time under safelights should also be restricted to as short a time as possible.

10.5 Intraoral Radiographic Techniques

Many x-ray generators have variable kilovoltage (kVp), milliampere (mA), and exposure time settings. Variation in milliampere and exposure time will affect only the density (overall blackness of the resulting image) and have no effect on visual contrast (shades of gray). They affect the image density equally so that doubling the milliampere setting on the unit will allow the x-ray exposure time to be cut in half and vice versa. Variation in the kilovoltage setting will, however, affect both density and contrast. As the kilovoltage is increased there will be an increase in density. However, there can also be an undesirable increase in shades of gray in the images with a loss of distinct blacks and whites, resulting in low visual contrast in the image (Table 10.1).

The projection geometry when exposing images of a specimen is simple and straightforward. The receptor should be placed parallel to the dentition and the beam should be directed perpendicular to the receptor plane (Figure 10.2). Ideal imaging in many cases may be improved by resecting the mandible and maxilla.

10.6 Biological Effects of X-rays

All human tissues are adversely affected by ionizing radiation, particularly cells with high mitotic rates such as those in developing embryos, blood-producing tissue, and reproductive organs. Were it not for this selective sensitivity, x-radiation as a treatment to kill rapidly growing, immature tissues of neoplasms while creating less damage in normal healthy tissues would be ineffective. Tissue and host damage occurs through two mechanisms: direct and indirect. When the damaging energy is delivered directly from the x-ray photon to the molecule, it is deemed a direct effect. An indirect

Forensic Dental Radiography

Table 10.1 Factors Affecting the Image Quality

Factor	Density	Contrast Scale (Shades of Gray)
mA	Increases	
Exposure time	Increases	
kVp	Increases	Increases (more shades of gray)
Aluminum	Decreases	Increases (more shades of gray)
Distance	Decreases	

Exposure time and mA control density; kVp alters film contrast.

Figure 10.2 Setup for exposing a postmortem radiograph. Proper projection geometry is critical whether using film or digital sensors.

effect occurs when the x-ray energy ionizes water into ion pairs or free radicals, which then produce damage to molecules within the host. Most damage from x-radiation does occur through the ionization or radiolysis of water. There are numerous biological molecules that can be adversely affected by x-ray energy, including nucleic acids and proteins. Damage at the cellular level affects the nucleus, cytoplasm, and chromosomes and may result in cell and eventually host death. Given enough ionizing radiation, damage will occur with whole body systems such as the gastrointestinal organs and hematopoetic system.

Predicting damage from x-radiation falls follows two distinct models: deterministic and stochastic. Deterministic effects occur with large doses that produce certain types of bodily damage for which a definite threshold may be determined and damage increases above that threshold. Stochastic effects typically do not have a known threshold, are all or none in effect, and typically relate to cancer and genetic effects. In lieu of a threshold with stochastic damage, it is a matter of continuing higher odds of damage as the dose is increased. Dental radiographic exposures are typically low to the point of not reaching threshold doses required to cause deterministic somatic damage such as skin burns, damage to the lens of the eye, or hair loss. However, the development of cancer does not have a defined dose threshold. One study determined that the risk of fatal cancers including all organs was 0.25 per 1,000,000 full-mouth radiographic exams.[7] Risks in this and other similar studies, however, are estimated using a linear nonthreshold model. Actual fatal cancer risk for this amount of radiation may be more, less, or none at all; it is scientifically unproven. For the purpose of this textbook, it is obvious that the greatest concern to the forensic dentist involves protection to the operator from tube head leakage and scatter radiation. However, any measures taken to reduce the exposure dose and scatter radiation to victims also reduce exposure and its ill effects to all personnel in the immediate work zone.

10.7 Radiation Protection

Radiation protection standards regarding x-ray machine safety and performance are promulgated by the Food and Drug Administration.[8] The National Council on Radiation Protection and Measurements (NCRP)[9] published new guidelines and recommendations relative to use of x-radiation in dentistry in 2004. The document containing these guidelines supersedes NCRP Report 35, which was published in 1970. The purpose of the NCRP reports is to give guidance to the dental profession in protecting the public and those occupationally exposed to the dangers of excessive x-radiation. The radiation safety guidelines proposed by the American Dental Association typically follow the NCRP reports. This section lists and describes the guidance provided by this document.

10.7.1 Maximum Permissible Dose

Agencies typically follow the more conservative stochastic model when determining safety limits for the population, dental patients, and dental personnel.

Limits to x-radiation occupational doses are known as maximum permissible doses (MPDs). The MPDs are considered to be at a level that the body can tolerate with little or no damage. The dose limits are established by comparing to estimated risks of occupational injury in other vocations that are considered to be safe, such as agriculture, construction, and manufacturing. The yearly maximum for occupationally exposed individuals is 50 mSv per year, and the public limit is one-tenth of that, at 5 mSv per year (Table 10.2). Note that the cumulative dose becomes more restrictive than the annual dose over time. For example, a person who begins receiving an annual dose of 50 mSv every year will exceed his or her cumulative maximum of 230 mSv at age twenty-three. No ill effects or injuries have been demonstrated as a result of anyone receiving exposures equal to established MPDs.[10]

Forensic Dental Radiography

Table 10.2 Maximum Permissible Doses for Dental Operators

Occupational	50 mSv annual effective dose
	10 mSv × age (y) maximum cumulative effective dose
Public	5 mSv annual effective dose for infrequent exposure
Embryo and fetus	0.5 mSv equivalent dose in a month from occupational exposure of the mother once pregnancy is known
Annual U.S. background	3.65 mSv per year

Public MPDs are one-tenth that of those who are occupationally exposed.

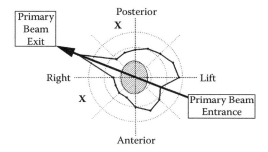

Figure 10.3 Graphic showing typical scatter pattern involved when exposing a specimen. The safer areas are behind, indicated by X. (Plotted from data from De Haan and Van Aken.[11])

Protection of x-ray operators and surrounding personnel is provided by the patient protection practices detailed above and by workplace wall shielding and the maintenance of proper operator distance. Operators must be able to stand at least 2 m away from the source of x-ray scatter (i.e., the patient's head) and out of the primary beam (Figure 10.3). If this is done, no barrier is required. If the operator cannot stand at this desired distance, a protective barrier or apron must then be provided. Most published research shows that x-ray doses in dentistry continue to decline. This is most likely due to the dramatic reduction in exposure times brought about by high-speed film and even more so by direct digital receptors.

10.7.2 Personnel Monitoring

The decision of whether dental personnel are monitored for x-ray exposure depends on state law and the expected exposure predicted on the basis of workload and protective measures. The NCRP recommends monitoring to all personnel who are likely to receive an effective dose greater than 1 mSv/year. Many states require monitoring only if the exposure will exceed one-tenth of the applicable MPD (5 mSv). Past studies of occupational exposure to a large sample of dental personnel have shown that the typical exposure is less than the minimum detectable dose, or less than 1 mSv/year.[12,13] These data suggest that it is *not* expected that dental personnel will reach the monitoring threshold of 1 mSv/year, and hence monitoring is not warranted. However, the MPD applicable for the fetus of an occupationally exposed pregnant worker of 0.5 mSv/month indicates that those individuals be monitored once pregnancy is discovered.

Also, special considerations may be necessary in a disaster morgue setting where numerous individuals are nearby the x-ray source in the dental section. This is based on recommendations by the NCRP concerning open-bay dental units that a patient in the room during a diagnostic exposure of another patient shall be treated as a member of the public. When portable x-ray machines are used, all individuals in the environs not directly involved with obtaining the radiographic image should be deemed as "public" when considering their MPD levels.

10.7.3 Handheld X-ray Devices

Handheld x-ray-generating devices have come into common use in forensic dentistry, particularly in multiple fatality incident morgues. In fact, the Aribex Nomad was used to produce all of the radiographic images in the Disaster Mortuary Operational Response Team's (DMORT) response to Hurricane Katrina. During that operation it was assumed that operators using the device were "safe" based on dose studies conducted by the manufacturer and the device's approval by the U.S. Food and Drug Administration. These studies demonstrated extremely small tube head leakage and backscatter to the operator. Subsequent research simulating morgue usage in a mass fatality setting has confirmed the safety of those working within the dental station. A study simulating doses to those comprising the dental team over a two-week deployment (5,760 exposures) showed that the team member receiving the highest dose at that position (60° to the side of the emanating beam) received an exposure of 0.253 mSv. This dose corresponds to 1/200th of the annual occupational limit of 50 mSv and approximately 3.5 weeks of the U.S. average background radiation.[14] Another study that simulated the dose received at numerous bodily locations revealed a dose of 0.454 mSv after 7,200 exposures, which equals 0.9% of the annual limit.[15]

10.8 Radiographic Equipment in the Forefront of Forensic Dentistry (CBCT)

Dental cone beam computed tomography (CBCT) is an advanced digital x-ray system that was released in 2001. A tomograph is a radiographic presentation of a thin slice of tissue selected from within the object being imaged. CBCT provides numerous new and traditional image projections related to various tasks related to dental diagnosis, such as panoramic, cephalometric, and TMJ images. The device also allows the creation of multiplanar, real-time reconstructions (MPRs) of the maxillofacial structures from three planes: axial, sagittal, and coronal, all from one scan. The three planes are also linked to each other so that any entity within the object may be observed from three different views in an interactive manner.

This CT system is termed cone beam because the beam is not thin or fan shaped such as that used in typical medical CT units. The beam is instead three-dimensional and produces three hundred or more full-size images of the selected tissue field from 360° of rotation about the patient. Each individual projection is similar to a lateral or posterior-anterior (PA) cephalometric radiograph. The transmitted energy from the tube is received by either an image-intensifying tube married to a charge coupled device (CCD) or a flat-panel receiver composed of amorphous silicon. The CBCT image is comprised of three-dimensional image units known as voxels. A voxel is similar to a pixel except that it includes the third dimension. Where a pixel is a lateral and vertical image unit, the voxel is in the shape of a cube (e.g., 0.2 × 0.2 × 0.2 mm). The equal-sided voxels are termed isotropic, unlike medical CT systems in which voxels are rectangular with an axial measurement of 2 mm and cannot reach the resolution provided by CBCT. The resolution of the images is determined by the voxel size and can range from 0.4 mm to as small as 0.125 mm in some units. Cone beam CT also produces fewer streak or *star* artifacts than medical CT when the beam strikes metallic objects such as dental restorations within the scan zone. The scatter artifact is also primarily confined to the horizontal plane.[16]

After the captured images are acquired, they are reconstructed together in all planes by the system's computer algorithm and displayed on a computer monitor. Other than that fundamental principle, CBCT units vary in size, shape, image receptor technology, and the volume size of the tissue field that may be captured. At least two machines currently available scan the subject in the supine position, which makes them suitable as morgue and mass disaster equipment.

The imaging possibilities with CBCT technology and its possible forensic applications are almost limitless. Any intraoral or extraoral film image view can be reconstructed from one scan, with the operator having the ability to select any desired slice location and orientation. Thus, it is possible to plot a plane within the confines of the dental arches and produce a panoramic image without the overlying ghost images and artifacts of conventional panoramic radiography. The thickness of the panoramic image layer may also be determined in an unlimited manner. Also, computer algorithms correct for the geometric distortions that are present within all plane film images. Thus,

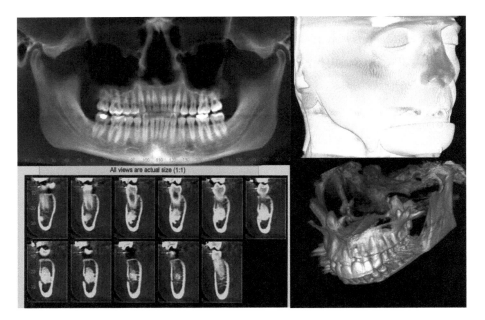

Figure 10.4 Multiple cone beam CT images. Several programs are available that provide three-dimensional rendering of the soft tissue. (See color insert following page 304.)

there is no magnification and the printed images are 1:1 representations of the maxillofacial structures (Figure 10.4).

It is only a matter of time before a victim identification case is solved using the image data from antemortem and postmortem CBCT scans, as these machines are growing rapidly in the private sector. Also, a single scan of the victim can later be compared to any possible variety of submitted antemortem plane film images (i.e., panoramic, periapical, occlusal, PA skull, Waters, etc.). The CBCT multiplanar views (particularly the axial) exhibit structures that have not been discernable previously. For example, the paranasal sinuses are clearly seen (Figure 10.5) and may possess enough unique bony architecture to allow for human identifications.[17] Three-dimensional images of the soft tissue of the face may also be re-created from scan data when manipulated by third-party software systems allowing comparison to photographic images of the victim.

Finally, there are other applications of this technology under investigation, including one in which postmortem dental and skull fragments scanned by CBCT might be compared to the three-dimensional antemortem image of a victim's skull image also obtained from a CT scan, as well as the possibility of aiding in artist reconstructions of soft tissue added to CBCT stereolithic models of missing persons' skulls.

Forensic Dental Radiography 199

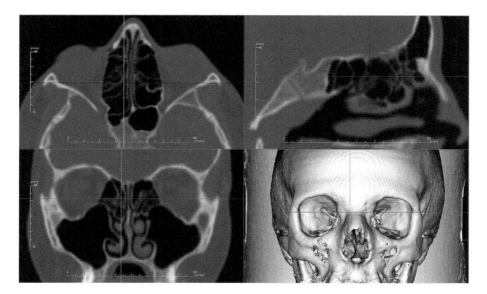

Figure 10.5 CBCT multiplanar images of the paranasal sinuses. May become useful in victim identification. (Courtesy of Dr. Douglas Chenin, Anatomage, Inc.)

10.9 Radiographic Pitfalls and Tips

10.9.1 Skeletal and Carbonized Remains

Resected maxillas and mandibles require about 50% of the beam exposure energy needed for normal tissue of the oral cavity. With skeletal and carbonized tissues, beam energy should be reduced to 30%.

10.9.2 Issues with Duplicated Radiographs

Many dentists in the United States orient dental films on the view box and in film mounts with the film's orientation dimple facing upward or out. This establishes the orientation of the film series as if the viewer is facing the patient. However, a very common concern in forensic dentistry is the orientation of duplicate antemortem radiographs where the dimple of the original film can be seen but not felt. This can confuse the issue of the patient's right from left. The most reliable method of determining if the dimple was up at the time of duplication is to consider the location of the bump relative to the corners of the individual films themselves. Whenever the dimple is placed upward or out on a film, its location will always be in the lower right or upper left corner of the film when it is oriented horizontally (Figure 10.6). The letters stamped on the film read correctly when the dimple is up or out when viewed on the view box.

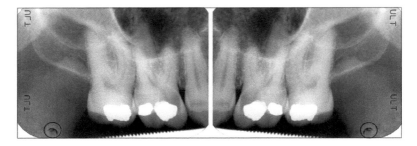

Figure 10.6 The film on the left is oriented dimple down (note bottom left position). The film on the right is dimple up (note bottom right position). Ultra™ film speed notation ULT reads correctly on right image.

10.9.3 Jaw Fragments

When taking radiographic images of jaw fragments, care must be taken that the beam and film orientation be the same as would be expected on a living patient. That is, the beam should enter from the buccal or facial with the film placed to the lingual. Should this not be taken into account, the film-orienting dimple will be positioned incorrectly and the teeth will appear to be from the opposite side of the arch and will be misidentified.

10.9.4 Panoramic

In panoramic radiography the beam direction is the opposite from intraoral radiography. That is, the beam is always directed upward and from lingual toward buccal or facial. Consequently, when attempting to emulate antemortem panoramic views with postmortem intraoral radiography, it may be helpful to reverse the beam direction by placing the film or sensor on the buccal or facial and directing the beam from the lingual upward. Also, the laws of the buccal rule will be reversed from what is normally true. As the panoramic beam is directed upward, lingual objects will be projected higher than buccal objects that are at equal heights.

References

1. McCoy, J.D. 1916. *Dental and oral radiography*. St. Louis, MO: C.V. Mosby Company.
2. Langland, O., Sippy, F., and Langlais, R. 1984. *Textbook of dental radiology*. 2nd ed. Springfield, IL: Charles C. Thomas.
3. Fry, W.K. 1943. The Baptist Church cellar case. *Br Dent J* 75:154.
4. Goaz, P., and White, S. 1994. *Oral radiology: Principles and interpretation*. 3rd ed. St. Louis, MO: Mosby.

5. Duncan, R., Heaven, T., Weems, R., Firestone, A., Greer, D., and Patel, R. 1995. Using computers to diagnose and plan treatment of approximal caries detected in radiographs. *J Am Dent Assoc* 126:873–82.
6. Dunn, S., and Kantor, M. 1993. Digital radiography: Facts and fiction. *J Am Dent Assoc* 124:38–47.
7. White, S.C. 1992. Assessment of radiation risk from dental radiography. *Dentomaxillofac Radiol* 21:118–26.
8. *Code of Federal Regulations*, Title 21, Vol. 8, April 2004.
9. Radiation Protection in Dentistry. 2004. NCRP Report 145. Bethesda, MD: National Council on Radiation Protection and Measurements.
10. Taylor, L.S. 1980. Let's keep our sense of humor in dealing with radiation hazards. *Paerspect Biol Med* 23:325–34.
11. De Haan, R., and Van Aken, J. 1990. Effective dose equivalent to the operator in intraoral dental radiography. *Dentomaxillofac Radiol* 19:113–18.
12. Kumazawa, S., Nelson, D., and Richardson, A. 1984. Occupational exposure to ionizing radiation in the United States: A comprehensive review for the year 1980 and a summary of trends for the years 1960–1985. *EPA* (National Technical Information Service) 520:1–84.
13. Weems, R. 1986. Radiation monitoring of dental school personnel. *J Dent Ed* 50:274–76.
14. Hermsen, K., Stanley, J., and Jaeger, M. 2008. Radiation safety for the Nomad™ portable x-ray system in a temporary morgue setting. *J For Sciences* 53:917–21.
15. Danforth, R., Herschaft, E., and Leonowich, J. 2009. Operator exposure to scatter radiation from a portable hand-held dental radiation emitting device (Aribex™ NOMAD™) while making 915 intraoral dental radiographs. *J For Sciences* 54:415–421.
16. Scarfe, W., Farman, A.G., and Sukovic, P. 2006. Clinical applications of cone-beam computed tomography in dental practice. *J Can Dent Assoc* 72:75–80.
17. Weems, R. 2008. Cone beam CT radiography for dental identifications. Paper presented at the annual meeting of the American Academy of Forensic Sciences, Washington, DC.

Forensic Dental Photography

11

FRANKLIN D. WRIGHT
GREGORY S. GOLDEN

Contents

11.1	Introduction	203
11.2	The Spectrum of Light	204
11.3	Properties of Illumination	205
11.4	Properties of Injured Skin: Inflammation and Repair	207
11.5	Forensic Dental Photography: Types and Techniques	211
11.6	Visible Light Photography	212
11.7	Film-Based Photography	214
11.8	Digital Photography	216
11.9	Alternate Light Imaging and Fluorescent Imaging Techniques	219
11.10	Nonvisible Light Photography	222
	11.10.1 Ultraviolet (UV) Light Sources	226
	11.10.2 Infrared (IR) Light Sources	226
11.11	Focus Shift	226
11.12	Reflective Long-Wave Ultraviolet Photography	229
11.13	Infrared Photography	232
11.14	Application and Use of Forensic Photography	234
11.15	Management of Forensic Photographic Evidence	237
References		243

11.1 Introduction

The concept of accurate forensic photography has established itself as a crucial part of forensic investigation as a means of documenting evidence. Historically, photography has been the most significant method of preserving the physical evidence of patterned injuries in skin. The need to accurately photograph injury patterns as they appear on skin is paramount to the odontologist, pathologist, law enforcement, and the legal system. Since vast amounts of time often elapse between the commission of crimes and the trial of the perpetrator, photographs frequently are the only permanent record of the injuries to the victims. Therefore, it is imperative that the forensic investigator be able to properly photograph injury patterns as a means of preserving such evidence.

This chapter is better understood if the reader has a good grasp of photographic terminology and the skills for operating basic camera equipment. It is not the authors' intent to attempt to cover basic photography. There are many publications that can provide the necessary background to improve one's understanding of the photographic principles described in this chapter. Two readily available and easy reading books are *Basic Photography* by Michael Langford[1] and *The Basic Book of Photography*, Fifth Edition, by Tom Grimm.[2]

This chapter will provide an overview of the various techniques utilized by the forensic dental photographer to properly record injuries using advanced photographic methods. It will also present the historical photographic techniques utilizing traditional film and the exploding era of digital imaging.

11.2 The Spectrum of Light

The process of photographically recording images on film, videotape, or magnetic media occurs through the capture of electromagnetic radiation (light) of specific wavelengths. These wavelengths are measured in millionths of millimeters, referred to as nanometers (nm). Photographic images can be recorded on film emulsions that are sensitive to light wavelengths anywhere between 250 and 900 nm. Visible light, which we see with our unaided human eye, comprises only from 400 to 760 nm. Most modern digital cameras and traditional photographic films are specifically designed to record images seen in the visible range of light as we see them However, it is also possible to record images we cannot see when specifically illuminated in the shorter ultraviolet range (210 to 400 nm) and longer infrared range (750 to 900 nm). Since ultraviolet and infrared radiations are outside the visible range of the spectrum, they are commonly referred to as nonvisible light. Recent generations of digital cameras have been designed to allow the recording of patterned injuries in skin using both visible light and nonvisible light. While the electronic transfer of light to magnetic recording media is very different than exposing photographic film, for the most part, the techniques utilized for image capture are basically the same. Photography using nonvisible light requires special techniques to record the injury, including an occasional minor focusing adjustment called focus shift[3] that provides correction for the optical properties of lenses that were designed primarily to be used for visible light photography (Table 11.1).

Forensic Dental Photography

Table 11.1 Spectrum of Electromagnetic Radiation

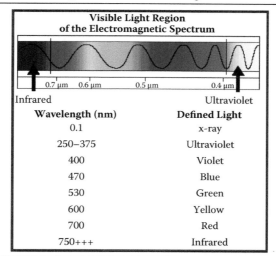

Wavelength (nm)	Defined Light
0.1	X-ray
250–375	Ultraviolet
400	Violet
470	Blue
530	Green
600	Yellow
700	Red
750+++	Infrared

Source: Data from Kochevar et al.,[4] with permission. Color spectrum image from www.science.hq.nasa.gov/kids/imagers/ems/visible.html.

11.3 Properties of Illumination

When an object is illuminated with white light, four phenomena take place.[4] *Reflection* occurs when electromagnetic radiation strikes and bounces back from the object. It is this reflection of visible light that accounts for the colors seen by the human eye. Not all light energy that strikes an object is reflected. Some of the light can be absorbed. It is the *absorption* of all colored light by an object that makes that object appear black. A third reaction that occurs, especially when light strikes human skin, is the *transmission* and scattering of the energy associated with the light through successive layers of cells until the energy of the light is spent and has dissipated. The final reaction, which occurs when light energy strikes an object, is a molecular excitation called *fluorescence* (Figure 11.1).

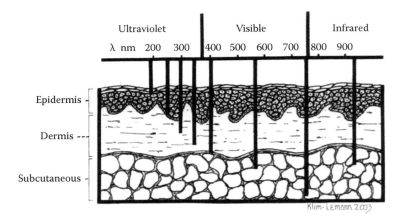

Figure 11.1 Diagram of levels of light penetration into skin when illuminated by varying wavelengths.

Molecules in tissue absorb the energy from light and release that energy as a fluorescent glow. It only lasts as long as the light's excitation energy is applied, usually about 100 nanoseconds (10^{-9} seconds).[5] Fluorescence of skin occurs at a very low level of energy and usually cannot be seen by the naked eye. Filters are generally required. Fluorescent inks and paint can be easily seen under "black light" (near UV), as well as latent blood and fingerprints, when treated with chemicals that facilitate viewing the level of excitation.

When light energy of various wavelengths strikes human skin, all four of the previously mentioned events can occur simultaneously. Depending on the wavelength of the source of the incident light and the configuration of the camera, lenses, and filters, it is possible to record, individually, any of the four reactions of skin to light energy (Figure 11.1).

Ultraviolet light only penetrates a few microns into skin, whereas infrared light can penetrate skin to a depth of up to 3 mm.[3] What is usually seen when visible light strikes the skin is reflected light energy. What isn't seen, however, is the light energy that is absorbed by the skin. By varying the wavelength of incident light used for illumination and setting up the appropriate configuration of the camera, lens, filters, and film, it is possible to photograph any of the four events that occur. This ability creates an opportunity for interesting pictures, especially when looking at bruises and other injuries to skin. Sharp surface details can be seen with ultraviolet light, while images well below the surface of the skin can be seen using infrared light. Images created using reflected visible light and fluorescence allow other potentially different appearances of the patterned injuries to be captured. The techniques and photographic protocols for documenting injuries to human skin in visible and nonvisible light using film and digital imaging are vastly different.

Forensic Dental Photography 207

11.4 Properties of Injured Skin: Inflammation and Repair

In forensic photography it is equally important to understand the physiological changes that occur in living tissue when skin is injured. These changes from normal to injured to healing states allow discriminatory recording of the contusions illuminated by light sources of various wavelengths.

Since the healing process is long term and ongoing, it is sometimes possible to photograph an injury after that tissue appears to have healed to the naked eye. Similarly, photographing the same injuries over time using both visible and nonvisible light techniques can yield different appearances of those injuries, which can sometimes add to their evidentiary value (Figures 11.2 to 11.10).

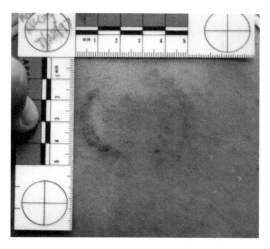

Figure 11.2 Back, color, day 1. (See color insert following page 304.)

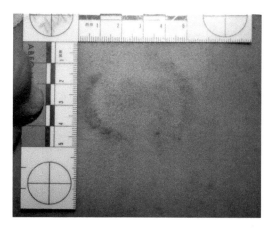

Figure 11.3 Back, color, day 8. (See color insert following page 304.)

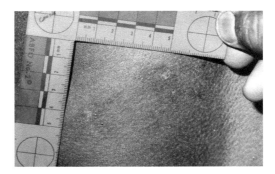

Figure 11.4 Back, UV, day 1.

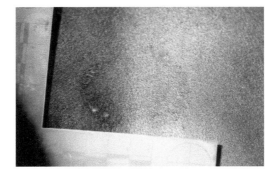

Figure 11.5 Back, UV, day 8.

Figures 11.2 to 11.5 show the same bitemark on the back over eight days. Figures 11.2 and 11.3 are color images and Figures 11.4 and 11.5 are UV images. This series of images indicate that day 8 was the best interval for UV imaging. Note the increased details of the bitemark as seen in the day 8 UV image.

Success in photographing healing bruises over time will depend on several variables, including the composition of the injured skin, the thickness of the skin, the wavelength and intensity of light used to photograph the damaged area, the equipment used, and the type of film used. Depending upon the specific injury, it may be necessary to photographically capture the injuries digitally or with film, in color and black and white using visible light, as well as nonvisible light. The injury may also vary in appearance in the photographs of each of these incident light sources and over time if photographed serially.

The location and type of skin injured has profound effects on the ability to photograph the injuries. For example, thick skin of the palm of the hand is usually much easier to photograph immediately after an injury than after it has partially healed. The thick, keratinized covering of the palm of the hand often exceeds the ability of most light energy to penetrate enough to record

Forensic Dental Photography

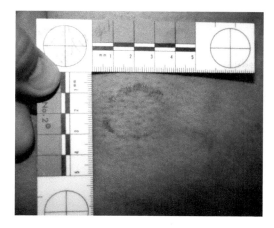

Figure 11.6 Hip, color, day 1. (See color insert following page 304.)

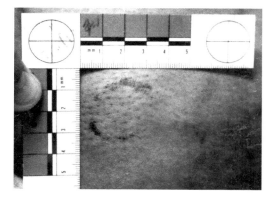

Figure 11.7 Hip, grayscale, day 1.

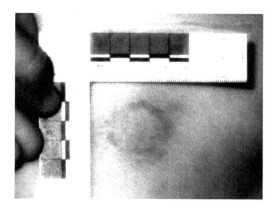

Figure 11.8 Hip, IR, day 1.

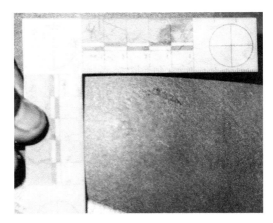

Figure 11.9 Hip, UV, day 1.

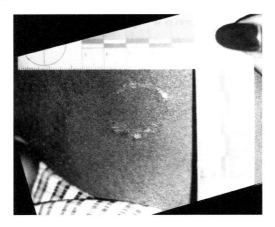

Figure 11.10 Hip, UV, day 8.

Series of photographs showing the appearance of a bitemark on the hip of a sexual assault victim. Figures 11.6 to 11.9 document the bitemark on the day of the injury in color, grayscale, infrared (IR), and ultraviolet (UV), respectively. Figure 11.10 is an image made on the eighth day after the attack. Note the significant increase in the detail of the bitemark discernable in this ultraviolet image. This injury was still clearly visible on day 28 in ultraviolet light images, the last time the injury was photographed by author FW.

the subepithelial injuries. Such cases require fluorescent photography due to the highly fluorescent nature of thick skin. In contrast, thin skin found in areas such as the face or female breast tissue can frequently lend itself to recording injury patterns long after the visible damage from the injury has faded when viewed under room lighting with the naked eye. Again, this is directly related to the ability of the specific wavelength light energy to react with the skin to a sufficient depth to record the injury.

It is important to understand a final concept about the physiology of injured skin. In a living victim, at any point in time after the initial response to injury occurs, the composition of the injured skin is different from all other points in time; the skin is continuously changing as the injury heals. These changes will affect the response of the tissues to the selected wavelength of light energy over time and may change the appearance of the injured tissue when photographed. This is especially true in a living victim but it is also true in a deceased victim, for the first few days after death. If the anticipated photographic evidence is not obtained with the first attempts, one should continue to rephotograph the injuries over time. If the victim is deceased, one should attempt to have the medical examiner or coroner delay any invasive autopsy procedures of the tissues associated with the injury until all attempts to photograph the injury have been exhausted.

11.5   Forensic Dental Photography: Types and Techniques

When presented with an injury, the forensic dentist or investigator must decide what information the injury may contain, the extent of the injury, and how best to photographically record it. As previously mentioned, preserving the detail of the injury with photographs may involve a combination of color and black-and-white visible light photographs as well as the use of the nonvisible ultraviolet and infrared photographs. The photographer should develop a *standard technique* that includes capturing orientation photographs showing where the injury occurred on the body. Additionally, this protocol should include taking closeup photographs with and without a scale. The scale serves as a reference to record the relative size of the injuries in the photographs. The evolution of scales for forensic photography resulted in the development of a two-legged (right-angled) scale, known as the ABFO 2® scale, which is used by modern crime scene photographers. This scale was developed by a photogrammetrist (Mr. William Hyzer) and a forensic dentist (Dr. Thomas Krauss) for the purpose of minimizing photographic distortion and ensuring accuracy in measurement (Figure 11.11). The ABFO 2 scale is now available through Amour Forensics, a branch of Lightning Powder Company, Inc. As this book goes to press, an ABFO 3 scale is being designed and developed in cooperation with the National Institute of Standards and Technology (NIST).

The photographer should retain the scale used in the photograph should enlargement to life-size reproductions become necessary at some future time. It is essential that the *standard technique* developed by the forensic photographer include exposing many photographs for each case. One should not be hesitant to use several rolls of film or create significant numbers of digital images for a photo shoot. Typically, most beginning

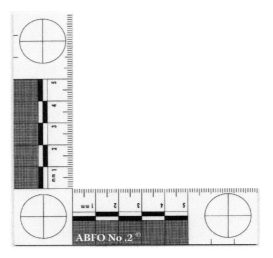

Figure 11.11 ABFO 2 scale.

forensic photographers do not take enough photographs in the first few cases in which they become involved.

11.6 Visible Light Photography

By far, the most common types of modern photography use visible light. Manufacturers of film-based and digital photographic equipment develop and market equipment and supplies that are specifically designed to have an optimal performance in the 400 to 760 nm range of the electromagnetic spectrum. For the photographer wishing to take pictures in this range of light energy, there is little practice required to ensure highly detailed and sharply focused photographs.

Many 35 mm film-based and digital SLR cameras are considered "automatic" point-and-shoot cameras. By definition, the object to be photographed is viewed through the lens and the camera automatically adjusts the focus and exposure variables before image capture. These types of cameras have been manufactured for optimal photography in the *visible light spectrum*. Part of that manufacturing process includes coated lenses and filtered flash units that block out the unwanted segments of the nonvisible light spectrum.

Most 35 mm cameras have serious size limitations when it comes to recording life-size images. This limitation comes from the restricted emulsion area exposed on the film (24 × 35 mm rectangle) or the small surface area used in the digital capture device. Since there are very few objects that will fit into that small of an area, considerable enlargement of the photographs may

Forensic Dental Photography

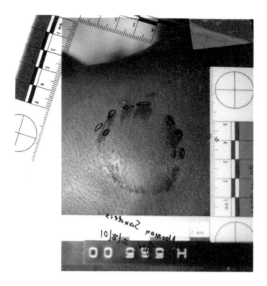

Figure 11.12 Deltoid with maxillary hollow-volume overlay (initial contact).

Figure 11.13 Deltoid with maxillary hollow-volume overlay (end position).

Maxillary overlay of hollow-volume exemplars of a suspected biter on the bitemark at the initial contact position (Figure 11.12) and the end position (Figure 11.13).

be necessary to see the injuries life size. Evaluations and comparisons of the injuries to the teeth, weapons, or tools that created them are often done in direct relation to the life-size object, creating the need for photographs that can be enlarged to life size without loss of the detail necessary for the comparison (Figures 11.12 and 11.13).

11.7 Film-Based Photography

Film manufacturers have designed photographic films that record light wavelengths from 250 to 700 nm. Special infrared films are available that can record photographs captured in light from 250 to 900 nm.[32,34] Choosing the proper film is critical for successfully recording the detail of an injury. The film must be sensitive to the wavelength of light being used to photograph the injury or no image will appear when the film is developed. There are many quality photographic negative films manufactured, in both color and black and white. Table 11.2 suggests some readily available films, processing techniques, and their potential uses in forensic photography.

In addition to the photosensitivity range of the film, the correct film speed must also be determined. Films come with a rating, referred to as the ASA/ISO number, which serves as an indicator of the amount of light energy necessary to properly expose the film. The higher the ASA/ISO number, the faster the film; in other words, less light is needed to expose an image. Films with high speed ratings (ASA 1600 or 3200) require very little light energy to expose, but caution must be exercised. The higher the ASA/ISO number, the lower the grain density on the film where the image is recorded, which translates into less versatility during enlarging. Large-grain fast films tend to produce prints that appear to lose focal sharpness and detail as they are enlarged toward their normal limits, i.e., life size or 1:1.

Just as there are good and bad attributes for high-speed films, slower-speed films can also have limitations. Using a film speed that is too slow for the amount of available light will result in an underexposed picture that may also lack clarity and detail. There are some situations where the photographer *does* need to underexpose for better detail, particularly during fluorescent photography. This will be discussed later in Section 11.9. The

Table 11.2 Summary of Nonvisible Light Techniques

	Ultraviolet	Infrared	ALI
ASA/ISO range	ASA 100–400	ASA 25–64	ISO 100–400
f-stop range	f-4.5 through f-11	f-4.5 through f-22	f-4.5 through f-11
Shutter speed range	1/90" through 1/125"	1/125" through 2"	1/60" through 2"
Filter (on lens)	Kodak Wratten 18A (film photography),[25] Baader *Venus* UV filter (digital photography)	Kodak 87 gel or glass[25]	Varies with subject matter. For injuries on skin, #15 yellow gel[25]
Illumination wavelength (nm)	200–390 nm	700–960 nm	Varies with subject. For bites and other injuries, 450 nm.
Films	Kodak TMAX 400 Kodak PhysX 125	Kodak Highspeed Infrared, ISO 100	Fuji Reala 100

basic recommendation is to use the slowest film speed that will have the most grain density for the lighting present. Problems caused by having the wrong film or improper lighting may be minimized by bracketing the exposures over a wide range of camera settings. The term *bracketing* means to expose individual photographs in a range of *f*-stops and shutter speeds.

As the digital era of photography has evolved, the availability of photographic film and film processing labs has significantly decreased. Even though many types of film are still available, fewer stores and camera shops carry the special films that are used in full-spectrum photography.

Previously it was possible to have film processed in one hour; it can now take a week or longer when the film must be sent to a commercial processing lab. Specialized films such as infrared film or even black-and-white film can take even longer, as fewer and fewer labs process these types of film.

The potential for a significant lag time between image capture and print delivery can create a potentially catastrophic problem for the photographer should the photographs need to be retaken due to fogged film, poor focus, or improper lighting. If the injuries are on a living victim, they may have healed and would no longer be viewable. In a deceased victim, postmortem degradation probably will have destroyed aspects of the injuries that were present when the photographs were taken. The net result would be no preservation of the evidence and no chance to recover it.

Advancements in design and manufacture of modern 35 mm cameras have greatly simplified film photography. These cameras have the capability to photograph objects with great accuracy and precise color detail. As discussed previously, the lenses have coatings and the flash units are filtered to direct only visible light to the film. When handled properly, photographic films record the images in remarkably sharp detail. The most critical variables to consider when taking film-based photographs are (1) the type of the film and (2) the intensity of the light present when the film is exposed. While the photographer has great latitude in film speed selection, films of ISO 100 are usually adequate for injury documentation with adequate light.

It may seem redundant to rerecord the injury with black-and-white film when color film photographs of the same injury were just taken—or is it? The human eye is very adept at seeing images in color. Because of the color information processed optically by the retina, other important details of the injury can be overlooked. When the injury is photographed in black and white, the eye is not distracted by the color composition of the injury and the normal surrounding areas. Consequently, this absence of color allows the viewer to see more detail of the injury. The duality of image capture available with digital cameras and imaging software such as Adobe Photoshop® allows the photographer to see both color and black-and-white images from one file, demonstrating one advantage of taking digital photographs instead of traditional photographic film.

11.8 Digital Photography

The modern era of photography is being redefined as the digital era. While film-based cameras are still around, most forensic photography is done with digital cameras. Manufacturers have created a wide range of digital cameras that vary in both capabilities and cost. The new generation of digital cameras has everything from the simplest point-and-shoot consumer camera to the most complex professional camera, and a combination of several "pro-sumer" cameras in between.

Digital image capture is unlike film in that a specialized computer chip in the camera reads the light coming through the lens and electronically saves the image on magnetic media, eliminating the need for film. Typical image capture devices used today include the charge coupled device (CCD) and complementary metal-oxide semiconductor (CMOS). CCD sensors are arranged with geometric green, red, and blue areas known as pixels that are sensitive to their corresponding colors of light. Typically there are two green pixels for every red or blue pixel in an arrangement known as the Bayer pattern image (Figure 11.14). Other manufacturers have developed other image capture technologies, including the Foveon layered model and the six-sided pixel technology from Fuji (Figures 11.15 and 11.16). The density of the pixels on the sensor and the firmware driving the electronics in the camera determine the quality of the image. Generally speaking, the more pixels there are, the sharper the image. For bitemark photography, the larger the image file size, the less pixilation (blurring) when enlarging to life-size proportions.

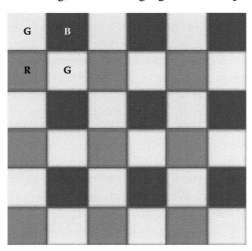

Figure 11.14 Bayer model. (Source: www.photo.net/learn/raw) RGB = red, blue, green. Each pixel in the sensor responds to either red, green, or blue light. There are two green sensitive pixels for each red and blue pixel because the human eye is more sensitive to green. (See color insert following page 304.)

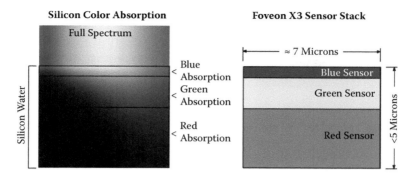

Figure 11.15 Foveon X3 CMOS sensor, method of capture and creation of a color digital image. (See color insert following page 304.)

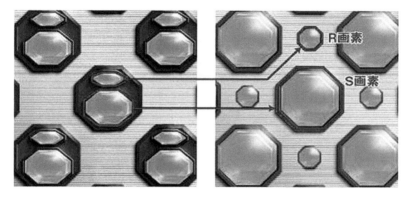

Figure 11.16 Illustration of the six-sided double-photodiode digital CCD sensor from Fuji. This sensor uses a different pixel shape and layout as opposed to the traditional Bayer patterned sensor. The two channels of the six-sided pixel create a sharper and smoother image that can be enlarged with less distortion (pixilation).

(See at www.dpreview.com/news/0608/6080904fujifilms3prouvir.asp.)

CMOS refers to both a particular style of digital circuitry design and the family of processes used to implement that circuitry on the associated electronic chips. CMOS circuitry dissipates less power and is denser than other implementations having the same functionality. As this advantage has grown and become more important, CMOS processes and variants have come to dominate, such that the majority of modern integrated circuit manufacturing is on CMOS processes.

Once the digital image is captured on the sensor, the digital camera's chip set applies the appropriate processing steps to create and save the image. These steps are usually directed by the photographer when choosing the camera settings before the image is photographed. Digital cameras can save the images in a number of file formats, each of which has advantages and

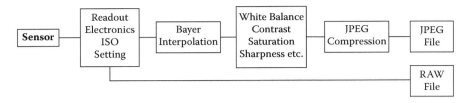

Figure 11.17 Schematic of how a captured digital image is processed from the time it is formed on the sensor until either a JPEG or RAW data file is created. (Source: www.photo.net/learn/raw/.)

disadvantages. The two most common file formats are JPEG and RAW. The simplest explanation of the differences between these two formats is that the JPEG file format is a compressed image of the RAW image data the sensor captured when the photograph was taken, which yields a smaller file. The RAW data file is quite large compared to the JPEG but still contains all the image data and detail, allowing more versatility when working with the image. RAW images require a separate proprietary "plug-in" program, usually supplied by the camera manufacturer, to open them. These files are then saved in a different non-RAW format when closing the altered file (Figure 11.17).

To understand the capabilities of any individual camera, the photographer should familiarize himself or herself with the operator's manual and perform several test runs of trial photographs in order to find the preferred camera settings and lighting methods.

While most digital cameras are automatic point and shoot, there are still some settings that must be applied before taking the photographs. One of the most significant would be setting the ISO for the environment where the photographs will be taken.

Other settings would include the file format that will be used to capture and process the digital image (such as RAW or JPEG, which were discussed previously) and the resolution of the resultant image, usually expressed in megapixels. The more resolution, the more pixels utilized and the larger the size of the file holding the digital data for the image.

When the digital camera settings have been applied, the photographer is ready to take the photographs. Unlike photographic film, which has to be processed before the images can be printed on paper, digital images can be immediately viewed, evaluated, and if necessary, retaken. Once again, the standard technique should be utilized.

Archiving digital image files must be addressed. Who among us hasn't had a computer crash, which involves the loss of all data on the affected computer? The risk of the loss of the data or the integrity of the data of the digital images requires the digital photographer to take additional steps to protect the data. The most common task is making duplicate backups frequently. This can involve writing the files to external removable hard drives, the use of CDs or DVDs, or even an Internet off-site data storage vender, if legally allowable.

The frequency of the backups and the suggestion to use multiple backup media cannot be emphasized enough. There will be an event in the digital photographer's lifetime when there will be a catastrophic loss of digital data via hard disk crash, electrical power surge during a storm, water or dust contamination, or some other incident. The only protection of the existing data is multiple frequent backups. While some have advocated doing backups on CDs or DVDs, it is important to understand that the surface integrity of the CD or DVD can environmentally degrade over time, rendering the data on the disks worthless. As part of the standard technique, the digital photographer must include routines for multiple frequent backups of the data files.

11.9 Alternate Light Imaging and Fluorescent Imaging Techniques

The field of forensic investigation has seen a tremendous growth in the utilization of alternate light imaging for both locating and photographing latent evidence. Fingerprints,[7,8] serological fluids left behind at a crime scene (blood, semen, saliva),[9] types of ink used to counterfeit or falsify documents,[10] and bruises or other pattern injuries left on human skin sustained during violent crimes can not only be more easily detected but also transformed into exciting and important exhibits with the utilization of fluorescence.[11] The application of this new technique has numerous titles. For simplicity, in this chapter, it will be referred to as alternate light imaging (ALI). The technique of photographing evidence with alternate light is called fluorescent photography. "*Fluorescence* is the stimulation and emission of radiation from a subject by the impact of higher energy radiation upon it. *Luminescence* is a general term for the emission of radiation that incorporates fluorescence and phosphorescence, as well as other electro-chemical phenomena like *bioluminescence*."[12]

All of the physical reactions that occur during illumination with full-spectrum visible light (reflection, absorption, transmission, and fluorescence) also occur for monochromatic light. Monochromatic light refers to the production of a single wavelength of light. Almost any object can be made to fluoresce, depending on the wavelength of light radiated upon it.[13] Although this filtered light is sometimes as much as 30 nm in width, it is called monochromatic (a misnomer) because very bright full-spectrum light is filtered to only allow one color of visible light to be predominant. This is accomplished with the use of bandpass filters, which are placed in the path of the light usually on the front of the lens.

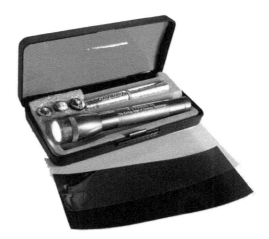

Figure 11.18 Personal handheld light source. (Photo courtesy of RC Forensic.)

The technique requires an alternate light source that is capable of producing the monochromatic beam. Most forensic light sources are capable of emitting several frequencies of visible light by using different filters, but they are limited in purity since generally each color band is 30 mm in width. Fortunately, there are a large number of less expensive, less complicated, and more portable light sources available that make photography at a remote location easier. Several manufacturers now produce what are called personal light sources that can be handheld, are reasonably priced, and are very portable (Figure 11.18). These allow the user to individually use several different frequencies of light, specific for each forensic application. The particular wavelength one uses depends upon what evidence the forensic investigator is seeking. There are optimal wavelengths for different applications; therefore, the color (frequency) of the light and blocking filters will vary. Research and investigation of pattern injuries on human skin has shown that peak fluorescence of the epidermis occurs at 430 to 460 nm[14,15] and is deep blue in coloration. Most light energy striking the surface of the skin is reflected.[16] Of the rest, about 30% penetrates below the surface. Some of it gets scattered, some is absorbed, and some is *remitted* as fluorescent light. When illuminated with an alternate light source, the electrons of these excited molecules return to their normal state by releasing energy in the form of light. The light that is emitted during this transition is of a lower frequency and weaker than the incident light. The phenomenon causes the tissue to appear to glow, or fluoresce. The scientific explanation for this phenomenon was described many years ago by Professor G.G. Stokes and is referred to as Stokes law or the Stokes shift.[35] Since the fluorescent light is always less bright than the incident light, one must observe the fluorescence of an object with the use of

Forensic Dental Photography

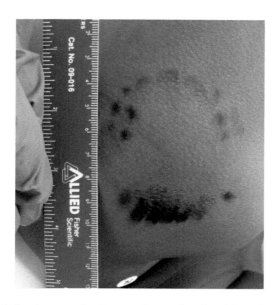

Figure 11.19 Color image of a bitemark on the shoulder of a black homicide victim. (See color insert following page 304.)

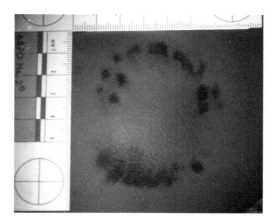

Figure 11.20 ALI image of same bitemark seen in Figure 11.19. *Note the enhancement of the bruise pattern in the fluorescent image acquired with ALI.* (See color insert following page 304.)

filters that allow only the fluorescent light through to the eye and block the more powerful reflected source light.

Light returning to the camera must be filtered to allow only the fluorescent image to be captured. In documenting injury patterns, this filtration is accomplished with a yellow filter such as the Kodak gelatin 15 filter, which blocks light transmission in the 400 to 500 nm range (Figures 11.19 and 11.20).

Fluorescent photography is best accomplished successfully in complete darkness, where all other sources of light are eliminated. One can imagine the difficulty in setting up and capturing this kind of photograph, especially when the exposure times can range up to 2 to 4 seconds in length and the subject is alive and moving. A high ISO setting and very bright alternate light source can maximize successful imaging. For weaker lights and low ISO settings, use of a tripod-mounted camera is recommended. The light source can be handheld to illuminate the injury from different angles. Experience has shown that slightly underexposing one to two *f*-stops will produce better results than the actual metered exposure. This is true because during longer exposures, even the fluorescent light coming back to the camera is still bright enough to wash out some of the fine detail in the injury at the metered "correct" exposure factor.

Several variables can influence the photographic protocol and parameters of exposure. Skin color (amount of melanin), skin thickness, wound healing response, light intensity, film speed, and location of the injury are but a few factors that affect the exposure times. Thick skin as found on the palm of the hand and sole of the foot fluoresces more than the thin skin covering the face. Darkly pigmented skin will require longer exposure times than lighter skin because more light is absorbed by the melanin pigmentation of the darker skin. Persons who bruise easily, such as the elderly, will produce injuries that may require shorter exposure times due to the thinness of the skin, but one can also expect longer exposures when greater hemorrhaging occurs beneath the skin since the blood absorbs light.

11.10 Nonvisible Light Photography

The photographic requirements for recording injuries using nonvisible light become somewhat more complex. The appearance of the injury using nonvisible light illumination cannot be seen by the naked eye. Therefore, special techniques must be employed to record the injury. Once recorded, the image must then be converted to a media from which it can be visualized. This would include photographic paper for film or a computer monitor for digital. Just as in ALI, these techniques require that bandpass filters be used. They are placed between the injury and the film or digital sensor, usually on the front of the lens of the camera. The filters allow *only* the selected wavelengths of light to pass to the film or digital sensor. It is important that several factors be considered when attempting to photograph injuries in nonvisible light (Figures 11.21 to 11.26).

For film-based nonvisible light photography, one must consider the type of film being used. The film's photoemulsion must be sensitive to the light wavelength the filter is allowing it to "see." Additionally, the light source must emit the appropriate wavelength and be strong enough to expose the

Figure 11.21 The Kodak Wratten 18-A filter used for film UV photography.

Figure 11.22 The Baader Venus UV filter used for either film-based or digital UV photography.

film after the bandpass filter is placed in front of the lens. The camera's exposure settings (*f*-stop and shutter speeds) must be set to properly bracket for the type of light being used. The camera's ASA/ISO value must be correctly set for the film being used, and the lens must be focused correctly for the type of nonvisible radiation being used. It will take some experimentation with any camera to find the optimal settings. The forensic photographer, as a rule, will practice using his or her camera and establishing techniques before photographing actual cases. Keep in mind that each camera is slightly different and these starting points may not work for every camera.

For digital nonvisible light photography, the photographer must ensure that the digital sensor is capable of recording the wavelengths of nonvisible light being used. Most commercially available digital cameras are designed to block the nonvisible ends of the spectrum.

There are two major problems encountered with nonvisible light photography. First, specialized light sources that emit enough of the desired wavelength to adequately illuminate the injury being photographed are somewhat

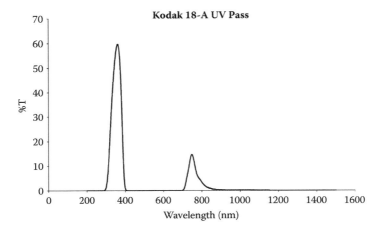

Figure 11.23 The transmission curve of the Kodak Wratten 18-A UV filter. Note the small peak in the IR range of light that contaminates the UV image, preventing this filter from being used (by itself) for digital UV photography. (Source: www.beyondvisible.com/BV3-filter.html.)

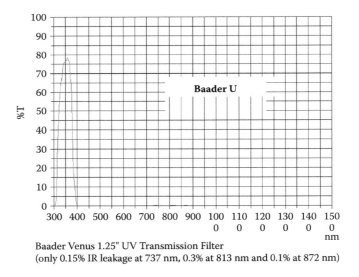

Figure 11.24 The transmission curve of the Baader Venus UV filter. Notice that predominantly UV light is able to pass through this filter with negligible IR transmission spikes.

unpredictable and vary to a great degree in intensity and quality. Second, the exact amount of focal shift to produce a sharp photograph must be determined. With viewing screens on the newer digital cameras designed specifically for nonvisible imaging, focal shift can be determined long before the image is acquired[36] (Figure 11.27). Developing confidence and getting predictable

Forensic Dental Photography

Figure 11.25 The Peca 87-A glass filter used for IR photography.

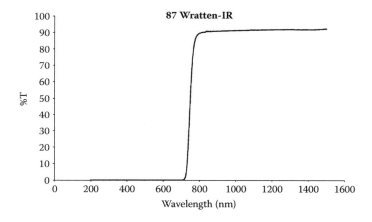

Figure 11.26 Graph showing the infrared transmission of Peca 87-A IR filter. (Source: www.ir-uv.com/peca_906 curve.htm.)

Figure 11.27 The Fujifilm IS-1 digital UVIR camera.

results in nonvisible light photography will require some trial-and-error experimentation. Available and predictable sources of nonvisible lighting are listed below for both ultraviolet and infrared photography. This list is by no means totally inclusive and is intended to be a potential resource. It is possible to find sources of adequate nonvisible light other than those listed here.

11.10.1   Ultraviolet (UV) Light Sources

- *Sunlight*: A good source of long-wave UV light but not practical for situations requiring indoor or nighttime exposures.[24]
- *Fluorescent tubes*: Produced in short, medium, and long UV wavelengths, these are routinely used for indoor lighting; some useful UV emission. One of these types of lights is known as a black light, which emits good UV radiation; the brighter, the better.
- *Mercury vapor lights*: Particularly useful in lighting small areas with intense UV light. Problems include long warm-up time for the light and limited availability.[24]
- *Flash units*: Many older units provide adequate UV light emission. Some newer units emit a measurable amount of UV but will require experimentation to determine the correct output.[24]
- *Combination fluorescent/black light*: This light combines the emission of the two light sources in one light fixture; commonly known as a Wood's lamp.
- *Special wavelength UV LED lights*[37] (Figures 11.28 and 11.29).

11.10.2   Infrared (IR) Light Sources

- *Flash units*: Most commercial flash units emit sufficient IR light to be adequate but require experimentation to determine their acceptability in infrared photography.
- *Tungsten lamps*: Used routinely in forensic investigation. The brighter the Kelvin value, generally the more IR output.
- *Quartz-halogen lamps*: Good source of IR radiation if unfiltered; more readily available and easy to use.
- *Special wavelength IR LED lights*.[37]

11.11   Focus Shift

After securing a source of predictable nonvisible light illumination, the problem of focus shift must be addressed. By definition, focus shift is "the distance between the visible focus and either the infrared or ultraviolet focus."[17]

Forensic Dental Photography

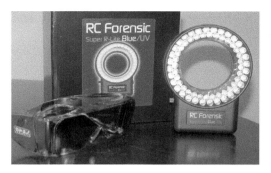

Figure 11.28 RC Forensic blue lite pack.

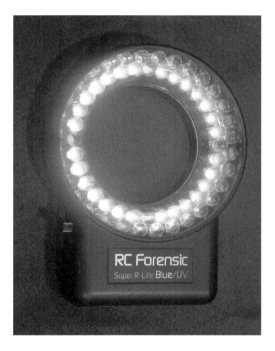

Figure 11.29 Example of a portable UV light source, manufactured by RC Forensic, Las Vegas, Nevada (www.rcforensic.com).

Focus shift is necessary because nonvisible wavelengths do not behave in the same way as visible light as they pass through a compound lens. The focal length of a lens is specific to a given wavelength of light. Most lenses are chromatically corrected to work within the 400 to 700 nm wavelengths (visible light). When the light energy falls outside of the visible spectrum, the optimal visual focus is no longer the optimally focused point for the nonvisible light energy used to expose the film.[17] While some manufacturers have developed achromatic lenses, which act to bring two different wavelengths to a single coincident focus, or apochromatic, which bring together three colors to a single focal point, many readily available chromatic lenses may require a focus shift for nonvisible light wavelengths. There are several published opinions concerning the correction of the focal point for nonvisible light ultraviolet photography. Kodak[19] has suggested a simple one, and this is the one the authors recommend you try first. It is Kodak's opinion that the focus shift required for ultraviolet photographs may be accounted for by simply increasing the depth of field.[29] The recommendation is to decrease the lens aperture at least two stops if shooting from wide open. Since the construction of compound lenses can be so different, Kodak suggests that test exposures at various aperture settings be performed to determine the exact change for an individual lens.[18,33] Cutignola and Bullough[19,28] support Kodak's recommendation in correcting the ultraviolet focus by changing the depth of field. The downside to this modification is that it may significantly alter exposure times, lighting, and film speed.

Other authors have suggested small focus shifts by turning the focusing ring slightly from the visible focus position. Arnold et al.,[20] Lunnon,[21] Nieuwenhuis,[22] and Williams and Williams[23] have suggested shifting focus for UV photography in the *same* direction and amount as is done in infrared photography. The majority of modern high-quality achromatic compound lenses have a focus color correction to achieve sharp photos.

The authors of this chapter have found that no focus shift is required for ultraviolet photographs when using lenses designed specifically for UV, such as lenses constructed of silicon-based (quartz) glass. Exposures using a silicon lens have produced very sharp ultraviolet photographs with no shift from the visible focus (Figure 11.30).

As forensic photography evolves, manufacturers are continuously modifying and upgrading equipment. For this reason, it is difficult to recommend any particular brand of equipment. There are digital cameras specifically designed for nonvisible light photography. Each application for recording the nonvisible ends of the light spectrum requires specific filters that allow only that portion of the spectrum to pass through the lens.

Finding the optimal camera setup, the correct focal point, and a dependable source of lighting takes some research and many sessions of experimental trials. The photographer should exercise patience and remember to record the exposures and f-stops with every trial photograph taken in order to determine the optimal parameters.

When doing infrared photography, a focus shift is required. This focus shift moves the focus point of the object being photographed *away* from the visible focus since the actual infrared focus in patterned injuries in skin is below the surface of the skin.[31] Many commercially available lenses have a small mark on the focus ring that represents the infrared focus (Figure 11.31).

Forensic Dental Photography

Figure 11.30 Nikon Nikkor UV 105 fused silicon (quartz) lens.

Figure 11.31 The infrared focus point on the focus ring of a Nikon Nikkor lens, as marked by the small dot adjacent to the visible focus.

11.12 Reflective Long-Wave Ultraviolet Photography

Ultraviolet photography is used by forensic odontologists for two reasons: The first is to visualize surface detail of the injury. Reflective ultraviolet photography helps to enhance surface detail (Figures 11.32 to 11.34).

The second reason is to attempt to record an injury after a period of time of healing when it is no longer visible to the unaided human eye. This second

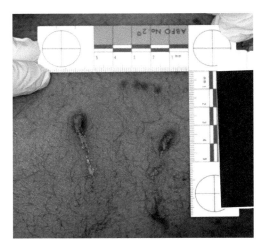

Figure 11.32 Dog bite on leg, full color spectrum. (See color insert following page 304.)

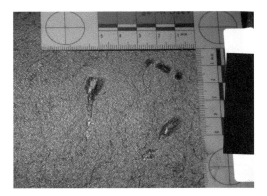

Figure 11.33 Same dog bite as seen in Figure 11.32, reflective UV image. Note enhanced surface detail.

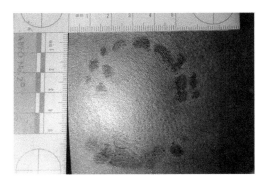

Figure 11.34 UV photo of same shoulder bite as seen in Figures 11.19 and 11.20.

Forensic Dental Photography 231

use occurs because ultraviolet light is strongly absorbed by pigment in the skin.[26] Any area of the healing injury that contains excess surface pigmentation compared to the surrounding normal tissue may sometimes be recorded with favorable results using reflective ultraviolet photography.[26] The optimum time for photographing injuries in the living skin using ultraviolet light is seven to eight days after the injury is inflicted (Figures 11.2 to 11.10). Case reports suggest that it is possible to photograph a healed injury up to several months after the injury. Such a case, reported by David and Sobel,[27] illustrated a five-month-old injury recaptured using reflective ultraviolet photography where no injury pattern was visible to the naked eye. Using UV photography, one author (Wright) has recaptured remnants of different bitemark injuries seven and twenty-two months after the victims were bitten (Figures 11.35 to 11.38).

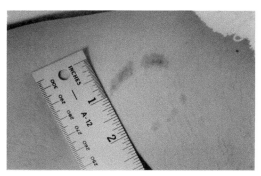

Figure 11.35 An alleged bitemark; grayscale image acquired within hours after injury inflicted (visible light).

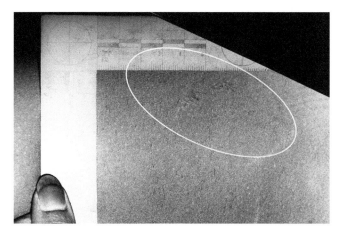

Figure 11.36 The same alleged bitemark as in Figure 11.35 (inside the white oval). UV image taken twenty-two months after the injury. The pattern was not visible to the human eye under normal lighting conditions.

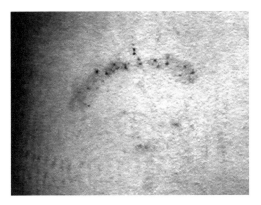

Figure 11.37 Bitemark on back, day 1, grayscale.

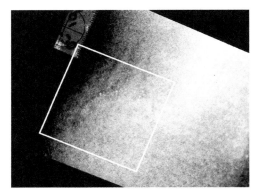

Figure 11.38 Bitemark on back, seven months later, UV.

Figures 11.37 and 11.38 are two images of the same bitemark on the back of an abuse victim. Figure 11.37 shows the bitemark as seen on the day of the injury. There is no scale in the photograph. Figure 11.38 shows the same bite in reflected UV seven months after the injury. It was not visible to the human eye in natural light. The UV image captured the surface disruption and photographically recorded it.

The UV-exposed film or digital CCD records the unseen surface damage contained in the affected area of the injured skin, which later becomes visible to the human eye on the photographic image, assuming proper UV photographic techniques were used as images were acquired.

11.13 Infrared Photography

Just as in reflective UV photography, infrared photography also requires special techniques. The infrared band of light is at the opposite end of the

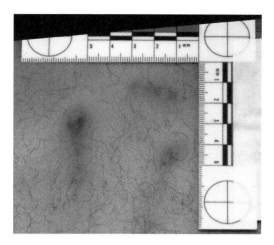

Figure 11.39 IR image of the same dog bite as seen in Figures 11.32 and 11.33. Note the deeper level of penetration of IR energy into the tissue.

light spectrum from the ultraviolet band, with ultraviolet light being about one-half of the wavelength of infrared light. Because infrared is longer in wavelength transmission, it penetrates up to 3 mm below the surface of the skin (Figure 11.39).

Since the depth of the injury that will be recorded with the infrared technique is below the surface, the infrared focus point will not be the same as the visible focus point, requiring a focus shift. The field of digital infrared forensic photography has grown to include documentation of gunshot residue, tattoo enhancement, questioned documents, blood detection, background deletion, wound tracking, and tumor detection. The injury documented with infrared technique will not appear the same as photographs taken using visible light. In Kodak Publication N-1, *Medical Infrared Photography*,[6] this difference is discussed (pp. 26–27): "It should not be overlooked that even when the lens is focused correctly, the infrared image is not as sharp as the panchromatic one. The reason is the lens aberrations have been corrected for panchromatic photography, so the anastigmatism is not as perfect in the infrared. The majority of biological infrared images are formed from details not on the outside of the subject.... This feature accounts for the misty appearance of many infrared reflection records."[32]

Successful infrared photography is a trial-and-error process, particularly when dealing with injury patterns. If the injury did not cause sufficient damage to the deeper skin tissues, i.e., no bleeding below the surface of the injured skin, or if the surface of the injured skin is too thick for the infrared light to penetrate to find the site of the bleeding, there may be no infrared detail recorded in the photographs. The advantage of digital photography is that the image can be either previewed before or immediately seen after exposure on an LCD screen on the back of the camera. If no image appears when employing the IR photographic technique, it should not be interpreted as a failure of the technique. Situations can occur wherein IR techniques are less effective than visible spectrum photography.

11.14 Application and Use of Forensic Photography

It bears repeating that even when nonvisible light photography is performed perfectly, images may not appear in the resultant photographs. This does not necessarily indicate a failure of the techniques. Rather, it may just mean that the injuries are not such that the incident wavelength of nonvisible light doesn't "see" the injuries based on the components in the injured skin. It must also be pointed out that even if the techniques work and images are captured, the resultant images may not add to the evidentiary value (Figures 11.40 to 11.47).

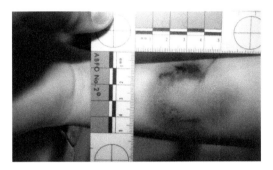

Figure 11.40 Grayscale.

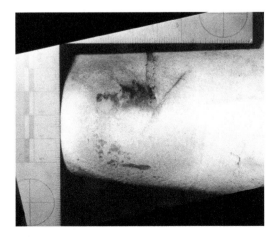

Figure 11.41 UV.

Forensic Dental Photography

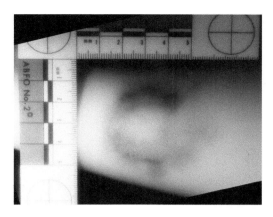

Figure 11.42 IR.

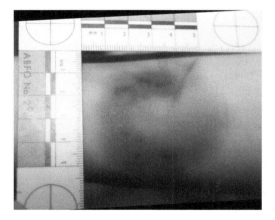

Figure 11.43 Full spectrum color image of patterned injury with very little forensic or evidentiary value. (See color insert following page 304.)

In this reported bitemark on the forearm of a homicide victim (Figures 11.40 to 11.43), the patterns are unclear in all imaging modes. Consequently, this mark has very little forensic or evidentiary value. The use of multiple photographic modalities failed to increase the forensic/evidentiary value. The images in Figures 11.40 to 11.43 are grayscale, UV, IR, and color, respectively.

Sometimes, nonvisible light photography can be used to help determine if the injuries represent human bitemarks or come from another source. In such cases, the use of digital full-spectrum photography benefits the investigator since the resultant images are instantly available for review (Figures 11.48 to 11.53).

This chapter has dealt with the photographic techniques that apply to collecting evidence of patterned injuries in skin, primarily human bitemarks.

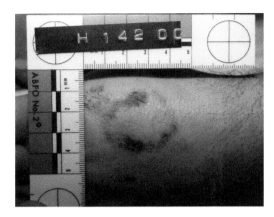

Figure 11.44 Hand, grayscale.

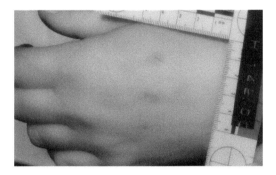

Figure 11.45 Hand, IR.

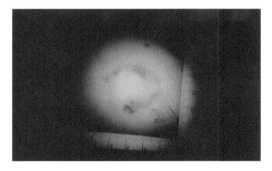

Figure 11.46 Hand, ALI.

It should be mentioned that these techniques work for other types of injuries in human skin. While this chapter's authors are forensic odontologists whose area of expertise is bitemark analysis, full-spectrum photographic documentation of injuries in skin not made by teeth can also be important and should be pursued by other criminal investigators (Figures 11.54 to 11.63).

Forensic Dental Photography

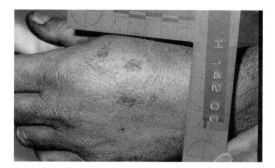

Figure 11.47 Hand, UV.

The appearance of a bitemark on the back of the left hand of a homicide victim (Figures 11.44 to 11.47): grayscale, infrared, alternate light, and ultraviolet photographs, respectively.

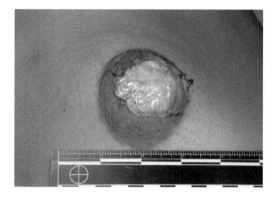

Figure 11.48 Excised breast, color. (See color insert following page 304.)

11.15 Management of Forensic Photographic Evidence

The photographs documenting a victim's injuries may become part of the legal system and, as such, are subject to *chain-of-evidence* rules. The appropriate protocols for evidence management are fully discussed in Chapter 17. The protocols require accountability as to who had possession of the evidence from the time it was collected until it is marked and introduced into the legal system. It is important to maintain the integrity of the evidence in terms of its original form and reproducibility.

Photography is one of the most important tools used in the practice of forensic dentistry. The demands on the photographer can be great, especially in situations where an injury is the only evidence tying a suspect to the crime. Time, patience, and preparation in forensic photography are requirements for successful pattern injury documentation. While often frustrating and

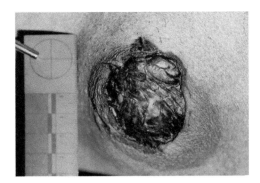

Figure 11.49 Excised breast, UV.

Injury to this breast was thought to have been the result of an avulsive bite when viewed in the color photograph (Figure 11.48). However, when photographed in UV light (Figure 11.49), the borders indicate that the tissue was more likely removed with a sharp instrument.

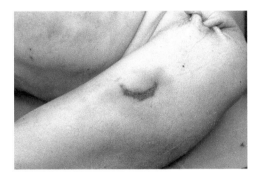

Figure 11.50 Orientation photograph of a patterned injury on the left arm that is suggestive of being a human bitemark, using ABFO terminology. (See color insert following page 304.)

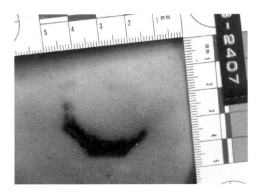

Figure 11.51 Arm injury, color. (See color insert following page 304.)

Forensic Dental Photography 239

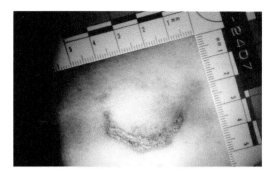

Figure 11.52 ALI to grayscale, enhanced.

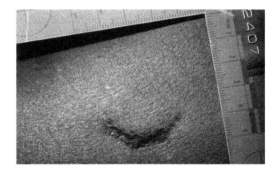

Figure 11.53 UV, enhanced.

Figures 11.50 to 11.53 depict a patterned injury on the upper left arm of a homicide victim described as a "possible bitemark" by investigators. Closeup photographs in color; alternate light imaging and UV show that the patterned injury is not a bitemark.

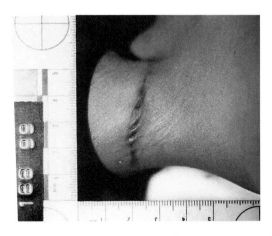

Figure 11.54 Neck, color. (See color insert following page 304.)

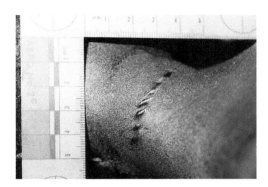

Figure 11.55 Neck, UV.

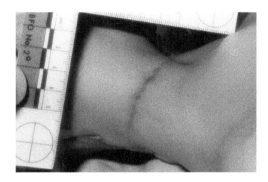

Figure 11.56 Neck, IR.

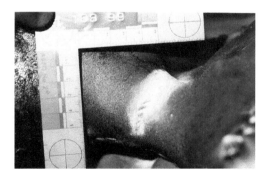

Figure 11.57 Neck, UV (Sirchie).

Figures 11.54 to 11.57 are photographs of the neck of a young homicide victim. The images show details indicating that a rope had been used to tie the neck of the bound victim to the headrest of an automobile. The images seen in Figures 11.54 through 11.57 are, respectively, color, standard UV, infrared, and UV using Sirchie® powder (a powder highly reactive to ultraviolet light). The details of the rope seen in the photographs indicated that the ligature marks on the neck were not caused by the border of the seat belt, as the defendant had claimed.

Forensic Dental Photography

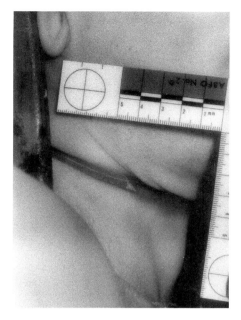

Figure 11.58 Strangulation with electric cord, IR.

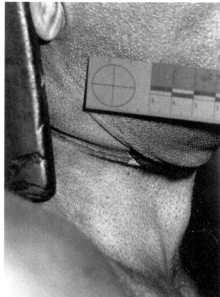

Figure 11.59 Strangulation with electric cord, UV.

Figures 11.58 and 11.59 show ligature marks on the neck made by an electric cord used in a death by strangulation case.

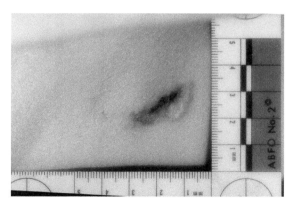

Figure 11.60 Knife wound, IR.

time-consuming, when properly performed, the results yield good evidence, bringing with it a sense of accomplishment and satisfaction that the forensic dentist has made a significant contribution to the case. Developing the skills necessary to competently document these injuries with visible and nonvisible light is one of the great challenges in forensic dentistry.

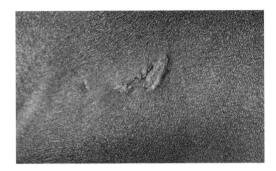

Figure 11.61 Knife wound, UV.

Figures 11.60 and 11.61 show the different appearance of a knife stab wound in the skin of the leg in infrared and ultraviolet light.

Figure 11.62 Tattoo, color. (See color insert following page 304.)

Figure 11.63 Tattoo, IR.

The color photograph (Figure 11.62) shows the tattooed mummified tissue of the back of a homicide victim, and the infrared photograph (Figure 11.63) of the same tissue clearly shows tattoo detail not visible with the naked eye. The tattooed area illustrated in both photographs is outlined.

References

1. Langford, M. 2000. *Basic photography*. 7th ed. http://www.amazon.com/gp/reader/0240515927/ref=sib_fs_top?ie=UTF8&p=S00I&checkSum=T0kc6clbFQMX0TOKe1FptuqNk3JTFQXTRJfWyVgDJ50%3D#reader-link.
2. Grimm, T., Grimm, M. *The basic book of photography*. 5th ed. http://www.amazon.com/gp/reader/0452284252/ref=sib_dp_pt#reader-link.
3. Williams, R., Williams, G. 2002. Infrared photography, medical and scientific photography: An online resource for doctors, scientists, and students. http://msp.rmit.edu.au/Article_03/02d.html.
4. Kochevar, I.E., Pathok, M.A., Parrish, J.A. Photophysics, photochemistry, photobiology. In *Dermatology in general medicine*, chap. 131, p. 1632. 4th ed. New York: McGraw-Hill.
5. Eastman Kodak Co. 1972. *Ultraviolet and fluorescence photography*, 12. Kodak Publication M-27. Rochester, NY: Eastman Kodak Co.
6. Eastman Kodak Co. 1973. *Medical infrared photography*, 6. 3rd ed. Kodak Publication N-1. Rochester, NY: Eastman Kodak Co.
7. Bramble, S.K., Creer, K.E., et al. 1993. Ultraviolet luminescence from latent fingerprints. *Forensic Science International* 59:3–14.
8. Ray, B. 1992. Use of alternate light sources for detection of body fluids. *Southwestern Association of Forensic Science Journal* 14:30.
9. Stoilovic, M. 1991. Detection of semen and blood stains using Polilight as a light source. *Forensic Science International* 51:289–96.
10. Masters, N., Shipp, E., Morgan, R. 1991. DFO, its usage and results—A study of various paper substrates and the resulting fluorescence under a variety of excitation wavelengths. *Journal of Forensic Identification* 41(1).
11. Golden, G. 1994. Use of alternative light source illumination in bite mark photography. *Journal of Forensic Sciences* 39(3).
12. Williams, A.R., Williams, G.F. 1994. The invisible image—A tutorial on photography with invisible radiation. 2. Fluorescence photography. *Journal of Biological Photography* 62(1), 3.
13. Guilbault, G. 1973. *Practical fluorescence*. New York: Marcel Dekker.
14. Devore, D. 1974. Ultraviolet absorption and fluorescence phenomena associated with wound healing. Thesis for doctor of philosophy, University of London, Oral Pathology, London Hospital Medical College.
15. Dawson, J.B. 1980. A theoretical and experimental study of light absorption and scattering by in vivo skin. *Physics in Medicine and Biology* 25(4).
16. Regan, J.D., Parrish, J.A. 1982. *The science of photomedicine*, 160. New York: Plenum Press.
17. Nieuwenhuis, G. 1991. Lens focus shift required for reflected ultraviolet and infrared photography. *Journal of Biological Photography* 17.
18. Eastman Kodak Co. 1972. *Ultraviolet and fluorescence photography*, 6–7. Kodak Publication M-27. Rochester, NY: Eastman Kodak Co.
19. Cutignola, L., Bullough, P.G. 1991. Photographic reproduction of anatomic specimens using ultraviolet illumination. *American Journal of Surgical Pathology* 15(11), 1097.
20. Arnold, C.R., Rolls, P.J., Stewart, J.C.J. 1971. *Applied photography*. London: Focal Press.

21. Lunnon, R. 1974. Reflected ultraviolet photography in medicine. MPhil, University of London.
22. Nieuwenhuis, G. 1991. Lens focus shift for reflected ultraviolet and infrared photography. *Journal of Biological Photography* 59(1).
23. Williams, A.R., Williams, G.F. 1993. The invisible image—A tutorial on photography with invisible radiation. 1. Introduction and reflected ultraviolet techniques. *Journal of Biological Photography* 61:124.
24. Eastman Kodak Co. 1972. *Ultraviolet and fluorescence photography*, 5. Kodak Publication M-27. Rochester, NY: Eastman Kodak Co.
25. Eastman Kodak Co. 1981. *Kodak filters*, 3. Publication B-3A KIC. Rochester, NY: Eastman Kodak Co.
26. Eastman Kodak Co. 1972. *Ultraviolet and fluorescence photography*, 9. Kodak Publication M-27. Rochester, NY: Eastman Kodak Co.
27. David, T.J., Sobel, M.N. 1994. Recapturing a five month old bite mark by means of reflective ultraviolet photography. *Journal of Forensic Science* 36:1560–67.
28. Cutignola, L., Bullough, P.G. 1991. Photographic reproduction of anatomic specimens using ultraviolet illumination. *American Journal of Surgical Pathology* 15(11), 1096–97.
29. Williams, A.R., Williams, G.F. 1993. The invisible image—A tutorial on photography with invisible radiation. 1. Introduction and reflected ultraviolet techniques. *Journal of Biological Photography* 61:119.
30. Williams, A.R., Williams, G.F. 1994. The invisible image—A tutorial on photography with invisible radiation. 2. Fluorescence photography. *Journal of Biological Photography* 62:4.
31. Eastman Kodak Co. 1973. *Medical infrared photography*, 26. 3rd ed. Kodak Publication N-1. Rochester, NY: Eastman Kodak Co.
32. Eastman Kodak Co. 1973. *Medical infrared photography*. 3rd ed. Kodak Publication N-1. Rochester, NY: Eastman Kodak Co.
33. Eastman Kodak Co. 1995. Personal communication with technical support department. Rochester, NY: Eastman Kodak Co.
34. Fuji Photo Film Co. 1994. *Professional Fujichrome, Fujicolor/Neopan data guide.* Ver. A. Tokyo: Fuji Photo Film Co.
35. Stokes, G.G. 1852. On the change of refrangibility of light. *Phil Trans R Soc London* 1853:385.
36. Fujifilm, USA. http://www.fujifilmusa.com/products/digital_cameras/is/finepix_ispro/index.html.
37. RC Forensic, Inc. http://rcforensic.com.

Dental Identification in Multiple Fatality Incidents

12

BRYAN CHRZ

Contents

12.1	Introduction	245
12.2	Multiple Fatality Incident/Mass Disaster	246
12.3	Types of Multiple Fatality Incidents/Mass Disasters	248
	12.3.1 Natural Disasters	248
	12.3.2 Transportation Accidents	248
	12.3.3 Terrorism	249
	12.3.3.1 Weapons of Mass Destruction	249
12.4	Disaster Site/Crime Scene Management	249
	12.4.1 Crime Scene	249
	12.4.2 Incident Perimeter	249
	12.4.3 Grid/GPS	250
	12.4.4 Security and Safety	250
12.5	Interagency Relations	251
12.6	Identification Section	252
12.7	Odontology Section	255
12.8	Computer Assistance Programs and Digital Information	258
12.9	Digital and Conventional Images	258
12.10	Preparedness Training and Planning	260
12.11	Conclusions	260
References		261

12.1 Introduction

Multiple fatality incident (MFI) is now the common term given to the fatal results of any mass disaster. In this chapter the management of an MFI will be discussed and specific issues will be addressed. The overall response includes a multitude of agencies, and their interaction is critical to the success of any MFI operation. The world has experienced several overwhelming MFIs in the recent past. The World Trade Center attack, the Pentagon attack, and the Pennsylvania crash all related to the 9/11 terrorist's attacks left the United

States in shock. The triple hurricane assaults on Florida were taxing on the systems in 2004. The world watched in astonishment as the tsunami of December 26, 2004, wiped out entire cites and islands around the Indian Ocean. How societies and nations respond to an MFI or mass disaster is determined by the plans in place and the methods chosen to handle the remains of victims. Cultural and religious differences must always be considered to create an MFI response appropriate for different geographical areas around the world. After the care of the injured, housing issues, and local infrastructures are addressed in a mass disaster, the remainder of the operation tends to be the disposition of human remains, hence the MFI. The identification process is very important to the family members of the deceased for legal and psychological reasons. Every phase of the identification process should lead toward an accurate and scientifically based identification. The forensic odontologist plays an integral role in this process. Those who read this chapter should have the ability to define a multiple fatality incident and know the types and causes of those incidents. He or she should understand the principles of site management, relationships with other agencies, and the role of the forensic odontologist in the response, including the latest technological advances in imaging and forensic dental identification software.[1]

12.2 Multiple Fatality Incident/Mass Disaster

By most definitions a mass disaster is a situation that overwhelms the infrastructure of a community or locale sufficiently to require assistance from outside sources to respond to the needs created by the disaster. For instance, a transportation accident in the New York City area with one to two hundred fatalities could be well within the abilities of the local emergency management system, fire and police departments, and medical examiner. Therefore, it would not be by the strictest terms a mass disaster. Conversely, in a jurisdictional area in the Midwest with sparse population and limited resources, a multicar accident with twenty fatalities could create a situation that would be beyond the capabilities of the existing system. This situation could be termed a mass disaster, and some sort of outside aid would be needed to assist the local responders.

Of course, the survivors of a mass disaster are the initial concern of the responders. The injured and displaced must be treated and housed properly. Once this priority is accomplished, the efforts intensify to locate, identify, and return to their families those victims who did not survive. A multiple fatality incident (MFI) is the forensic extension of the mass disaster. The response efforts to multiple fatalities include recovery, autopsy as required, identification, and release of bodies to families or to their country of origin.

Embalming may be a consideration, but in respect to religious concerns, should be done only after identification and legal release. In some situations the overwhelming numbers of fatalities or other extenuating circumstances may require mass management and no individual processing. This type of body disposition must be defined in planning and training sessions and implemented when the situation warrants.

The MFI affects many levels of society. The direct family members have suffered a tremendous loss with the death in their family, but also could have suffered injuries or sustained significant personal property losses. Dealing with the family member is the primary concern of the Family Assistance Center, an important part of the MFI response structure. In cases like the 9/11 World Trade Center attacks, we found that the responders who were in the line of fire and the same people depended upon to supply fire and police protection and assistance became part of the missing persons list. Local, state, and national resources are tapped, and the effects of an MFI are felt across an entire nation.

An MFI requires the work of many forensic experts.[2] Who are they and from where do they come? Normally local agencies have some emergency management resources in place to begin the initial operation. A large majority of responders may be volunteers who feel the need to give aid to families and victims in a time of need. Organizing and choosing these volunteers is a complicated job. Often there are imposters who try to cross the security lines just to look at the scene. This type of security problem must be prevented, and a good protocol for accepting volunteers must be in place. Many governments supply either salaried employees or activated employees to work at a disaster site. These are trained individuals who tend to be able to stay at a scene longer than volunteers due to the fact they are not losing income while working at an MFI. There are also independent contractors who provide services for an MFI. These contractors work for an entity such as an air carrier or government and provide the requested services. The final decision on who works an MFI rests with the person who has jurisdictional authority. This individual is usually a medical examiner, coroner, or assigned emergency management director. It is recommended that all agencies involved have daily meetings to assess the progress of the operation and to determine how each agency is functioning. This also helps solidify this conglomerate of groups into a single team.

Many people pose the question, "Why do you do this type of work?" Normally the volunteer has a feeling of responsibility to provide his or her services to those in need. The employee working an incident of course sees it as a job, but still will have that deep conviction to serve. Perhaps the simplest way to explain the worker's need to serve is as a basic human desire to assist those who have no other source of aid. Whatever the reason the responders come, the people to whom the services are being rendered should be, and most often are, aware of the great sacrifices these responders are making.

12.3 Types of Multiple Fatality Incidents/Mass Disasters

Mass disasters can result from multiple causes. The forces of nature are very powerful and can easily destroy man-made objects and take human lives. Transportation of large numbers of people presents the possibility for large-scale injuries and death if an accident occurs. Terrorism has become another disturbing source for mass disasters.

12.3.1 Natural Disasters

Natural disasters include hurricanes, tornadoes, floods, volcanoes, earthquakes, tsunamis, and any other natural phenomena that cause destruction. Most of nature's disasters come very quickly and surprise the unsuspecting population. Weather and seismic predictors have been employed and improved as early warning systems that have the potential to save countless lives. Other than warning and evacuation to safer areas, there are no dependable ways and means to prevent natural disasters.

12.3.2 Transportation Accidents

Transportation modes that have experienced accidents include airplanes, trains, and passenger liners on lakes, seas, and oceans.[3] Any mode of transportation that moves large numbers of people is at risk. Within the United States and its protectorates, the National Transportation Safety Board (NTSB) is the agency in charge of the site of an incident and for the investigation and management of the search and recovery element of the event. The Family Assistance Act of 1996 gives the authority to the NTSB to ensure that the families of victims associated with a multiple fatality incident are properly served by the responders. If the local response fails to accomplish this task within the requirements of the NTSB, the NTSB is allowed to replace the local response teams with an entity that can provide the needed services. One of the services deemed most needed for the families is a Family Assistance Center to provide the family members with things necessary for their comfort and well-being. Many times transportation accidents involve fuels, toxic container rupture, or other complications creating extreme hazards for the responders. A safety and hazmat team is used to maximize responders' safety and protect the health of the public.

12.3.3 Terrorism

Terrorism is the use of threats and actions to create fear within a target population. The most common form of terrorism seen in current times involves some sort of explosive device used to kill and maim large numbers of people.

12.3.3.1 Weapons of Mass Destruction

Weapons of mass destruction (WMD) is a term used to define these mechanisms that go beyond the conventional improvised explosive devices or suicide car bombs and escalate to mechanisms introducing radiation and pathogens that increase the fear-producing powers of the devices. A bomb similar to the one used in the 1995 Oklahoma City bombing of the Murrah Federal Building that killed 168 individuals and injured nearly 800 others could have injured or killed most of the population downwind of the explosion if it had contained certain WMD elements. The response to terrorism must always consider such WMD usage. Protection of everyone, including the responders, is important in an MFI, and decontamination may be necessary before the normal forensic process can begin.

12.4 Disaster Site/Crime Scene Management

12.4.1 Crime Scene

All disaster sites should be managed and processed like a crime scene. Control of the scene and maintaining appropriate chain of evidence is of utmost importance. There should be an assigned position for scene security to ensure the integrity of the investigation. In transportation accidents the NTSB will have authority in these matters. In other mass disaster scenarios law enforcement officers may be utilized to fill the position due to the nature of their formal training. Each team member should be instructed on the importance of following the instructions of the site security officer.

12.4.2 Incident Perimeter

The incident perimeter should be established by the authorities of the site as soon as possible. This will facilitate control of the scene. Whether it involves a single building or half a city, the barriers controlling the perimeter should be well defined and visible to workers and the public. Once the perimeter is established it can be changed. Experience has shown that it is easier to decrease the area of control than to increase it, making it wise to establish a generous perimeter in the opening phases of the operation.

12.4.3 Grid/GPS

Once the perimeter is established, a system to define the different areas of the crime scene is needed to allow recovery teams to easily record the exact location of the evidence they find. In the past it was found that simple grid systems worked well with flat scenes, similar to an airplane crash on land. Grid numbers on maps could be marked as evidence was recovered. It has been found in recent incidents that when dealing with collapsed buildings and water settings, simple grids fail to give adequate information about elevation and return points of reference. Global positioning system (GPS) technology has been employed in some situations, allowing not only a positional record, but also a very accurate way to return to a specific area and, if necessary, establish and document elevations of points.

12.4.4 Security and Safety

Security of the site is always a concern. An established perimeter will aid with this. A control system needs to be established early in the incident to allow access to the crime scene to only authorized personnel. Badge systems, periodic changes of codes, and photo identification have all been used to prevent unauthorized access. Fingerprint scans are now easy to implement and monitor with computer systems. These systems could be used for access control as well as computer system logins. A security officer and an established security protocol should be the focus areas for all security-related questions. The protocol should be required reading for all workers, who should avow that they have read and understood the policies in a signed document as a part of their indoctrination into the operation of the incident.

Safety should be among the most important aspects in the management of a mass disaster site. There may be a large number of injured and dead individuals as a result of the disaster. There is no justification for adding to those numbers with responders who do not follow safe practices. A safety office should be created, and as in the security section, there should be a safety protocol developed, understood, and adhered to by the workers to prevent more injuries and loss of life. The debris field of an airplane crash or the resulting devastation of a detonation of a bomb in an urban setting yields situations and materials that are physically dangerous and potentially contaminated with dangerous substances. The safety officer and the person responsible for hazardous materials (hazmat) should coordinate and establish the safety of the site before responders enter the area. The safety office should address the problem of proper decontamination of materials being sent to the morgue area.

12.5 Interagency Relations

The initial response to any mass disaster is with local assets. There are, in most cases, emergency plans of action in place by the local emergency response office to activate police, fire protection, and medical services. Local agencies may also include hazmat responders to address safety of the site. These local agencies should be in close communication and would be best organized if one person is assigned to supervise the operation. There should also be contingency plans for contacting outside resources for assistance if the situation becomes too complex for the local responders to handle.

Within the United States the National Disaster Medical System (NDMS) can be activated if needed. Included in the NDMS structure is the Disaster Mortuary Operational Response Team (DMORT). DMORT has the ability to aid local response teams in recovering, storing, identifying, and casketing of victims. DMORT can provide and manage a stand-alone morgue facility and supply the human resources for an extremely large MFI. DMORT has been used effectively in responses in 2001 to the combined World Trade Center/American Airlines Flight 587 incidents and in 2005 to the aftermaths of Hurricanes Katrina and Rita. DMORT will tailor its response to the needs of the local authorities. DMORT has three disaster-portable morgue units (DPMUs) stored in California, Texas, and Maryland for deployment at any time of need. In times of natural disaster the state agencies work with their state governor to seek this aid through a presidential declaration. This declaration sets the National Response Plan (NRP) into action, activates the NDMS and DMORT, and releases federal funding for operations.

In the early and middle 1990s several complaints were lodged by family members of victims killed in airplane crashes. The complaints revolved around poor treatment and lack of communication with the families. Congress felt strong enough about these complaints that they enacted the Family Assistance Act of 1996. This allows the National Transportation Board to evaluate the performance of local teams to care for these family members and, if the response is deemed inadequate, to activate DMORT to provide family and identification services.

In other parts of the world disaster victim identification (DVI) teams provide services much like DMORT. They use Interpol systems for charting and record keeping. DVI teams have been used in situations such as the Bali bombings and the December 26, 2004 tsunami in the Indian Ocean. Most of these units act as a resource for each country but work together in an international response when needed.

12.6 Identification Section

The first priority for responders to an MFI is to establish who is in charge. Through training and establishment of response manuals this person is usually defined for most local jurisdictions. The problem arises when many state and federal agencies come together in a response role and then try to work as independent agencies. Most response models are developed around the Incident Command System (ICS).[2] This system identifies a hierarchy of command with well-defined assignments for support groups. The overall responsibility for the operation lies with the incident commander. Any assignments within the command framework not delegated to others revert to the incident commander. In more complicated situations with many agencies involved, a modified system can be implemented with a unified command replacing a single incident commander. The unified command is composed of agency representatives who have full authority to make policy decisions for their respective agencies. There is an agreement made before the unified command is activated that it will work as a democratic body, with each agency represented committed to follow the decisions of the entire group. By following the ICS chain of command all team members can identify the appropriate individual to whom they can address concerns.

Early in an MFI identification operation the decision of how to accomplish the identification of unidentified victims must be established. Although deciding on a victim and missing person numbering system may seem uncomplicated, in the past there have been almost as many different numbering systems as operations done. In this author's opinion, simple is better. With the use of computers so prevalent, numbering systems need only be tailored for estimated numbers and some obvious differentiation between antemortem and postmortem records. Always begin the antemortem or postmortem records with a variation of the number 1. For purposes of electronic database storage and order, zeros can be added before the 1 to approximate the total number of records expected. For example, if there are three hundred expected fatalities, the first number would be 001. This would not only accommodate the 300 expected fatalities, but also could be used if the incident increased up to 999 victims. Antemortem records can easily be numbered in the same way, with the addition of A to appear before the number to make an obvious difference between antemortem (A001) and postmortem (P001). Never allow the name of missing persons to be used for unique record identifiers. There is too great a chance that two people could share the same name. As far as using the numbering system to show recovery areas, operation names, or any other bits of information, digital databases replace this function with record-specific data cells that can be ordered or searched as necessary. If numbering systems are changed for any reason in the middle of an operation, many problems will ensue. Some computer systems use the numbering system for unique identification within a particular software application, and in addition might use the number to connect or bridge to another software program to accomplish a different task. These types of systems do not allow quick and easy updates of initial numbering systems. In fact, it could require many hours of work and coordination to establish a new numbering system. Therefore, establish the exact numbering system before operations begin and stick with it.

Most MFI responses are represented by several specialty sections.[4] The anthropology section is concerned with skeletal identifiers and is very important in sex and age determination of unidentified victims. They are also very active in reassociating fragmented and comingled body parts. Forensic anthropologists who have a doctorate degree and are board certified by the American Board of Forensic Anthropology should be used if available. The dental section is responsible for antemortem and postmortem dental records, the dental postmortem examination, and the comparison of dental records for identification. The dental section will be explained more fully later in this chapter. The DNA section collects samples from the unidentified victims and stores then transfers them for sequencing to a certified DNA laboratory. Most DNA laboratories will send their own employees experienced in sampling to staff or augment this station. The DNA section is also responsible for collecting and analyzing samples collected that relate to missing persons and their family members. The fingerprint section retrieves all fingerprint information from the victims. This section is also charged with obtaining latent prints from missing persons' premises that may lead to identification of missing persons with no antemortem fingerprint records. This section is staffed with experienced fingerprint experts from local, state, or federal agencies. The personal effects section is usually the first physical station in the morgue area. Personal effects collects those items associated with each victim, documents and stores those items, and prepares them for return to the families of the victims. This section usually confirms proper numbering of victims and begins the sequential processing of the morgue operation. The forensic pathology section is usually headed by the medical examiner in charge of the morgue. This section provides the forensic pathology examinations and autopsy services. The radiology section provides medical radiographic documentation for each unidentified victim. They may then compare antemortem radiographs from missing persons to those of unknown victims for identification. Other sections may be added or any of the sections listed above can be modified, expanded, combined, or deleted to fit the needs of the operation.

The morgue sections are set up in such a manner as to allow easy flow of the unidentified victims from station to station, along with all pertinent records. A tracker is assigned to stay with each individual as he or she moves from beginning to end through the morgue. The tracker ensures that all records stay with the proper individual and that all sections sign off on the activity done at each station.

The flow of the morgue also requires the movement of victims from the storage area to the morgue facility.[5] This storage area needs to be cooled and monitored to ensure proper preservation. Multiple refrigeration trucks are normally used for the purpose. This allows for flexibility with changing numbers of victims. It also allows the storage team more choices in where to position their facility for the easiest and most secure transport of the unidentified bodies to and from the morgue. Computerized databases are now commonly used to document each truck's contents, maintenance records, fueling, temperatures, and release records.

The final release of the identified victim to family members is the main function of the MFI morgue. The entity in charge of the morgue will usually verify identification and release with the issuance of a death certificate. The quality control of this process is extremely important. Once released, all evidence is compromised. There must be a very explicit protocol for the release to be allowed. Normally the section or sections responsible for the identification will be called either at a release station or back to the morgue to verify the remains being released are in fact that of the individual they associated to the case. After the final check of the identification is complete the remains may be released to the family's representative from the funeral home they have chosen. Mistakes at this level could compromise the overall operation and destroy confidence of families in the system on site. A thousand accurate positive identifications cannot undo the harm and anguish of one misidentification!

Another protocol that is important to an MFI is the release of information to the public and media. It has been found through many past experiences that only one person should be designated as the information control officer. This person reports directly to the medical examiner or MFI director and releases only that which is approved. The best advice for any other person involved with the operation is to not make comments to anyone. This prevents any ambiguity from occurring. It has also been suggested that the information office give reports to family members before releasing any information to the media. This procedure takes families into consideration and prevents the possibility of close family members hearing sensitive material in the media before they hear it from official sources. It also prevents accidentally erroneous information from reaching their ears. The media has been known to make mistakes.

The final point that is important to any identification section of an MFI is an exit plan. All personnel need to be aware that the operation will be completed or reach an endpoint at some time in the future. All too often this is not addressed and workers are left with the empty and desolate feeling of no longer being needed. Education of the workforce about expectations of the actual time of service helps with this transition and is essential to the well-being of workers.

12.7 Odontology Section

The classic dental section model included antemortem, postmortem, and comparison sections, but response to large incidents in recent years has resulted in more modern dental morgue operations.[6] These operations have taken advantage of technological advances, and that the classic model has changed somewhat. The details of those changes will be discussed below. The tasks, in general, have not changed. Postmortem section members still examine, photograph, and take radiographs of the victims. Antemortem section members remain responsible for locating, interpreting, and archiving antemortem records. With the advent of instantaneous digital information from the postmortem section, these same antemortem team members will often also compare antemortem and postmortem data. Commonly, the postmortem section will be very busy for the first part of the operation. As postmortem examination rates slow or are completed on all victims, the postmortem team members can be transferred to the antemortem section to assist with antemortem duties and make comparisons. The overall odontology section is headed by the chief odontologist. This position can rotate among designated individuals over extended operations. Operations in the odontology section are under the authority of the chief odontologist and all dental identifications are reviewed and approved by the chief.

Prior planning is essential. The postmortem team requires autopsy instruments, personal protective equipment, radiography equipment, clerical equipment, and sundries. A cache of equipment must be assembled along with an accurate inventory. After each activity the inventory must be refreshed and repacked. Any adjustments to the cache can be done during operations, but need to be noted and corrected in the inventory. The antemortem team will also require a cache of equipment to allow them to accomplish their assignments. These items have been listed many times in several publications. The reference listing at the end of this chapter will have most of the articles and equipment lists. There are many suggested equipment lists, but the most important thing is that there is one that is consistent with the training and needs of that team. A manual prepared by the U.S. Department of Justice entitled *Mass Fatality Incidents: A Guide for Human Forensic Identification* (www.ojp.usdoj.gov/nij) details the needs of each part of an MFI operation.[2] Each operation has its unique needs, and adjustments to these required inventories can be made. It is also wise to list all outside resources available for help with disposables and other sundries.[7]

In recent years the classic three-section team has become somewhat dated. With the widespread use of computerized radiography and record handling, the postmortem and radiology sections have merged into one. The entire dental section now runs as a real-time, integrated, and networked system that allows all sections immediate access to all information. Postmortem team members now can review and do quality control on radiographs as they are taken. Once the postmortem record is obtained, if antemortem information is already on site, the antemortem section can begin to work on comparisons. This integration and access allows for accurate construction of victim records, timely review of possible identifications, and more rapid information transfer to the waiting family members of the victims.

The tracker brings a postmortem case to the dental section. The first step is always digital photography to document the unique number and the appearance of the victim. The initial and last photographs for the case show the case number assigned to the unidentified victim. These two photographs bracket all other photographs of the case on the camera's storage media for easy distinction and separation when they are entered into the digital case file. Once the photographs are finished the decision for surgical or nonsurgical dental examination is made. The decision to resect the jaws is dependent on the viewablity of the victim's body. For viewable bodies, all methods to break rigor and access the dental structures for examination and radiographs are attempted short of surgical access. There should be a rigid protocol available to address this question. Sometimes the medical examiner or MFI commander will request to review all cases before surgical access is used. The chief odontologist may be empowered to make this decision. The protocol used should document all actions on each case and note who approved surgical access. Once the mode of access is determined, the postmortem dental examination begins with digital radiographs. A full mouth series is taken. In some forensic cases a single arch bitewing projection image is used on some periapical radiographs to allow comparisons to antemortem bitewings. Once the digital radiographs are completed the postmortem visual examination is carried out, with the radiographs available for review whenever necessary. The postmortem examination team is made up of three individuals. The leader performs the visual examination and places the digital sensor or film to assist with radiographs. The assistant operates and aims the handheld x-ray source and assists the leader with the visual examination. The third member of the team is the computer operator, who verifies correct radiographic images, controls the progress through the visual examination and radiographs, and enters the dental data into the electronic chart as the examination progresses. Figure 12.1 shows the DMORT dental morgue with workers in full personal protective equipment (PPE) and all three workstations in operation in Louisiana following Hurricane Katrina.

Figure 12.1 Dental Morgue, St. Gabriel, Louisiana, after Hurricane Katrina. At each station the leader and assistant work together in a postmortem *pas de deux* as the computer operator choreographs each step. An assistant using an Aribex™ Nomad™ handheld x-ray generator can be seen on the far left. The body trackers in the foreground are waiting to move each gurney to the next station.

Once the examination is complete, the computer operator reads back all the recorded data for the leader and assistant to verify as quality control. Quality control is a team effort and all checks should lead to the production of a pristine postmortem record with photographs, radiographs, and charting. If done properly, there should be no need to reexamine the case until final release verification after a positive identification has been made.

The antemortem section has three main goals: (1) obtain dental information on all missing persons from the MFI, (2) enter this information into computers to create the antemortem records on each missing person, and (3) perform comparisons and make dental identifications using the information that they and the postmortem section have gathered. The antemortem section, perhaps, has the most difficult job in the odontology section. They are responsible for gathering information on dentists of record, working with investigators to assess needs for further information searches, and translating the records of treating dentists into a standard form for comparisons. When all of the postmortem work is completed, the antemortem section is still working and busy up to the close of the overall operation. There has been some concern about Heath Insurance and Portability Accountability Act of 1996 (HIPAA) rulings for dental records. According to 45 CFR 164.512(g), "HIPAA Exemption for Medical Examiners and Coroners," dental records needed for identification purposes do not fall under HIPAA regulations and can be released and sent with no repercussions for the attending dentist.

Selection of workers in the odontology section is important. Hopefully all the workers have been trained in similar fashion and are aware of protocols in use. For extended operations, new workers are needed to replace those that have finished their tour of duty. The chief odontologist should have a standard procedure for reviewing applicants and verifying the accuracy of their credentials. Each new member needs to be closely monitored to ensure he or she understands what his or her role is to be and how important accuracy is to the identification process. Experienced and new workers must be monitored for physical and emotional stress and each case handled appropriately.

The chief odontologist will establish how each case will be signed out. Most use the American Board of Forensic Odontology (ABFO) guidelines and terminology.[8] Whether identifications are reviewed by the chief or a committee should be decided at the beginning of an MFI operation. These identification records are then given to the overall team leader or control center for processing. Most operations then require the identifying section to review the identification and all records before final release of the victim's body to the family. This gives one last point of quality control before the case leaves the chain-of-evidence control of the MFI command.

12.8 Computer Assistance Programs and Digital Information

There has been a natural evolution of computer-assisted identification programs to aid in comparing records of postmortem and antemortem cases.[9] The first DOS-based program was Computer-Assisted Postmortem Identification (CAPMI), developed by the Army and used many times with great success. With the development of MS Windows, Dr. James McGivney developed WinID. WinID uses an MS Access database and a Windows format to enter and store antemortem and postmortem record information. It assigns a unique number to each case and will not allow duplication of numbers. There are other programs under development and being tested, but most teams currently train with WinID.

12.9 Digital and Conventional Images

Digital images are now widely accepted and used for identification procedures.[10] Digital photographs and radiographs are very similar, with the main difference being the fact that photographs are created by the detection of light by the sensor, while the radiographic sensors detect radiation from an x-ray source. Digital imaging eliminates darkrooms and chemical processors and instead allows immediate images for viewing and comparison. Details on digital radiography are discussed in Chapter 10 of this book. Digital images have been accepted as evidence in state and federal courts. Digital imaging requires imaging software to capture, enhance, and store the images. The DEXIS™ digital radiography system has been used in mass disaster situations since the late 1990s. The system captures the radiographic image from a wired sensor and stores these images for viewing. DEXIS software can also capture and store photographic images.

A working relationship has been established to allow the WinID identification program and the DEXIS dental digital radiography system to work together seamlessly. This relationship allows WinID to create antemortem and postmortem records while ensuring unique number assignments. WinID then, in turn, assigns the corresponding case number to the digital image file and allows all the digital images (radiographic and photographic) to be captured and stored for retrieval and review anytime the case is being viewed in the WinID program. Dexis also creates a screening image to be used by WinID for initial radiological review of case comparisons. What this means in the field is that duplicate numbers are prevented by WinID before they can ever be introduced to the operations data. It has also been seen that this setup combined with three dental postmortem bays can handle large numbers of cases in each twelve-hour shift. This system prevents past bottlenecks in the dental identification morgue when conventional film radiography and paper records would invariably impede the flow. Dual-screen computer technology that enables gurney-side and computer input team members to view the information simultaneously is already possible and has been tested in the laboratory.

WinID uses a Microsoft Access database. This allows the mail merge features of Microsoft Word to be used to create a range of reports that may be needed by local authorities or other groups. The final records from an operation in the form of a database can be given to the responsible representative of the agency in charge of the operation in digital form for their use or retrieval at a later date. This will also allow for additional records to be added by the local authorities in charge of postaction operations. All the dental records, digital images, and digital radiographs from the Hurricanes Katrina and Rita response were contained in 20 gigabytes of information and presented to the Louisiana authorities in a removable computer drive. No paper postmortem records were created or stored from that operation. All antemortem records were scanned into digital form and then were sealed and boxed for storage.

The keyword in digital imaging and record keeping is *backup*. Be sure the system backs up regularly and gives a dependable and accessible backup copy of the data. Check frequently to ensure the backup will work to restore the information in case the system fails. Many backup systems are available. Choose one before an actual disaster occurs.

A response can still use conventional radiography and paper records, but it has been found that in the aspect of convenience and efficiency, digital materials are now much preferred. If paper records are to be used, copies of all antemortem and postmortem documents are available in the WinID program files. In training sessions the odontology section must study modes and methods available to their particular settings. The medical examiner or MFI commander also needs to approve the methods to be used. Do not expect to be able to switch from conventional to digital systems on the fly. The transition is time-consuming and could be prevented if all of this was approved and covered in training sessions. In the opinion of this author, there is a reasonable cutoff for the use of conventional methods. Once the MFI estimated number exceeds one hundred individuals, the operation enters the level where digital creation and storage of records is necessary.

12.10 Preparedness Training and Planning

All persons involved in an MFI response need to be trained in the job they do as well as about the entire operation and structure.[1] Protocols for each section need to be studied and understood before the individual is allowed to enter the work areas. Training before the incident is best, but on long deployments many personnel changes lead to new workers entering the process. These new workers must be trained and assessed for their abilities before they begin to work in sections. A designated training area is needed in this case, and a standardized training schedule needs to be developed to make sure the proper objectives are reached by each new worker. Training periods are also good to introduce the new workers to the response team environment. If regular training sessions are used by the various sections, hopefully the responders will have a fairly good level of knowledge and be familiar with others on the team. Going into the situations seen during an MFI is not the best time to build trust, and at least annual training sessions will allow team members to meet and learn about each other as well as learn how to handle new technology and techniques. Federal, state, and local agencies have some funding for training, and your team leaders can investigate and use available funds to help team members get to training sessions.

12.11 Conclusions

There is a large knowledge base now available from past operations involving MFI.[11] As leaders and workers on an identification team, we need to study history, assemble information, and endeavor to not repeat the mistakes made in every deployment. Training is the most valuable step in this process.[3] Individuals in a well-trained organization tend to experience fewer glitches in the operational phase of the identification process and also have better morale due to the fact that all go into these situations with a general idea of what to expect. The identification process is a huge team effort. No section stands alone.

One thing that is really important for the identification team to remember is that we are working for the families to return their loved ones to them. Team members should not do this work for personal recognition and gain. Once the egos are tamed, the operations run more smoothly and groups of individuals become teams.

Control information release by designating only one individual to represent the MFI operational staff. No worker should ever talk to the media for any reason unless asked by the person in charge of public information or the incident commander. This tight control of information will protect the workers from the stress of media pressure and improve the quality of the information reported to the public.

The last word in any identification operation is *accuracy*.[12] Use the utmost care in creating antemortem and postmortem records. Use multiple individuals to review and approve comparisons. Treat every case you work in an MFI as if it was your first. Speed is never as important as accuracy. Never bypass or shortcut protocols, as this will lead to failure. Take pride in your work and remember that the goal and purpose of your work is to ease grief and help families who have suffered tremendous losses.

References

1. American Dental Association. 1996. *Second National Symposium on Dentistry's Role and Responsibility in Mass Disaster Identification.* Chicago: American Dental Association.
2. National Institute for Justice. 2005. *Mass fatality incidents: A guide for human identification.* Washington, DC: USDOJ.
3. Levinson, J., and H. Granot. 2002. *Transportation disaster response handbook,* xiv. San Diego: Academic Press.
4. Eckert, W.G. 1997. *Introduction to forensic sciences,* xi. 2nd ed. Boca Raton, FL: CRC Press.
5. World Health Organization. 2004. *Management of dead bodies in disaster situations.* Washington, DC: Pan American Health Organization.
6. Herschaft, E.E., et al., ed. 2006. *Manual of forensic odontology.* 4th ed. American Society of Forensic Odontology.
7. Stimson, P.G., and C.A. Mertz. 1997. *Forensic dentistry.* Boca Raton, FL: CRC Press.
8. American Board of Forensic Odontology. 2008. *Diplomates reference manual.* http://www.abfo.org.

9. Cottone, J.A., and S.M. Standish. 1982. *Outline of forensic dentistry*. Chicago: Year Book Medical Publishers.
10. Bowers, C.M. 2004. *Forensic dental evidence: An investigator's handbook*. 1st ed. San Diego: Elsevier Academic Press.
11. Fixott, R.H., ed. 2001. *Forensic odontology, the dental clinics of North America*. Vol. 45, No. 2. Philadelphia: W.B. Saunders.
12. Luntz, L.L., and P. Luntz. 1973. *Handbook for dental identification: Techniques in forensic dentistry*. Philadelphia: Lippincott.

Age Estimation from Oral and Dental Structures

13

EDWARD F. HARRIS
HARRY H. MINCER
KENNETH M. ANDERSON
DAVID R. SENN

Contents

13.1	Introduction	263
13.2	Tooth Eruption/Tooth Emergence	264
	13.2.1 Statistical Concerns	269
13.3	Tooth Mineralization	269
	13.3.1 Primary Tooth Formation	270
	13.3.2 Permanent Tooth Formation	271
	13.3.3 Dental Age	275
13.4	Age Estimates for Adolescents and Adults	279
	13.4.1 Gustafson Method	281
	13.4.2 Root Transparency	284
	13.4.3 Cementum Annulations	284
	13.4.4 Tooth Wear	286
	13.4.5 Third Molar Formation	288
	13.4.6 Aspartic Acid Racemization	291
	13.4.7 Enamel Uptake of Radioactive Carbon-14	292
13.5	Summary, Conclusions, and Recommendations	293
References		295

13.1 Introduction

Here we provide an overview of the common dental methods used to estimate the age of an individual. Our intent is to discuss several of the many dental methods, but we make no claim of completeness in this overview. More detailed information is available in other textbooks as well as in the primary literature.[1-4] The methods reviewed here are applicable to a range of topics in the fields of physical anthropology, growth and development, and forensic dentistry.

For age estimation, the investigator is concerned with the person's degree of maturity: How far along the pathway from wholly immature to wholly mature has the person traveled? Investigators depend on useful landmarks along this pathway, like emergence of the first baby tooth or mineralization of the third molar. Commonly the attainment of specific biological events, such as crown completion of a particular tooth, is used to compare against the person's chronological age to gauge his or her tempo of development. Unfortunately, there are a number of confounding issues, like the person's sex (males develop slower for many traits), socioeconomic status (well-off kids tend to develop faster), health history (illness and poor nutrition both slow development), and race (some combinations of genes promote the tempo of growth; others slow it down). It is unusual that the investigator would know most, let alone all, of these important modifying factors. We also commonly need to assume that the person is (or was) growing near the average for his group and that we can apply appropriate norms for "his group" since the range of population growth patterns far exceeds the available published standards for any method.

A key issue *not* comprehensively covered here is that investigators should use as much information—and as many methods—as practical. *Composite* age estimates based on multiple kinds of data typically are more accurate than any one alone.[5–8] Similarly, while our expertise is with the dentition, some questions are better answered using other sorts of information. For example, craniometrics is much more informative about racial affiliation.[9,10]

13.2 Tooth Eruption/Tooth Emergence

The timing and pattern of tooth eruption are fairly well buffered from the external environment, and recording tooth eruption status can be a rapid, useful, and convenient way to estimate a person's age *if* the subject is in the primary or mixed dentition. There is a diversity of applications here. A person's age may be undocumented or, in preliterate settings, simply unknown, or the person may wish to falsify his age.[11–14] It may be of interest to ask whether groups of people differ in their tempos of growth or whether normative standards can accurately be applied to another group.[15–19] Most commonly for the forensic odontologist, the issue is to establish or estimate age (or age at death) in a contemporary, recent, or archaeological setting.[2,4]

In fact, tooth *eruption* is the process of a tooth's migration from its initial position in its bony crypt into occlusion. The proper term to be used here is tooth *emergence*,[20] and there are at least three operational definitions of emergence in the literature: In one, *emergence* commonly is defined as the appearance of some portion of the tooth's crown piercing the gingival mucosa (gingival eruption). This is the conventional clinical definition. But

Table 13.1 Mean Ages (in Months) for Emergence of Primary Teeth (Sexes Combined)

Tooth	Maxilla	SD	Mandible	SD
i1	9.03	2.20	7.18	2.46
i2	10.19	3.28	12.13	3.45
c	18.04	3.47	18.34	3.40
m1	15.13	2.69	15.01	2.80
m2	27.48	4.87	26.40	4.73

Source: Data from Tanguay et al., based on a sample of 314 French Canadian children.

what if you are dealing with macerated or skeletal material—with no mucosa? Similarly, what is one to do with radiographs where the gingiva is difficult or impossible to visualize? In such situations, one needs to find normative data where *emergence* is defined as the most occlusal portion of the crown emerging above the alveolar bone (alveolar emergence). Yet a third, occasional definition of emergence is when the tooth is fully erupted, so it is in functional occlusion with its antagonist in the opposing arch.[21,22] Obviously these three definitions yield different ages, with emergence above the alveolar bone preceding exposure through the gingiva, and eruption into functional occlusion occurring several weeks later.[23]

The twenty teeth of the primary dentition erupt (and emerge) when the infant is between about six and thirty months of age.[24,25] Tanguay and coworkers provide some of the more detailed statistics in this regard (Table 13.1). The modal eruption sequence is i1-i2-m1-c-m2 in both arches, defining a simple mesial-to-distal gradient, though m1 normally precedes canine emergence, so the canine is the one primary tooth that emerges into a confined space. Tanguay et al. document a slight but statistically significant male precedence of about one month in their sample of French Canadians, which agrees with the bulk of studies, but they also make the case that the amount, and even the direction, of sexual dimorphism exhibits ethnic and racial variability.[24,25]

Few studies have compared deciduous tooth emergence in American blacks and whites, but there seems to be trivial racial variation in the primary dentition, especially given the confounding effects of lower socioeconomic status in most studies of American blacks, which is likely to slow the tempos of development.[25,26]

Exfoliation of the primary teeth affords another means of estimating a child's age, and Moorrees' landmark publications are still considered the most reliable sets of data in this area.[27,28] Some other publications are available for more restricted groups.[29,30] Moorrees et al. reported on the three deciduous buccal teeth (c, m1, m2) in the mandible. These teeth exfoliated between about nine and twelve years of age in their sample of normal, white children in Ohio. The authors report the median ages of three stages of root

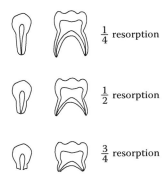

Figure 13.1 The three stages of root resorption of primary teeth used by Moorrees et al.[27] A fourth stage is exfoliation of the tooth. A representative single-rooted tooth is shown on the left, a multirooted tooth on the right.

resorption (Figure 13.1) plus age at exfoliation. The degree of tooth resorption can be scored directly or from radiographs. Obvious evidence of root tissue destruction (i.e., loss of a quarter of root length) occurs by five years of age on c and m1 in girls, and a bit later in boys. Given the ages covered by these events (Table 13.2), they generally can be combined with ongoing tooth formation in the permanent dentition, depending on the completeness of the dental remains.

Emergence of the permanent teeth exhibits fairly consistent sexual dimorphism across groups, meaning that girls' teeth typically erupt at earlier chronological ages than boys—though the degree of sex difference often is greater for bony than dental events.[31–33] There are racial differences in the timing of tooth emergence, though the studies are too sparse for most non-European groups to make definitive claims.[31,34] Broadly, it appears that peoples from Sub-Saharan Africa are developmentally advanced dentally, sometimes to striking extents.[35] On the other hand, there seems to be no clear-cut differences between Asians and Caucasians, and little is known about other racial groups.[36] Maki et al. found that American-Chinese and American-Japanese children were significantly delayed in their dental development compared to American whites.[17] Variation within populations, variation among the traits studied (such as dental vs. bony maturation), as well as the delaying effects of poor nutrition, low socioeconomic status, and childhood illnesses further muddy the issue of race differences.

For Americans with Western European backgrounds, emergence of the permanent teeth (defined as a cusp tip breaking through the gingiva) occurs in the range of roughly six to twelve years of age.[34,37] The most comprehensive data for the United States are those for blacks and whites from the Ten-State Nutrition Survey of 1968–1970; unfortunately, the standard deviations for these averages are not available.[38] Instead, we illustrate the eruption ages from a large contemporary cohort of London white children.[39] These data

Table 13.2 Average Ages at Mandibular Root Resorption Stages of Three Primary Teeth

	Females		Males	
Stage	Mean	SD	Mean	SD
Deciduous Canine				
Resorption 1/4	4.9	0.54	6.1	0.67
Resorption 1/2	7.3	0.78	8.4	0.89
Resorption 3/4	8.7	0.92	9.8	1.02
Exfoliation	9.5	1.00	10.6	1.10
Deciduous First Molar: Mesial Root				
Resorption 1/4	4.9	0.54	5.4	0.60
Resorption 1/2	7.2	0.78	7.6	0.82
Resorption 3/4	8.7	0.93	9.4	0.98
Exfoliation	9.5	1.05	10.7	1.12
Distal Root				
Resorption 1/4	5.1	0.58	6.4	0.69
Resorption 1/2	7.7	0.82	8.3	0.88
Resorption 3/4	9.3	0.97	10.0	1.04
Exfoliation	10.1	1.05	10.7	1.12
Deciduous Second Molar: Mesial Root				
Resorption 1/4	6.1	0.67	6.6	0.72
Resorption 1/2	8.3	0.88	8.5	0.90
Resorption 3/4	10.0	1.05	10.4	1.08
Exfoliation	11.1	1.15	11.6	1.20
Distal Root				
Resorption 1/4	6.9	0.74	6.6	0.79
Resorption 1/2	8.6	0.91	8.5	0.99
Resorption 3/4	9.9	1.04	10.4	1.14
Exfoliation	11.1	1.16	11.6	1.20

Source: Data derived from Moorrees et al.[28]

(Table 13.3) show the typical two waves of tooth emergence.[40] There is one group between six and eight years of age, when the first molar and the incisors emerge. Then there is a quiescent interval of about two years, followed by emergence of the canines, premolars, and second molar between about ten and twelve years of age. The characteristic female precedence also is evident in these averages (Figure 13.2), where girls, on average, emerge their teeth earlier than boys (mandibular M2 is the one exception in these data, where mean age is 6.3 for both sexes).

Numerous studies have attempted to describe the effect of primary tooth extraction on the eruption rate of the succedaneous tooth, though most are

Table 13.3 Median Emergence Times (Years) of the Permanent Teeth in a Contemporary Cohort of London White Children

Tooth	Males	Females	Sexes Pooled	Pooled SD
		Maxilla		
I1	7.19	7.07	7.14	0.57
I2	8.56	8.02	8.29	0.83
C	11.71	11.10	11.41	1.37
P1	10.82	10.60	10.70	1.39
P2	11.71	11.52	11.61	1.50
M1	6.28	6.38	6.34	0.61
M2	12.51	12.24	12.37	1.32
		Mandible		
I1	6.17	6.06	6.13	0.55
I2	7.47	7.20	7.34	0.64
C	10.79	9.94	10.37	1.15
P1	10.95	10.45	10.70	1.30
P2	11.98	11.58	11.78	1.35
M1	6.25	6.31	6.28	0.51
M2	12.07	11.67	11.86	1.27

Source: Data from Smith et al.[39]

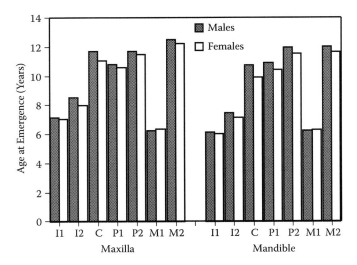

Figure 13.2 Plot of mean emergence age (years) in a contemporary cohort of London white children. Most teeth exhibit some sexual dimorphism, with girls ahead of boys. (Data from Smith et al.[39])

Age Estimation from Oral and Dental Structures

limited by small, varied sample sizes and weak study designs. The early work by Fanning sheds some light here. She showed that the effect of extraction depends on the eruption status of the successor, specifically, how developed the root is and how much overlying bone there is. If extraction was done when the successor was deep in its crypt with appreciable overlying bone and less than about one-fourth of the root formed, emergence is delayed. Conversely, if extraction was done when the successor was already close to the alveolar crest and there is considerable root formation, emergence is accelerated.[41]

13.2.1 Statistical Concerns

While tangential to this review, investigators and researchers need to be aware of the diversity of statistical techniques used to calculate the "average" age of a tooth's emergence (and of other developmental events). One issue is that the data can be derived from longitudinal or cross-sectional data. With longitudinal data, how frequently were the exams made? The precision of timing is one-half the examination interval, so broader-spaced examinations reduce the precision of the estimate of when a tooth emerged. When a tooth is unerupted at one exam but is erupted at the next, the convention is to assume that the tooth actually erupted halfway between the two exams.[42] With cross-sectional data, where children of various chronological ages are examined just once, various statistical methods have been applied to estimate the average age, but these don't all produce the same results because they are based on different assumptions. Methods range from calculating the arithmetic mean of the chronological ages of children exhibiting the tooth[43,44] to calculating probits or logits.[45] Analysts have applied other methods that are purported to better fit the nature of the data.[46,47]

Shortcomings of using tooth emergence to gauge a person's age are (1) tooth emergence is a fleeting event that occurs at an instant in time—a tooth can only be unemerged or emerged—and (2) permanent tooth eruption occurs within a span of only about 6 years (ignoring the highly variable M3), so the method is of no use outside the range of about six to twelve years of age.

13.3 Tooth Mineralization

Tooth mineralization—the development of a tooth's hard tissues—spans a much longer age range, making it more broadly applicable than emergence. So too, the extent of mineralization can be gauged radiographically so it is noninvasive. (We prefer *tooth mineralization* to the older term *tooth calcification* because little of a tooth's substance is calcium per se.)

Table 13.4 Normative Ages for Key Events in the Development of the Primary Teeth

Primary Tooth	Initiation of Cusp Mineralization (weeks in utero)	Amount of Crown Formed at Birth	Postnatal Age at Crown Completion (months postpartum)	Age at Root Completion (years)
		Maxilla		
i1	14	5/6	1.5	1.5
i2	16	2/3	2.5	2.0
c	17	1/3	9	3.25
m1	15.5	3/4	6	2.5
m2	19	1/5	11	3.0
		Mandible		
i1	14	3/5	1.5	1.5
i2	16	3/5	2.5	2.0
c	17	1/3	9	3.25
m1	15.5	Occlusal table complete	5.5	2.25
m2	18	Occlusal table incomplete	10.0	3.0

Source: Lunt and Law.[51]

13.3.1 Primary Tooth Formation

Tooth mineralization progresses in an invariant sequence from cusp tips, through lengthening of the crown's cervical loop, to formation of the cemento-enamel junction. Once the crown has formed, root development progresses, terminating with completion of the root apex or apices.[48–50] Dentine formation precedes and stays ahead of amelogenesis throughout formation.

Most development of the primary teeth occurs *in utero*. The synthesis by Lunt and Law[51] continues to be the most comprehensive data available (also see Kraus and Jordan[52]). They recorded four development landmarks for each tooth (Table 13.4): (1) the age at initiation of cusp mineralization, (2) how much of the tooth crown is mineralized at term, (3) the age after birth when the crown has finished mineralization, and (4) the age when the root is complete (root apexification). As shown in Table 13.4, all ten primary tooth types begin dentin and enamel deposition at their coronal tips early in the second trimester, and none of the crowns is completely mineralized at birth. Consequently, most primary teeth possess a neonatal line if the infant survives birth.[53–55] The maxillary molars are somewhat more advanced than those in the mandible at term. Mandibular m1 typically has the crown apices mineralized and united, with the occlusal table complete and evidence of enamel-dentine deposition along the crown's neck.

Mandibular m2 is less developed, with the individual cusp tips mineralized and some enamel bridging across the cusps, but the overall occlusal table is incomplete. Root growth progresses throughout the first year of life, completing for the incisors at about 1½ years of age, but not until about 3¼ years for the canines.

Problems with the use of formation stages are (1) there are few distinguishable grades and (2) there are "gaps" between one stage and the next. And yet, it is evident that crown-root mineralization is a continuous, seamless process. Liversidge et al.[56] suggest that finer age discrimination can be obtained by measuring tooth length and estimating age at death from predictive equations that they developed from regressing tooth length on age. Liversidge and coworkers provide predictive equations for the deciduous (and some permanent) teeth. Their data show that, between birth and about five years of age, tooth length increases on the order of 0.6 to 0.9 mm per year. They studied the eighteenth- and nineteenth-century internments of known age from Christ Church, London. The method shows promise of providing finer age estimation[57] than when using ordinally spaced morphological grades. Liversidge and Molleson[58] showed that the method can be applied to radiographs of the teeth, so extracted elements are unnecessary.

13.3.2 Permanent Tooth Formation

While tooth mineralization is a seamless developmental process, it can be partitioned into visually distinguishable stages or grades.[59] These artificial ordinal grades allow the researcher to visually identify the extent of crown-root development based on morphological features (so measurement and concerns about magnification are irrelevant). Several researchers have devised ordinal schemes for gauging tooth formation. Some are more elaborate than others.[28,60,61] The competing issues are that (1) there should be as many grades as possible so the extent of tooth formation can be gauged precisely, but (2) stages that are too close together create ambiguities and misclassifications between them. Currently, the two most popular grading schemes are those of Moorrees et al.,[28] with thirteen grades for single-rooted teeth and fourteen for multirooted teeth (Figure 13.3), and Demirjian's method,[62] which uses just eight grades (Figure 13.4).

An important convention when using these schemes is that the examiner should score the highest grade that has been attained (Moorrees, personal communication), because this improves repeatability. If a tooth's morphology places it between two grades, one should pick the higher grade actually achieved, not the "closer" grade, nor should one try to interpolate between grades.

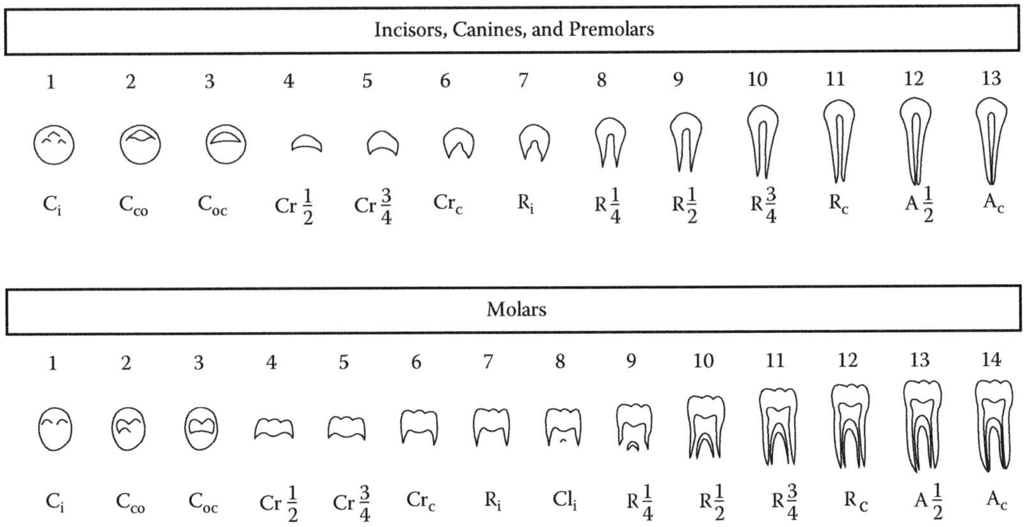

Figure 13.3 Schematic drawings of the stages of crown-root mineralization used by Moorrees et al.[28] There are thirteen stages for single-rooted teeth and fourteen stages for multi-rooted teeth, the difference being the addition of cleft initiation for the multirooted molars. The first three grades illustrate the mineralizing coronal aspects of a tooth within its bony crypt. Although the stages normally are numbered, it needs to be remembered that these ordinal grades do not represent equally spaced gradations of formation. Instead, they are different enough to permit visual discrimination between them. (Based on figures in Moorrees et al.[28])

Age Estimation from Oral and Dental Structures

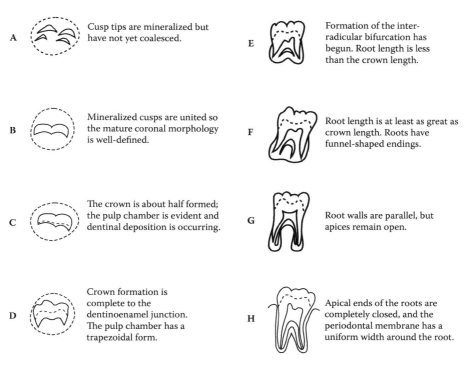

Figure 13.4 Schematic drawings of the eight ordinal grades used in the Demirjian system of dental age estimation. The dotted circles in stages A through D depict the encapsulating bony crypt. (Modified from Demirjian et al.[62])

Most grading schemes have been developed for assessing radiographic images of crown-root formation. With appropriate consideration, though, one can use actual teeth from recently deceased or archaeological material.[63–65] If histological sections of teeth are studied, it needs to be remembered that the soft tissue growth and differentiation of a tooth precedes the mineralization phase.[49] Kronfeld suggested that histological processes were two to six months ahead of radiological findings.[66]

The first large-scale study of permanent tooth mineralization in children followed longitudinally was the well-known work of Moorrees et al.,[28] which actually combined two growth studies. Records of young children (less than about ten years of age) were examined from the Boston area. These records had been collected by Harold C. Stuart, but World War II put an unexpected end to Stuart's study.[24a] Moorrees consequently combined these Boston data with those from older children enrolled in the Fels Longitudinal Study in Yellow Springs, Ohio.[67] The resulting data are somewhat confounded because the tooth development data for younger and older children are from different populations—and the Fels children had a faster tempo of growth than those in Boston.[68]

Table 13.5 Definitions of Tooth Formation Stages

Single-Rooted Teeth	Definition	Multiple-Rooted Teeth
1	Initial cusp formation: Amelogenesis has begun on the individual cusp tips.	1
2	Coalescence of cusps: Centers of calcification are merged but the entire border is not radiopaque.	2
3	Cusp outline complete: The coronal outline of the tooth is mineralized.	3
4	Crown 1/2 formed: Amelogenesis has proceeded halfway to the crown-root as judged from the morphology of the radiopaque portion.	4
5	Crown 3/4 complete.	5
6	Crown complete: Morphologically, the crown has mineralized but root formation has not begun.	6
7	Initial root formation: There is just a trace of root radiopacity below the crown outline.	7
—	Initial cleft formation: Mineralization is evident in the interradicular area.	8
8	Root length 1/4: The radiographic morphology of the root is 1/4 of its projected final size.	9
9	Root length 1/2 complete.	10
10	Root length 1/4 complete.	11
11	Root length complete.	12
12	Apex half closed: The lateral borders of the root tip become convex rather than tapered as earlier.	13
13	Apical closure complete: Size of the apical foramen is reduced to its mature size.	14

Source: Adapted from Moorrees et al.[27]

Moorrees et al.[28] developed the ordinal grading scheme shown in Figure 13.3, and descriptions of these stages are given in Table 13.5. This is a simplification of the sixteen-grade scheme developed by Gleiser and Hunt[50] and of the nineteen-grade scheme constructed by Fanning.[60] Moorrees and coworkers presented their data in graphical form with the idea that each scorable tooth from a child's examination would be marked on the graph, and the resulting cluster of tooth-specific dental ages would serve as that child's range of dental ages. In practice, these graphs proved time-consuming and ineffective (but they are commonly reproduced in the literature). Harris and Buck "reverse engineered" these graphs to provide the data in a more usable format (Tables 13.6 and 13.7).[68]

Other normative data for tooth mineralization are available for North America,[15,43] Europe,[69–74] Eastern Europe,[75] and the Mideast.[76]

Harris and McKee[15] developed formation standards for American blacks and whites separately, since blacks tend to form their teeth faster. The data were derived from the cross-sectional study of children attending a dental school in Tennessee for routine dental care (Tables 13.8 and 13.9).

13.3.3 Dental Age

Investigators should score all formative teeth in order to maximize the information about a case. An exception would seem to be dysmorphic teeth, including microform and pegged teeth, where the formative status is questionable. In Westernized settings, where the subject may have been exposed to chemotherapy or irradiation during growth, one needs to be alert to abnormal crown-root forms.[77-79] Alternatively, no one has found any consistent left-right asymmetry in tooth formation,[20] though mandibular teeth tend to form and erupt slightly ahead of their maxillary counterparts.[80-82]

In concept, the formative status of up to thirty-two permanent teeth could be scored for a given individual, though the incisors may have completed root apexification before M3 mineralization begins.[15] The question is how best to synthesize the age estimate. It is unlikely that multiple teeth will yield the same age estimate. The simplest suggestion for combining the individual tooth age estimates is to average across all scorable teeth.[61,83] The shortcomings are that (1) much of the data are redundant (such as estimates from homologous teeth on the two sides of the arcade), and (2) some teeth are more precise and reliable estimates than others (though this is a complex mix of tooth type and stage of formation). The third molar is notorious in this regard, with unusually high variability, and it is omitted from some scoring systems.

Demirjian et al.[62] developed a weighting scheme to overcome these issues. With his system, the mandibular seven tooth types are scored (omitting M3). Just the mandibular teeth are used because of their greater clarity on radiographs (whereas several maxillary teeth are obscured by the complex bony architecture), and because there is considerable statistical redundancy among the teeth in the two arches. Likewise, teeth from just the left (or from the clearer, better preserved) side are used because of the duplication of information between sides. There are four steps in the Demirjian system: (1) the extent of crown-root development of the seven teeth is scored (Figure 13.4), (2) a table of values is used to weight each tooth's score, (3) these weights are summed, and (4) the sum is checked against another table that provides the person's dental age. Several researchers have computerized this sequence of events to minimize the arithmetic.[62] Demirjian's system has been applied to a variety of ethnic samples.[57,84] Results have been varied, and those where dental age prediction has been poor occur because the target sample has a different tempo of growth during childhood or adolescence than the French Canadians used to develop the pattern.[18,85-87] That is, the dental age curve developed mathematically from Demirjian's French Canadian children is not uniformly applicable to other human groups. Deviations commonly are greatest in adolescence,

Table 13.6 Age at Attainment (Years) of Stages of Crown-Root Formation of Permanent Incisors[a]

Grade	UI1 $\bar{x}$	UI1 SD	UI2 $\bar{x}$	UI2 SD	LI1 $\bar{x}$	LI1 SD	LI2 $\bar{x}$	LI2 SD
Girls								
Ci	•	•	•	•	•	•	•	•
Cco	•	•	•	•	•	•	•	•
Coc	•	•	•	•	•	•	•	•
Cr 1/2	•	•	•	•	•	•	•	•
Cr 2/3	•	•	4.6	0.51	•	•	•	•
Cr 3/4	•	•	•	•	•	•	•	•
Cr c	4.9	0.54	5.7	0.62	•	•	•	•
R 1/4	6.0	0.66	6.6	0.71	4.5	0.51	4.7	0.53
R 1/3	•	•	•	•	•	•	5.2	0.57
R 1/2	6.6	0.71	7.2	0.76	5.1	0.57	5.9	0.65
R 2/3	7.1	0.76	7.7	0.82	5.6	0.62	6.3	0.68
R 3/4	7.6	0.81	8.3	0.87	6.1	0.66	6.7	0.72
R c	8.2	0.86	9.1	0.95	6.6	0.72	7.6	0.80
A 1/2	8.9	0.93	9.6	0.99	7.4	0.79	8.1	0.86
A c	•	•	•	•	7.7	0.82	8.5	0.89
Boys								
C i	•	•	•	•	•	•	•	•
C co	•	•	•	•	•	•	•	•
C oc	•	•	•	•	•	•	•	•
Cr 1/2	•	•	•	•	•	•	•	•
Cr 3/4	•	•	•	•	•	•	•	•
Cr c	5.3	0.59	5.9	0.64	•	•	•	•
R 1/4	6.3	0.68	6.9	0.75	•	•	5.3	0.60
R 1/3	•	•	•	•	•	•	5.6	0.62
R 1/2	6.9	0.74	7.6	0.80	5.2	0.59	6.2	0.68
R 2/3	7.6	0.80	8.1	0.86	5.8	0.64	6.8	0.74
R 3/4	8.1	0.85	8.7	0.91	6.4	0.70	7.4	0.78
R c	8.6	0.90	9.6	1.01	7.0	0.75	8.0	0.84
A 1/2	•	•	•	•	7.7	0.81	8.5	0.90
A c	•	•	•	•	8.1	0.85	9.3	0.98

[a] Codes: cusp (C), crown (Cr), root (R), apex (A). Stages: initiation (i), coalescence (co), cusp outline complete (co), complete (c), cleft (cl).

when the French Canadian "norms" deviate from the target samples they are applied to. The solution is to produce group-appropriate norms that accurately reflect the target group's tempo of growth.

Demirjian reflected that the statistical information among the seven tooth types was largely redundant since the teeth were developing synchronously.[88]

Table 13.7 Age at Attainment of Stages of Crown-Root Formation for the Permanent Mandibular Buccal Teeth

	C		P1		P2		M1		M2		M3	
Grade	x̄	SD	x̄	SD	x̄	SD	x̄	SD	x̄	SD	x̄	SD
					Girls							
C i	0.5	0.12	1.7	0.24	2.9	0.35	0.1	0.05	3.5	0.41	9.6	1.00
C co	0.7	0.15	2.2	0.28	3.5	0.40	0.2	0.09	3.8	0.43	10.1	1.05
C oc	1.2	0.18	2.9	0.35	4.1	0.47	0.7	0.14	4.3	0.49	10.7	1.11
Cr 1/2	1.9	0.25	3.5	0.41	4.7	0.53	1.0	0.17	4.8	0.54	11.3	1.17
Cr 3/4	2.9	0.35	4.2	0.49	5.3	0.59	1.4	0.20	5.4	0.59	11.7	1.20
Cr c	3.9	0.45	5.0	0.56	6.2	0.66	2.2	0.28	6.2	0.68	12.3	1.27
R i	4.7	0.52	5.7	0.63	6.7	0.73	2.6	0.32	7.0	0.75	12.9	1.32
R cl	•		•	•	•	•	3.5	0.41	7.8	0.83	13.5	1.39
R 1/4	5.3	0.57	6.5	0.69	7.5	0.79	4.6	0.52	9.1	0.96	14.9	1.53
R 1/2	7.1	0.75	8.1	0.86	8.7	0.92	5.1	0.57	9.8	1.01	15.8	1.62
R 3/4	8.3	0.88	8.8	0.97	10.0	1.05	5.5	0.60	10.5	1.09	16.4	1.67
R c	8.8	0.93	9.9	1.03	10.6	1.12	5.9	0.63	11.0	1.13	17.0	1.71
A 1/2	9.9	1.03	11.0	1.15	12.0	1.24	6.5	0.71	12.0	1.23	18.0	1.82
A c	11.3	1.18	12.1	1.26	13.6	1.40	8.0	0.85	13.8	1.43	20.1	2.01
					Boys							
C i	0.5	0.11	1.8	0.24	3.0	0.37	0.0	0.09	3.7	0.42	9.2	0.98
C co	0.8	0.15	2.3	0.31	3.5	0.42	0.2	0.11	4.0	0.44	9.7	1.01
C oc	1.2	0.19	2.9	0.36	4.2	0.48	0.5	0.11	4.8	0.52	10.3	1.07
Cr 1/2	2.1	0.27	3.6	0.43	4.7	0.53	1.0	0.17	5.1	0.56	10.9	1.14
Cr 3/4	2.9	0.35	4.4	0.52	5.3	0.59	1.5	0.21	5.7	0.61	11.6	1.20
Cr c	4.0	0.46	5.2	0.58	6.2	0.69	2.1	0.29	6.5	0.69	12.0	1.24
R i	4.8	0.55	5.8	0.64	6.9	0.74	2.7	0.34	7.1	0.76	12.7	1.32
R cl	•	•	•	•	•	•	3.5	0.41	8.1	0.84	13.6	1.41
R 1/4	5.7	0.63	6.8	0.74	7.8	0.83	4.7	0.53	9.3	0.98	14.6	1.50
R 1/2	8.0	0.86	8.5	0.91	9.4	0.99	5.1	0.57	10.1	1.04	15.1	1.54
R 3/4	9.6	1.00	9.9	1.04	10.8	1.13	5.4	0.61	10.8	1.12	15.9	1.62
R c	10.2	1.06	10.3	1.09	11.5	1.21	5.8	0.64	11.3	1.16	16.3	1.67
A 1/2	11.8	1.23	11.9	1.24	12.7	1.30	6.9	0.75	12.2	1.25	17.6	1.79
A c	13.0	1.35	13.3	1.38	14.2	1.46	8.5	0.91	14.2	1.46	19.2	1.95

He simplified the system further, so dental age can be estimated from just four mandibular teeth with little loss of accuracy.[89] Of note, Haavikko et al. had developed similarly simplified versions a few years earlier.[70] The Demirjian methods require that all of the required tooth types be scored, so missing data create a problem. Nystrom et al. provide prediction models for missing teeth based on Finnish children.[90]

Table 13.8 Age of Achievement of Mineralization Stages of Permanent Maxillary Teeth[a]

Stage	Subjects	Incisors			Premolars		Molars		
		Central	Lateral	Canine	First	Second	First	Second	Third
1	WM	...	...	...	...	...	...	3.9	9.3
	BM	...	...	...	...	...	...	4.3	8.6
	WF	...	...	...	...	...	...	4.0	8.9
	BF	...	...	...	...	...	...	3.3	9.2
	(SD)	...	...	...	...	...	...	(0.49)	(0.99)
2	WM	...	...	...	...	4.0	...	4.5	9.7
	BM	...	...	...	...	4.3	...	4.0	9.2
	WF	...	...	...	...	4.6	...	4.4	10.0
	BF	...	...	...	...	3.9	...	3.6	9.0
	(SD)	...	...	...	...	(0.83)	...	(0.55)	(1.32)
3	WM	...	...	...	4.1	4.8	...	4.8	10.8
	BM	...	...	...	3.3	4.2	...	4.4	9.9
	WF	...	...	...	3.9	4.4	...	5.1	10.7
	BF	...	...	...	3.3	4.7	...	4.6	9.2
	(SD)	...	...	...	(0.70)	(0.67)	...	(0.77)	(1.06)
4	WM	...	3.6	3.8	4.7	5.6	...	5.7	11.5
	BM	...	3.5	3.6	4.2	5.0	...	5.3	9.9
	WF	...	4.3	4.2	4.7	5.8	...	5.6	10.6
	BF	...	3.8	3.5	4.2	5.3	...	4.8	9.7
	(SD)	...	(0.70)	(0.80)	(0.79)	(0.78)	...	(0.65)	(1.03)
5	WM	3.6	4.5	4.8	5.8	6.4	3.4	6.7	11.9
	BM	3.7	4.6	4.1	5.7	6.7	2.7	6.3	11.1
	WF	4.3	4.5	4.6	6.0	6.4	3.5	6.5	11.9
	BF	3.8	3.8	4.1	5.2	6.0	3.0	6.0	9.9
	(SD)	(0.77)	(0.78)	(0.83)	(0.77)	(0.92)	(0.53)	(0.93)	(1.17)
6	WM	4.3	5.3	6.0	7.3	7.7	4.0	7.4	12.4
	BM	3.9	4.2	5.5	6.8	7.9	3.4	7.3	10.8
	WF	4.6	4.9	5.6	6.5	7.3	3.8	7.2	11.6
	BF	4.1	4.8	4.7	6.2	7.0	3.7	6.8	11.3
	(SD)	(0.79)	(0.90)	(0.93)	(1.14)	(1.04)	(0.71)	(0.99)	(0.70)
7	WM	5.3	6.5	6.7	8.0	8.5	4.6	8.5	13.2
	BM	5.4	5.8	6.1	8.2	8.4	4.2	8.6	13.1
	WF	5.4	5.9	6.1	7.7	8.4	4.3	8.4	13.4
	BF	5.1	5.4	5.9	7.2	8.2	4.3	7.9	12.1
	(SD)	(0.90)	(1.22)	(1.14)	(0.90)	(0.98)	(0.87)	(0.99)	(1.37)
8	WM	6.3	7.1	7.8	9.0	9.4	4.3	9.7	...
	BM	5.9	7.0	7.9	8.6	9.5	3.9	9.1	...
	WF	6.1	6.7	7.3	8.7	9.0	4.7	9.1	...
	BF	6.0	6.4	6.9	7.9	9.1	4.2	8.9	...
	(SD)	(0.75)	(0.83)	(0.90)	(0.95)	(0.98)	(0.58)	(1.13)	...

Table 13.8 (continued) Age of Achievement of Mineralization Stages of Permanent Maxillary Teeth[a]

Stage	Subjects	Incisors			Premolars		Molars		
		Central	Lateral	Canine	First	Second	First	Second	Third
9	WM	7.5	8.1	8.9	9.2	10.1	5.3	10.5	...
	BM	7.2	7.8	8.3	9.7	9.9	5.3	9.7	...
	WF	6.9	7.4	8.1	9.5	10.1	5.5	10.0	...
	BF	6.4	6.9	7.9	8.9	9.3	4.8	9.5	...
	(SD)	(0.82)	(0.90)	(0.88)	(0.85)	(1.12)	(0.78)	(0.81)	...
10	WM	8.1	8.5	10.2	10.7	11.4	6.4	11.8	...
	BM	7.6	8.3	9.7	9.7	10.4	6.9	11.5	...
	WF	7.5	8.2	9.4	10.0	10.5	6.3	11.3	...
	BF	7.1	8.5	9.1	10.3	10.0	6.0	10.8	...
	(SD)	(1.04)	(0.75)	(0.92)	(0.96)	(1.06)	(0.74)	(1.10)	...
11	WM	8.8	9.6	11.9	12.3	12.6	7.5	12.6	...
	BM	8.9	9.7	11.3	12.1	12.3	7.5	12.5	...
	WF	8.1	9.1	11.0	11.2	11.2	7.3	11.5	...
	BF	8.5	9.1	10.1	10.5	11.4	6.6	11.4	...
	(SD)	(0.86)	(0.84)	(1.07)	(1.02)	(0.81)	(1.03)	(0.93)	...
12	WM	9.7	10.5	12.5	12.7	12.3	8.5	12.4	...
	BM	9.3	9.6	12.7	11.9	12.8	8.5	12.8	...
	WF	9.1	9.7	11.8	11.6	12.0	8.0	12.1	...
	BF	8.8	9.6	11.5	11.1	12.2	8.4	12.2	...
	(SD)	(0.91)	(1.00)	(0.95)	(0.85)	(0.77)	(1.05)	(1.06)	...
13	WM	...	...	...	...	...	9.5	12.5	...
	BM	...	...	...	...	...	9.3	13.0	...
	WF	...	...	...	...	...	9.2	12.9	...
	BF	...	...	...	...	...	8.8	11.8	...
	(SD)	...	...	...	...	...	(1.07)	(1.43)	...

[a] Data are unreported (...) when the sample size is less than five and for the first and last mineralization stages, which cross-sectional data do not accurately define. The values are the arithmetic mean and standard deviation (SD) weighted for the four race-sex subgroups of each tooth and stage.

13.4 Age Estimates for Adolescents and Adults

The several methods discussed so far are unavailable once the subject is at least in his mid-teens. After this, the variable third molar is the only tooth that has not yet completed root formation.[69,91,92] A more comprehensive discussion of age estimation for adolescents and young adults from third molar development is provided below (Section 13.4.5). Focus then turns to age estimation based upon the aging and, oftentimes, degenerative processes associated with adulthood, or to techniques that look at histological, biochemical, or special changes in teeth.

Table 13.9 Age of Achievement of Mineralization Stages of Permanent Mandibular Teeth[a]

		Incisors			Premolars		Molars		
Stage	Subjects	Central	Lateral	Canine	First	Second	First	Second	Third
1	WM	...	...	...	...	3.9	...	4.1	9.0
	BM	...	...	...	...	3.2	...	3.7	8.2
	WF	...	...	...	...	5.0	...	3.6	9.6
	BF	...	...	...	...	3.7	...	3.5	8.4
	(SD)	...	...	...	...	(1.17)	...	(0.51)	(1.23)
2	WM	...	...	...	...	4.6	...	4.2	9.9
	BM	...	...	...	...	3.8	...	3.9	9.0
	WF	...	...	...	...	4.3	...	4.7	10.0
	BF	...	...	...	...	3.6	...	3.6	9.0
	(SD)	...	...	...	...	(1.02)	...	(0.58)	(1.21)
3	WM	...	...	...	4.0	5.3	...	5.0	11.0
	BM	...	...	...	3.4	4.4	...	4.7	9.6
	WF	...	...	...	4.0	4.8	...	5.2	10.6
	BF	...	...	...	3.4	4.8	...	4.5	9.4
	(SD)	...	...	...	(0.63)	(0.94)	...	(0.81)	(1.10)
4	WM	...	...	3.5	4.5	5.4	...	5.9	11.5
	BM	...	...	3.7	4.3	5.1	...	5.5	10.4
	WF	...	...	4.2	4.6	5.7	...	5.6	11.2
	BF	...	...	3.2	4.2	4.9	...	5.2	9.8
	(SD)	...	...	(0.62)	(0.63)	(0.73)	...	(0.74)	(1.03)
5	WM	...	3.8	4.3	5.8	6.1	...	6.3	12.5
	BM	...	3.6	4.2	5.8	6.7	...	6.4	11.3
	WF	...	4.1	4.2	5.5	6.4	...	6.4	12.0
	BF	...	3.5	4.4	5.2	5.7	...	6.0	10.7
	(SD)	...	(0.79)	(0.64)	(0.73)	(0.90)	...	(1.03)	(1.22)
6	WM	3.7	4.1	5.6	6.8	7.6	3.5	8.0	12.6
	BM	4.0	4.4	5.5	6.3	7.2	3.0	7.3	12.2
	WF	3.9	4.5	5.0	6.4	7.3	3.5	7.4	11.7
	BF	3.4	4.1	4.9	6.2	6.6	3.2	6.8	12.2
	(SD)	(0.57)	(0.73)	(0.75)	(1.14)	(1.12)	(0.29)	(1.04)	(0.74)
7	WM	4.7	5.3	6.4	7.7	8.6	4.3	8.1	13.0
	BM	4.1	4.9	6.6	7.8	8.5	4.0	8.7	13.2
	WF	4.4	4.7	5.8	7.3	8.0	4.2	8.1	13.5
	BF	4.0	5.1	6.1	7.0	7.6	3.5	7.6	12.6
	(SD)	(0.75)	(0.77)	(1.21)	(0.84)	(0.89)	(0.64)	(0.95)	(1.34)
8	WM	5.5	6.1	7.6	9.0	9.8	4.3	9.0	...
	BM	5.4	6.0	8.3	9.0	9.4	4.0	8.7	...
	WF	5.2	5.8	6.9	8.5	8.8	4.5	8.9	...
	BF	5.4	5.7	6.8	8.4	8.9	4.1	8.6	...
	(SD)	(0.70)	(0.76)	(0.94)	(0.86)	(0.96)	(0.68)	(0.92)	...

Table 13.9 (continued) Age of Achievement of Mineralization Stages of Permanent Mandibular Teeth[a]

		Incisors			Premolars		Molars		
Stage	Subjects	Central	Lateral	Canine	First	Second	First	Second	Third
9	WM	6.4	6.9	8.8	9.4	10.2	5.2	10.2	...
	BM	5.8	6.7	8.8	9.7	10.0	5.2	9.7	...
	WF	6.2	6.3	8.0	9.0	9.7	5.2	9.8	...
	BF	5.5	5.9	7.7	9.0	9.5	4.8	9.5	...
	(SD)	(0.72)	(0.73)	(0.99)	(0.90)	(1.16)	(0.70)	(1.03)	...
10	WM	6.6	7.7	9.7	10.8	11.3	6.1	11.5	...
	BM	6.5	7.7	9.9	9.8	10.7	6.6	11.2	...
	WF	6.3	7.3	9.1	10.0	10.7	6.3	10.8	...
	BF	5.9	6.9	9.3	9.6	9.7	5.8	10.7	...
	(SD)	(0.74)	(0.86)	(0.95)	(0.91)	(1.03)	(0.71)	(1.02)	...
11	WM	8.0	8.2	11.5	11.7	12.4	7.5	12.2	...
	BM	7.4	8.4	11.3	11.4	11.9	7.4	12.1	...
	WF	7.4	8.0	10.1	11.1	11.7	7.0	11.6	...
	BF	6.5	8.0	9.4	11.2	11.7	6.6	12.0	...
	(SD)	(0.85)	(0.99)	(1.04)	(1.00)	(0.99)	(1.05)	(0.93)	...
12	WM	8.6	9.2	12.4	12.5	12.5	8.4	12.6	...
	BM	8.6	9.4	11.6	12.0	12.8	8.4	12.5	...
	WF	8.3	8.9	11.3	11.5	11.8	8.1	12.2	...
	BF	7.8	8.6	10.7	11.7	11.4	7.8	11.5	...
	(SD)	(0.97)	(1.01)	(0.91)	(0.95)	(0.97)	(0.87)	(1.01)	...
13	WM	...	...	...	...	...	9.2	12.8	...
	BM	...	...	...	...	...	9.5	13.0	...
	WF	...	...	...	...	...	9.0	13.2	...
	BF	...	...	...	...	...	8.4	13.2	...
	(SD)	...	...	...	...	...	(1.01)	(1.21)	...

[a] Codes are the same as in Table 13.8.

13.4.1 Gustafson Method

One of the most broadly referenced studies in forensic odontology is the work by Gustafson on age estimation from teeth.[93] Gustafson made a multifaceted attack on this subject by assessing six age-progressive changes:

1. *Occlusal attrition.* This is the wearing down of the occlusal surface, predominantly from grit in the diet (ignoring hyperfunctional issues in individuals with bruxism). The major issues that Gustafson did not deal with are (1) the virtual absence of grit in the modern, Westernized diet, so rates of wear can be inconsequential, and conversely, (2) that rates of wear differ widely depending on the group

under study.[4,94] Rates of wear developed for one cultural group may have no relevance to others.

2. *Periodontosis.* Two unrelated issues are confounded here. Destruction of the gingival, periodontal, and alveolar tissues may develop due to pathogens—acute infectious diseases. These often rapid destructive processes have to be distinguished from "continued eruption," in which teeth continue to erupt, albeit slowly, throughout much of life.[95,96] Danenberg et al. have shown that the distance from the cementoenamel junction (CEJ) to the bony alveolar crest increases as people age.[97] Continued eruption seems to be an accommodation to occlusal wear that, over time, reduces crown height. Continued eruption helps to extend a tooth's functional longevity. In contemporary peoples, with trivial wear, the effect is simply to increase lower face height.[98]

3. *Secondary dentin.* The slow age-progressive deposition of secondary dentin diminishes and ultimately occludes the pulp chamber.[99] Secondary dentin needs to be distinguished from tertiary dentin that is thought to accumulate in response to caries and trauma.[100] Pulp dimensions can be assessed radiographically, especially if computer-assisted methods for enlargement and measurement are available. Precision is required if teeth are sectioned so as to find and preserve the maximum dimensions of the pulp chamber. Kvall et al. reported a method in 1995 that allows estimation based on morphological measurements of two-dimensional radiographic features of individual teeth.[101] Earlier similar studies had reported unsatisfactory results.[102] The Kvaal method is less discriminatory than other methods but has the important advantages of being noninvasive, not requiring extraction of teeth, and being useful for examination and regression analysis of all data performed, with age as the dependent variable. The measurements include comparisons of pulp and root length, pulp and tooth length, tooth and root length, and pulp and root widths at three defined levels. Vandevoort et al. reported a morphometric method pilot study using microfocused computed tomography to compare pulp-tooth ratios.[103] Cameriere et al. proposed a method in 2004 using pulp-tooth ratios of second molars.[104] In 2007 Berner et al. reported a larger clinical trial of the Kvall method showing similar results.[105] Adetona et al. presented in 2008 a three-dimensional method using cone beam computed tomography (CBCT).[106]

4. *Cementum apposition.* The thickness of cementum often increases on roots with age, and cementum generally is thinnest near the CEJ and thickest on the apical third of the root. Recent attention has focused on cementum annulations—the deposition of a new layer of cementum onto roots in a manner analogous to the growth of tree

rings. This new emphasis tends to overshadow simply measuring the *thickness* of enamel.[10] Gustafson,[92] Johanson,[93] Solheim,[108] and several others have documented a significant positive relationship between cementum thickness and age (but with appreciable inaccuracy). This method might be of some use when the tooth cannot be used to count annulations, though one needs to control for differences among tooth types. Researchers have also found systematic differences when using impacted teeth (cementum is thicker[92]) or periodontally involved teeth (cementum is thicker[109]). Johanson (p. 52) claimed that "extensive attrition is always followed by extensive increase in thickness of the cementum."[110]

5. *Root resorption.* External apical root resorption (EARR) occasionally is seen to be age progressive. More accurately, the incidence of resorption is somewhat more common in older-aged groups in cross-sectional studies. Woods et al.[111] found only a few cases of EARR (less than 5%) in contemporary American adults who had not experienced orthodontic treatment, and these cases characteristically are the result of compromised dentitions (missing teeth, periodontal involvement) that led to lack of tooth support.[112] This perspective is in concert with Maples' 1978 findings that root resorption is the least dependable predictor of those suggested by Gustafson. In modern populations, a very common cause of EARR is orthodontic treatment.[113]

6. *Root transparency.* Root transparency is due to the age-progressive occlusion of dentin tubules leading to sclerotic dentin. This change can occur in the crown and root of a tooth, but changes in the crown are in consequence to attrition, trauma, caries, and other noxious stimuli.[93,100] Root transparency commences in the apical region, progressing coronally with age.

Various researchers have adapted and refined Gustafson's method.[114–117] In contrast to his considerable biological and clinical expertise, Gustafson's method failed statistically. He assumed that all six of the parameters could be visually graded on a four-step scale. This assumed that (1) these four (unequally spaced) ordinal stages informatively reflect the trait distributions; (2) all six parameters are equally effective at estimating age, so they can just be added together; (3) the *rates* of change are equivalent among all of the parameters, so they can just be added together; and (4) the imprecision (variability of true and predicted ages) is the same for all parameters, so the variances can be assumed to be equal. Gustafson's scoring method also assumes that the six pieces of age information are statistically independent, which is far from true. These various shortcomings were overcome in subsequent studies using multiple linear regression methods and similar statistical techniques that more appropriately account for the nature of the data.[92,114]

Gustafson's legacy is considerable for the forensic odontologist. But, rather than pursuing his collective mix of six dental changes, most researchers have elaborated on the scientific bases of one or another of these processes. We revisit some of these topics below.

13.4.2   Root Transparency

Beginning with Bodecker,[93] followed by Gustafson[118] and others, several researchers have described the age-progressive increase in root transparency. The procedure is destructive in that the tooth has to be extracted, sectioned, and polished before measurement. Root transparency develops due to progressive sclerosing of the tubules, first at the root apex, then advancing coronally.[119] De Jonge reported that the average width of dentinal tubules is about 3.2 μ in young individuals, narrowing to about 1.5 μ at fifty, and down to 1.2 μ at seventy years of age.[120] Opaqueness of young dentin is due to differences in the refractive indexes between the crystalline and the intratubular organic components. Aging causes the refractive indexes to converge, making the dentin transparent to transmitted light.[121] Whittaker and Bakri[122] note that there are diverse opinions as to the cause of the dentin changes; factors causing deposition of sclerotic dentin seem to be toxins from diseased periodontal tissues, absence of functional stimuli, and the diminished diameters of the dentinal tubules due to increased mineralization of the intratubular matrix, especially at the root apex. Changes may be seen as early as the later teens, though typically starting in adulthood.[123]

Whittaker and Bakri[122] showed that staining with methylene blue in a partial vacuum significantly enhanced contrast of the sclerotic and nonsclerotic root areas. They also suggest that the rate of sclerosis differs among populations (possibly due to dietary differences), and that the rate may not be linear throughout adulthood (possibly slowing in older adults). Sognnaes et al. describe a method of quantifying the sclerotic portion of dentin using intact teeth.[124]

Drusini et al.[125] reported a correlation of 0.84 between the proportion of transparent dentin and chronological age, which is fairly low. These authors found that at least 55% of their age estimates deviated by more than five years.

13.4.3   Cementum Annulations

Wildlife biologists have noted that in mammals living in temperate and polar regions (where climate, activity, and food availability vary seasonally), layers of cementum are laid down on tooth roots as alternating light and dark bands, with one layer corresponding to (approximately) one year of life.[126,127] Stott et al. showed that humans also possess these layers (now termed cementum annulations) and suggested they would be useful for age estimation (adjusting for when the tooth erupted).[128] Etiology of these annulations probably is due to alternating cycles of mineralization of the cementum that is deposited as dense bundles of collagen that are subsequently replaced (mineralized) by hydroxyapatite crystals. Differences in the alternating light and dark bands are due to their different crystal orientations.[129,130]

Counting cementum annulations is destructive, requiring histological examination of thin sections through a root,[131–133] and the method has some limitations, among these are: different teeth from the same individual and different sections from the same tooth yield different counts because cementum is not deposited uniformly; there may be some remodeling after deposition; some teeth exhibit thin or hard-to-read annulations; and annulations may not necessarily be deposited annually.

Interestingly, the annuli show evidence of certain life history events that negatively influence calcium deposition, such as pregnancies, and renal and skeletal disease.[134] Such qualitative findings may be valuable in forensic identification cases.

Charles et al.[135] and Condon et al.[136] reported on a well-structured study of teeth of known age. They found the correlation between true and predicted age was just 0.78, and there tended to be fewer lines than one per year, but variably so. A few specimens exhibited doubling, where there were about twice as many lines as anticipated from the person's known age. These authors made several observations: (1) archaeological specimens tend to have fainter annulations (and should be decalcified more gently), (2) cementum deposition may be affected by periodontal disease, (3) a few teeth possess no cementum annulations, (4) accuracy of the method diminishes with the person's age, which is common to most aging techniques, and (5) population differences need to be studied.[137] In spite of these several caveats, Condon et al. noted that this method probably is at least as accurate as any other skeletodental technique.

Wittwer-Backofen et al. readdressed this issue of accuracy, providing updated preparation methods and observational techniques (including scanning electron microscopy).[138] They found that the error was just 2.5 years or less (gauged from the 95% confidence limits), making the method more accurate than many other approaches. This accuracy, however, was achieved after about 16% of the sample was omitted where the cementum pattern was irregular—"where the cementum band partly surrounds artifacts or overlays itself in undulations"—or where image quality was poor after tissue processing. These authors made the points that, in their hands, (1) males and females could be aged with equal precision and (2) periodontal disease did not affect accuracy.[134] They concluded, "This provides a strong argument for the application of [this] method in archaeological skeletal samples in which most of the individuals suffered from extreme dental disease."[139]

13.4.4 Tooth Wear

Occlusal attrition has been a characteristic feature of adults' teeth throughout time; it is only recently in industrialized countries that grit (and grit size) has been reduced to the point that attrition is now trivial. Rates of tooth wear are intimately associated with diet, particularly food preparation processes.[140,141] The amount of occlusal attrition is proportional to a person's age (i.e., the duration of time a tooth has been functioning), but the rate of wear varies tremendously among cultures, so age needs to be tied to the appropriate food processing context.

Murphy[141] was among the first to develop a tooth wear grading system that accounted for morphologies of the individual tooth types as well as using enough grades to approximate a continuous scale (also see Molnar, 1971[142]; Scott, 1979[143]). Figure 13.5 illustrates the tooth-specific scheme developed by Smith. Smith's scheme is individualized for the different tooth types, and ranges from traces of wear (polished or small facets on the enamel) to loss

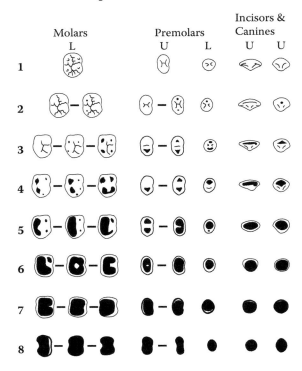

Figure 13.5 The tooth wear ordinal scheme developed by Smith.[94] The system accounts for different tooth types. Grade 0 (not shown) would be a pristine, completely unworn tooth. Multiple examples connected by horizontal lines show the ranges of variation within a grade. While the grades are numbered, this is an ordinal, morphological system; one cannot assume the grades are equally spaced.

of most of or the entire crown, leaving just the root stumps functioning at the gum line.[94]

In the broad view, tooth wear can be useful in interpreting three topics: cultural modifications (such as caused by crafts and food preparation), dietary reconstruction (for instance, hunter-gatherers wore their teeth slower than prehistoric agriculturalists), and of course, age estimation (see Rose and Ungar[144] for an in-depth review of attrition studies). In addition, there is a growing body of research using microscopic wear patterns to disclose dietary adaptations.[144,145]

Grading schemes as in Figure 13.5 are of little use in contemporary populations where the rates of wear are slower, so attrition is only moderate even in old age. In modern, Westernized countries there have been important demographic and oral health shifts. With people living longer—and retaining more of their teeth—the occlusal wear in the older age grades has been increasing (because teeth are less often extracted or decayed), and this trend is expected to continue. Moreover, and especially when abrasion and erosion are included in wear, it becomes relevant to consider all tooth surfaces, not just the occlusal.[146] Smith and Knight[147] developed the tooth wear index (TWI) as an epidemiological tool (Table 13.10), and various other indexes also are available.[148,149] Donachi and Walls provide a critical review of tooth wear indexes, and make helpful suggestions for extending the TWI to more situations encountered in the elderly.[150]

Table 13.10 Tooth Wear Index (TWI) Scoring Criteria

Score	Surface		
	Buccal, Lingual, and Occlusal	Incisal	Cervical
0	No loss of enamel surface characteristics	No loss of enamel surface characteristics	No change in contour
1	Loss of enamel surface characteristics	Loss of enamel surface characteristics	Minimal loss of contour
2	Loss of enamel exposing dentine for less than 1/3 of the surface	Loss of enamel just exposing dentine	Defect <1 mm deep
3	Loss of enamel exposing dentine for more than 1/3 of the surface	Loss of enamel and substantial loss of dentine, but not exposing pulp or secondary dentine	Defect 1–2 mm deep
4	Complete loss of enamel, or pulp exposure or exposure of secondary dentine	Pulp exposure or exposure of secondary dentine	Defect >2 mm deep, or pulp exposure or exposure of secondary dentine

Source: Adapted from Donachi and Walls.[150]

It does not appear that anyone has yet used the TWI for age estimation, but the use of simpler systems continues to show that, while attrition is age progressive in a population, the wide ranges of variability within and among people make the precision of age estimates poor.[149,151]

13.4.5 Third Molar Formation

Although the third molar is the most variable tooth in the dentition with respect to developmental chronology, it is sometimes used to estimate age during late adolescence and early adulthood.[13,92,152] Arguably, there are no other, more reliable biological indicators available during this period, and third molar development is readily assessable from dental radiographs.

In most jurisdictions the attainment of a specific calendar age marks adulthood and legal implications change significantly. In the United States this is at eighteen years. Recently, the chronology of third molar development has been used extensively to judge whether an individual is a juvenile or an adult. Such estimations are often requested by immigration authorities in cases involving foreign nationals.

Several studies have indicated that although actual age estimation using third molar development is relatively inaccurate because of large variation among individuals, a reasonable evaluation as to whether a subject has reached adulthood can be made by this method.[13,153–155] An early prototype study was done under the auspices of the American Board of Forensic Odontology.[13] Data submitted by ABFO diplomates from the United States and Canada were taken from dental radiographs, principally panoramic radiographs. Subjects (n = 823) were between 14.1 and 24.9 years of age, and 80% were categorized as Caucasians. The contributing dentists scored the stage of third molar development using the eight-grade scheme developed by Demirjian and coworkers[62] (Figure 13.4).

Only 54% of cases showed the same stage of crown-root development in both the maxilla and mandible. Left-right symmetry was present in 78%.

The sex-specific mean age at each formative stage was calculated (Table 13.11). Only data from white subjects are shown in these computations because too few from other racial groups were available to yield reliable estimates. Stages A and B did not occur in the ages under study, and Stage C was present in only 1%.

As reported by others, staging of third molar development was shown to be inaccurate for prediction of chronological age. Standard deviation (SD) for each grade ranged from 1.5 to 3.4 years, with an average of about 2 years. This means that age predictability within each stage consists of an interval of about eight years.

Table 13.11 Mean Ages at Attainment of Stages of Third Molar Crown-Root Formation in U.S./Canadian Caucasians

Grouping	Statistic	\multicolumn{5}{c}{Grade of Formation}

Grouping	Statistic	D	E	F	G	H
			Maxilla			
Males	Mean	16.0	16.6	17.7	18.2	20.2
	SD	1.97	2.38	2.28	1.91	2.09
Females	Mean	16.0	16.9	18.0	18.8	20.6
	SD	1.55	1.85	1.95	2.27	2.09
			Mandible			
Males	Mean	15.5	17.3	17.5	18.3	20.5
	SD	1.59	2.47	2.14	1.93	1.97
Females	Mean	16.0	16.9	17.7	19.1	20.9
	SD	1.64	1.75	1.80	2.18	2.01

There was significant dimorphism between the sexes, with third molars developing earlier in males than in females. This is opposite from the pattern observed in the other, earlier-forming teeth.

Table 13.12 shows the empirical probability of an individual being at least age eighteen—that is, legally an adult—based on stage of third molar formation. As with the computation of chronological age, the relationship between third molar development and attainment of legal adulthood is quite variable. However, a forensically useful observation was made. It was found that if third molar root formation was complete with closed apices and uniform

Table 13.12 Empirical Probabilities (%) of an Individual Being at Least 18 Years of Age Based on the Grade of Third Molar Formation[a]

Group	D	E	F	G	H
		Maxilla			
Males	15.9	27.8	44.0	46.8	85.3
Females	9.7	28.4	50.4	63.3	89.6
		Mandible			
Males	6.1	69.4	40.5	56.0	90.1
Females	11.3	27.4	43.2	69.8	92.2

[a] Values are based just on whites from the United States and Canada. Probabilities for the terminal grade (H) presume that, based on other criteria, the subject is less than twenty-five years of age.

Table 13.13 Comparison of Empirical Probabilities (%) of Individuals with Complete Third Molar Root Formation (Grade H) from Different Population Groups Being at Least 18 Years of Age

Study	Year	Subjects	Number	Results
Mincer et al.	1993	U.S. and Canadian whites	657	Max F 89.6%
				Max M 85.3%
				Mand F 92.2%
				Mand M 90.1%
Solari and Abramovitch	2002	Hispanics (Houston, TX)	679	Max F 79–84%[a]
				Max M 75–76%
				Mand F 91–92%
				Mand M 85–89%
Gunst et al.	2003	Belgian whites	2,513	Max F 88.2%
				Max M 82.7%
				Mand F 94.0%
				Mand M 96.6%
Arany et al.	2004	Japanese	1,282	Max F 99%
				Max M 97%
				Mand F 99%
				Mand M 98%

[a] Contralateral third molar values.
Max = Maxilla
Mand = Mandible
M = Male
F = Female

width of periodontal ligament (Stage H), there is a high probability that an individual is at least eighteen years old. The probability that an individual with complete root formation of the mandibular third molars is eighteen or older is 90.1% for white males and 92.2% for white females.

Subsequent to this ABFO study, several investigations of the relationship of third molar development to chronological age have been performed using similar methods and parameters with other specific population groups.[153–157] In virtually every instance a wide range of ages associated with each stage of tooth development was confirmed. In a large multinational study carried out on three samples, it was found that at each stage of third molar development Japanese subjects were on average one to two years older than corresponding German subjects, and South Africans were on average one to two years younger.[156]

As for the probability that an individual with third molars showing complete root formation is at least eighteen years old, results among the comparative studies have also differed somewhat with the ethnic group studied (Table 13.13). This indicates that if an odontologist uses this technique to assess legal adulthood of an individual, he should consider ethnicity of the subject and refer to appropriate studies for data.

Because of overall inaccuracy, possible differences in interpretation among evaluators, ethnic variation, and the fact that ethnicity is a complex matter with poorly defined boundaries, several authors have cautioned against relying too heavily on third molar formation alone for estimation of age in late-adolescent/early-adult subjects as well as for determination of attainment of adulthood.[58,159] Various parameters of skeletal development are also related to chronological age but, again, without sufficient accuracy to ensure forensic certainty. Some studies indicate that using a combination of dental and skeletal developmental findings results in increased accuracy of age estimation. Schmeling et al., for example, recommend that age estimates should be based on, in addition to dental radiological findings, general physical examination and radiological examination to determine stage of development of the hand, wrist, and the clavicle.[159]

Cameriere and coworkers[160] analyzed panoramic radiographs of Italian subjects and found that the values resulting from the pulp-tooth area ratio of the second molar and the stage of the development of the third molar computed in combination resulted in better assessment of whether an individual was eighteen or older than either method alone. Comparably, Chaillet and Demirjian found that the addition of the third molar to Demirjian's original seven-tooth method increased accuracy of the estimate.[161]

13.4.6 Aspartic Acid Racemization

Racemization of amino acids of certain proteins occurs *in vivo*.[162] This process, the conversion of the L-isomer to the D-isomer toward equilibrium, proceeds slowly throughout life, so it can be related to chronological age of the host and therefore can be used forensically to determine age at death. Racemization continues at a markedly reduced rate after death, so it is also useful for aging archaeological samples.[163] It has been demonstrated in a variety of proteins such as those in the eye lens,[163] brain, vertebral discs,[164] and dental tissues.[165] Many amino acids have been shown to racemize, and the process has been described in all dental hard tissues.[162] Because of postmortem preservation, a relatively high rate of racemization, and other technical factors, the compound most often analyzed has been aspartic acid in dentine collagen.

Helfman and Bada first showed that age can be determined from the enamel or dentin of a tooth by quantifying the relative amounts of the D- and L-forms of aspartic acid.[166,167] With time, the L-form undergoes racemization to the D-form. Age estimation using racemization depends on the assumptions that (1) the aspartic acid has not been replaced by remodeling or by diagenesis since the tooth was formed, and (2) the temperature has been constant (as in the human mouth at 37°C). According to the claim put forth by Helfman and Bada in their original publication[166] (p. 2893): "Under suitable conditions, where there has been no racemization since death, i.e., relatively cold temperatures, or within a few hundred years of the burial, the age of an individual at death could be deduced using the extent of aspartic acid in tooth enamel from skeletons in cemetery populations."

Mörnstad et al. showed that high-performance liquid chromatography (HPLC) can be used instead of amino acid analyzers, making determinations faster and less expensive.[168]

Reported accuracy of this technique for estimation of chronological age in adults varies. Carolan and coworkers concluded that age estimation based on racemization is similar in reliability to other dental methods.[162] Waite et al., by contrast, indicated that with proper sampling and analytic techniques, this method provides a simple, cost-effective solution to age estimation in adults, and can achieve accuracy within the range of ±3 years.[169] Seitz et al. have reported success in detecting ratios using whole teeth and HPLC coupled with mass spectroscopy.[170]

13.4.7 Enamel Uptake of Radioactive Carbon-14

Spalding et al. reported in 2005 the development of a new method of age estimation from dental enamel.[171] Aboveground nuclear testing from 1955 to 1963 produced greatly increased atmospheric levels of the radioactive carbon-14 isotope (^{14}C). The increased ^{14}C isotope levels rapidly generalized around the globe. The atmospheric levels of this isotope had been stable at much lower levels prior to 1955. Since the cessation of testing in 1963 the levels have been decreasing exponentially. The half-life of ^{14}C is 5,730 years. Atmospheric ^{14}C reacts with oxygen to produce CO^2 with the radioactive signature. That radioactive CO^2 is taken up by plants worldwide. Eating those plants and animals that eat those plants causes ^{14}C uptake in the metabolically stable tooth enamel as it is formed. By calculating the levels of ^{14}C with consideration of known tooth development data for the specific tooth tested, an estimated date of birth can be calculated. Initial tests on twenty-two individuals of known age resulted in estimates that were reported to be accurate within about 1.5 years. The technique is useful only for individuals born after 1943 since the formation of the enamel of the third molars is completed at approximately age twelve. Third molar enamel ^{14}C levels of 0 would indicate only that the individual was born before 1944, but even this could be useful forensic information. Teeth samples were collected and prepared at Karolinska Institute, Stockholm, Sweden, and ^{14}C levels were determined by accelerator mass spectrometry at Lawrence Livermore National Laboratory, Livermore, California.[171] In 2009 Druid et al.[172] proposed combining aspartic acid racemization analysis and ^{14}C studies to provide investigators of unidentified body cases with more useful information. The age at death estimate from aspartic acid racemization analysis combined with the date of birth from ^{14}C levels could, by calculation, provide investigators with the estimated date of death.

13.5 Summary, Conclusions, and Recommendations

Teeth are common and they preserve well in traumatic settings (protected by thick surrounding bones and tissues), and they undergo a variety of age-progressive changes. These features make them of value to the forensic odontologist. We have reviewed some of the common methods used for age estimation. Collectively, the methods can be applied from before birth (all primary teeth begin formation in the second trimester) into very old age. Improved oral health—fewer caries and less periodontal disease—means that more forensic cases will have teeth (and more teeth) than in the past.

A predictable conclusion from this review is that more studies are needed, but not just more of the same. We identify two areas of need: First, Westernized countries are quickly becoming the homes of diverse peoples from around the globe, and the forensic specialist is ill-equipped to account for the range of racial variation. Few dental standards for age or sex are available for nonwhites. The problem becomes more complex when a person grows up in one cultural setting, then immigrates to quite another. More studies of ethnically-racially diverse segments of our population are needed. Second, our opinion is that forensic odontology research in age estimation has been lacking in scope and execution, with numerous reports suffering from small sample sizes and restricted analysis. Larger studies with more in-depth analysis would improve the precision of age estimations due to a better understanding of the nature of the differences and recognition of cases where normative standards do not apply.

For practical forensic casework the specific method or methods that are appropriate vary. While the capabilities and resources of investigators differ, we offer these guidelines for choosing the method or methods with the most potential for each case.

For living persons, the techniques utilizing radiological assessment are most applicable. In children up to the age of puberty, those techniques assessing dental development are more accurate but more complex and time-consuming than those using eruption/emergence schedules. From puberty to the time that growth and development have ceased, around age twenty-one, the most useful dental techniques involve analyzing the development or mineralization of the teeth, including the third molar. It is well known that third molar development variations are greater than for any other human tooth, but those are likely the only teeth still developing in the target age group. There is a need for more and larger population studies for specific geographic and ethnic groups. For adults after the age of around twenty-one years, the applicable dental techniques are those that look at gross, histological, biochemical, and trace element changes to teeth. In living adults these are limited to the radiological and visual examination techniques unless a valid clinical reason for removing all or part of a tooth exists. In those cases, aspartic acid racemization analysis or ^{14}C analysis may be performed.

The same radiographic and visual procedures used for living individuals can be used for the deceased. In addition to those methods, and with permission from coroners or medical examiners, teeth can be removed for age estimation studies. All of the methods that consider postformation changes in teeth can be used. Tooth cementum annulation (TCA) analysis is becoming relatively common for age estimation with other mammals. With proper consideration of the limitations of the technique, there is no reason that it could not be routinely used in human age estimation. Although work remains to be performed to validate the determination of the ratios in amino acid racemization, that method is a promising technique for improving dental age estimation in all age groups. Both tooth cementum annulation and aspartic acid racemization can be used, with associated limitations, for individuals of any age. Radioactive ^{14}C analysis from tooth enamel is a new and promising technique, potentially offering the most accurate and precise information on estimated date of birth for those individuals born after 1943.

Using human teeth for age estimation is well established in past and recent literature. Multiple studies have demonstrated varying accuracy, reliability, and precision. Reproducible and reliable results are possible when the appropriate techniques for a given situation are properly understood and applied. Soomer et al. correctly stated in their 2003 paper that "forensic odontologists should evaluate each age estimation case and, in addition to their visual age assessment, choose one or more methods that would best serve their particular case, keeping in mind that accuracy and precision are the main requirements."[14] Willems et al. in 2002 concluded that "an important aspect in dental age estimation is that the investigator should apply a number of different techniques available and perform repetitive measurements and calculations in order to improve reproducibility and reliability of the age estimation."[173]

Linear regression, confidence levels, and standard deviations relate to means for the populations studied. Specific individuals within a population may live at either end of the range or fall outside the normal limits. Error exists in every system of age estimation. The limitations of current methods and the paucity of population data available mean those methods are not adequate to allow precise age estimation results for every case. Age estimation reports must clearly convey that the data reported are based on mean ages derived from the features studied for a specific population and should include realistic ranges. Specific casework may require combining methods to arrive at the most accurate conclusions.

When possible, more than one dental technique or a combination of dental and skeletal or other techniques should be used. Since research into age estimation is ongoing, forensic dentists performing age estimation must continually monitor the scientific journals that report new developments and validate or challenge existing techniques.

References

1. Gustafson, G. 1966. *Forensic odontology*. New York: American Elsevier Pub. Co.
2. Ubelaker, D.H. 1999. *Human skeletal remains: Excavation, analysis, interpretation*, x. 3rd ed., Manuals on Archeology 2. Washington, DC: Taraxacum.
3. Scheur, L., and S. Black. 2000. *Developmental juvenile osteology*. San Diego: Academic Press.
4. Brothwell, D.R. 1981. *Digging up bones: The excavation, treatment, and study of human skeletal remains*. 3rd ed. Ithaca, NY: Cornell University Press.
5. Acsádi, G., and J. Nemeskéri. 1970. *History of human life span and mortality*. Budapest: Akadémiai Kiadó.
6. Lovejoy, C.O., et al. 1985. Multifactorial determination of skeletal age at death: A method and blind tests of its accuracy. *Am J Phys Anthropol* 68:1–14.
7. Meindl, R.S., K.F. Russell, and C.O. Lovejoy. 1990. Reliability of age at death in the Hamann-Todd collection: Validity of subselection procedures used in blind tests of the summary age technique. *Am J Phys Anthropol* 83:349–57.
8. Bedford, M.E., et al. 1993. Test of the multifactorial aging method using skeletons with known ages-at-death from the Grant Collection. *Am J Phys Anthropol* 91:287–97.
9. Howells, W. 1973. *Cranial variation in man*. Papers of Peabody Museum Archaeology Ethnology. Cambridge, MA: Harvard University.
10. Key, P., and R.L. Jantz. 1981. A multivariate analysis of temporal change in Arikara craniometrics: A methodological approach. *Am J Phys Anthropol* 55:247–59.
11. Voors, A.W. 1957. The use of dental age in studies of nutrition in children. *Doc Med Geogr Trop* 9:137–48.
12. Voors, A.W., and D. Metselaar. 1958. The reliability of dental age as a yard-stick to assess the unknown calendar age. *Trop Geogr Med* 10:175–80.
13. Mincer, H.H., E.F. Harris, and H.E. Berryman. 1993. The A.B.F.O. study of third molar development and its use as an estimator of chronological age. *J Forensic Sci* 38:379–90.
14. Soomer, H., et al. 2003. Reliability and validity of eight dental age estimation methods for adults. *J Forensic Sci* 48:149–52.
15. Harris, E.F., and J.H. McKee. 1990. Tooth mineralization standards for blacks and whites from the middle southern United States. *J Forensic Sci* 35:859–72.
16. Mappes, M.S., E.F. Harris, and R.G. Behrents. 1992. An example of regional variation in the tempos of tooth mineralization and hand-wrist ossification. *Am J Orthod Dentofacial Orthop* 101:145–51.
17. Maki, K., et al. 1999. The impact of race on tooth formation. *ASDC J Dent Child* 66:294–95, 353–56.
18. Farah, C.S., D.R. Booth, and S.C. Knott. 1999. Dental maturity of children in Perth, Western Australia, and its application in forensic age estimation. *J Clin Forensic Med* 6:14–18.

19. Hegde, R.J., and P.B. Sood. 2002. Dental maturity as an indicator of chronological age: Radiographic evaluation of dental age in 6 to 13 years children of Belgaum using Demirjian methods. *J Indian Soc Pedod Prev Dent* 20:132–38.
20. Demirjian, A. 1978. [Dental development: Index of physiologic maturation]. *Med Hyg* (Geneve) 36:3154–59.
21. Garn, S.M., and A.B. Lewis. 1957. Relationship between the sequence of calcification and the sequence of eruption of the mandibular molar and premolar teeth. *J Dent Res* 36:992–95.
22. Garn, S., A.B. Lewis, and B. Bonne. 1962. Third molar formation and its development course. In *Angle orthodontics*, 32(4), 270–79.
23. Sato, S., and P. Parsons. 1990. *Eruption of permanent teeth: A color atlas.* St. Louis, MO: Ishiyaku EuroAmerica.
24. Stuart, H.C., and H.V. Meredith. 1946. Use of body measurements in the school health program. *Am J Public Health Nations Health* 36:1365–86.
24a. Stuart, H.C. 1939. Studies from the center for research in child health and development, School of Public Health, Harvard University. Monograph. *Society for Research Child Development.* Series No. 20, 1–261.
25. Infante, P.F. 1974. Sex differences in the chronology of deciduous tooth emergence in white and black children. *J Dent Res* 53:418–21.
26. Ferguson, A.D., R.B. Scott, and H. Bakwin. 1957. Growth and development of Negro infants. VIII. Comparison of the deciduous dentition in Negro and white infants: A preliminary study. *J Pediatr* 50:327–31.
27. Moorrees, C.F., E.A. Fanning, and E.E. Hunt Jr. 1963. Formation and resorption of three deciduous teeth in children. *Am J Phys Anthropol* 21:205–13.
28. Moorrees, C.F., E.A. Fanning, and E.E. Hunt Jr. 1963. Age variation of formation stages for ten permanent teeth. *J Dent Res* 42:1490–502.
29. Saka, H., A. Kikuchi, and Y. Ide. 1996. A morphological study of root resorption of the maxillary first deciduous molars. *Bull Tokyo Dent Coll* 37:137–44.
30. Peterka, M., R. Peterkova, and Z. Likovsky. 1996. Timing of exchange of the maxillary deciduous and permanent teeth in boys with three types of orofacial clefts. *Cleft Palate Craniofac J* 33:318–23.
31. Garn, S.M., et al. 1958. The sex difference in tooth calcification. *J Dent Res* 37:561–67.
32. Anderson, D.L., and G.W. Thompson. 1973. Interrelationships and sex differences of dental and skeletal measurements. *J Dent Res* 52:431–38.
33. Thompson, G.W., D.L. Anderson, and F. Popovich. 1975. Sexual dimorphism in dentition mineralization. *Growth* 39:289–301.
34. Liversidge, H.M., F. Lyons, and M.P. Hector. 2003. The accuracy of three methods of age estimation using radiographic measurements of developing teeth. *Forensic Sci Int* 131:22–29.
35. Hassanali, J. 1985. The third permanent molar eruption in Kenyan Africans and Asians. *Ann Hum Biol* 12:517–23.
36. Eveleth, P.B., and J.M. Tanner. 1976. *Worldwide variation in human growth*, xiv. International Biological Programme 8. New York: Cambridge University Press.
37. Hurme, V.O. 1949. Ranges of normalcy in the eruption of permanent teeth. *J Dent Child* 16:11–15.
38. Garn, S.M., et al. 1973. Negro-Caucasoid differences in permanent tooth emergence at a constant income level. *Arch Oral Biol* 18:609–15.

39. Smith, J., R.N. Smith, A.H. Brook, and C. Elcock. 1999. Timing of permanent tooth eruption in London school children. In *Dental morphology*, ed. J. Mayhall and T. Heikkinen, 187–91. Oulu, Finland: University of Oulu Press.
40. Van der Linden FPGM, D.H. 1976. *Development of the human dentition: An atlas*. Hagerstown, MD: Harper and Row.
41. Fanning, E.A. 1963. Effect of extraction of deciduous molars on the formation and eruption of their successors. *Angle Orthod* 32:44–53.
42. Dahlberg, A.A., and R.M. Menegaz-Bock. 1958. Emergence of the permanent teeth in Pima Indian children: A critical analysis of method and an estimate of population parameters. *J Dent Res* 37:1123–40.
43. Anderson, D.L., G.W. Thompson, and F. Popovich. 1976. Age of attainment of mineralization stages of the permanent dentition. *J Forensic Sci* 21:191–200.
44. Moslemi, M. 2004. An epidemiological survey of the time and sequence of eruption of permanent teeth in 4–15-year-olds in Tehran, Iran. *Int J Paediatr Dent* 14:432–38.
45. Heidmann, J. 1986. Comparison of different methods for estimating human tooth-eruption time on one set of Danish national data. *Arch Oral Biol* 31:815–17.
46. Holman, D.J., and R.E. Jones. 1998. Longitudinal analysis of deciduous tooth emergence. II. Parametric survival analysis in Bangladeshi, Guatemalan, Japanese, and Javanese children. *Am J Phys Anthropol* 105:209–30.
47. Korhonen, M., T. Kakilehto, and M. Larmas. 2003. Tooth-by-tooth survival analysis of the first caries attack in different age cohorts and health centers in Finland. *Acta Odontol Scand* 61:1–5.
48. Logan, W., and R. Kronfeld. 1933. Development of the human jaws and surrounding structures from birth to the age of fifteen years. *J Am Dent Assoc* 20:379–427.
49. Schour, I., and M. Massler. 1940. Studies in tooth development: The growth pattern of human teeth. *J Am Dent Assoc* 27:1778–92, 1918–31.
50. Gleiser, I., and E.E. Hunt Jr. 1955. The permanent mandibular first molar: Its calcification, eruption and decay. *Am J Phys Anthropol* 13:253–83.
51. Lunt, R.C., and D.B. Law. 1974. A review of the chronology of calcification of deciduous teeth. *J Am Dent Assoc* 89:599–606.
52. Kraus, B.S., and R.E. Jordan. 1965. *The human dentition before birth*. Philadelphia: Lea & Febiger.
53. Szpringer-Nodzak, M. 1984. The location of the neonatal line in human enamel. *J Int Assoc Dent Child* 15:1–6.
54. Skinner, M., and T. Dupras. 1993. Variation in birth timing and location of the neonatal line in human enamel. *J Forensic Sci* 38:1383–90.
55. Kodaka, T., T. Sano, and S. Higashi. 1996. Structural and calcification patterns of the neonatal line in the enamel of human deciduous teeth. *Scanning Microsc* 10:737–43; discussion, 743–44.
56. Liversidge, H.M., M.C. Dean, and T.I. Molleson. 1993. Increasing human tooth length between birth and 5.4 years. *Am J Phys Anthropol* 90:307–13.
57. Liversidge, H.M., and T.I. Molleson. 1999. Deciduous tooth size and morphogenetic fields in children from Christ Church, Spitalfields. *Arch Oral Biol* 44:7–13.

58. Liversidge, H.M., and T.I. Molleson. 1999. Developing permanent tooth length as an estimate of age. *J Forensic Sci* 44:917–20.
59. Smith, B.H. 1991. Standards of human tooth formation and dental age assessment. In *Advances in dental anthropology*, ed. M.A. Kelley et al., 141–68. New York: Wiley-Liss.
60. Fanning, E.A. 1961. A longitudinal study of tooth formation and root resorption. *NZ Dent J* 57:202–17.
61. Nolla, C.M. 1960. Development of the permanent teeth. *ASDC J Dent Child* 27:254–66.
62. Demirjian, A., H. Goldstein, and J.M. Tanner. 1973. A new system of dental age assessment. *Hum Biol* 45:211–27.
63. Owsley, D.W., and R.L. Jantz. 1983. Formation of the permanent dentition in Arikara Indians: Timing differences that affect dental age assessments. *Am J Phys Anthropol* 61:467–71.
64. Saunders, S., et al. 1993. Accuracy tests of tooth formation age estimations for human skeletal remains. *Am J Phys Anthropol* 92:173–88.
65. Tompkins, R.L. 1996. Human population variability in relative dental development. *Am J Phys Anthropol* 99:79–102.
66. Kronfeld, R. 1935. Development and calcification of the human deciduous and permanent dentitions. *Bur* 15:18–25.
67. Roche, A.F. 1992. *Growth, maturation, and body composition: The Fels Longitudinal Study, 1929–1991*, xiii. Cambridge Studies in Biological Anthropology 9. New York: Cambridge University Press.
68. Harris, E.G., and A. Buck. 2002. Tooth mineralization: A technical note on the Moorrees-Fanning-Hunt standards. *Dental Anthropol* 16:15–20.
69. Haavikko, K. 1970. The formation and the alveolar and clinical eruption of the permanent teeth: An orthopantomographic study. *Suom Hammaslaak Toim* 66:103–70.
70. Haavikko, K. 1974. Tooth formation age estimated on a few selected teeth: A simple method for clinical use. *Proc Finn Dent Soc* 70:15–19.
71. Mornstad, H., M. Reventlid, and A. Teivens. 1995. The validity of four methods for age determination by teeth in Swedish children: A multicentre study. *Swed Dent J* 19:121–30.
72. Nystrom, M., and H. Ranta. 2003. Tooth formation and the mandibular symphysis during the first five postnatal months. *J Forensic Sci* 48:373–78.
73. Liversidge, H.M. 2000. Crown formation times of human permanent anterior teeth. *Arch Oral Biol* 45:713–21.
74. Nystrom, M., E. Kilpinen, and E. Kleemola-Kujala. 1977. A radiographic study of the formation of some teeth from 0.5 to 3.0 years of age. *Proc Finn Dent Soc* 73:167–72.
75. Alimskii, A.V., K.Z. Shalabaeva, and A. Dolgoarshinnykh. 1999. [The time periods for the formation of the permanent occlusion in children born and permanently living in a region close to a former nuclear test range]. *Stomatologiia* (Mosk) 78:53–56.
76. Uysal, T., et al. 2004. Relationships between dental and skeletal maturity in Turkish subjects. *Angle Orthod* 74:657–64.
77. Piloni, M.J., and A.M. Ubios. 1996. Impairment of molar tooth eruption caused by x-radiation. *Acta Odontol Latinoam* 9:87–92.

78. Kaste, S.C., K.P. Hopkins, and J.J. Jenkins 3rd. 1994. Abnormal odontogenesis in children treated with radiation and chemotherapy: Imaging findings. *AJR Am J Roentgenol* 162:1407–11.
79. Kaste, S.C., et al. 1997. Dental abnormalities in children treated for acute lymphoblastic leukemia. *Leukemia* 11:792–96.
80. Cohen, J.T. 1928. The dates of eruption of the permanent teeth in a group of Minneapolis children: A preliminary report. *J Am Dent Assoc* 15:2337–41.
81. Hagg, U., and J. Taranger. 1985. Dental development, dental age and tooth counts. *Angle Orthod* 55:93–107.
82. Hagg, U., and J. Taranger. 1986. Timing of tooth emergence. A prospective longitudinal study of Swedish urban children from birth to 18 years. *Swed Dent J* 10:195–206.
83. Liliequist, B., and M. Lundberg. 1971. Skeletal and tooth development: A methodologic investigation. *Acta Radiol Diagn* (Stockh) 11:97–112.
84. Liversidge, H.M., and T. Speechly. 2001. Growth of permanent mandibular teeth of British children aged 4 to 9 years. *Ann Hum Biol* 28:256–62.
85. Davis, P.J., and U. Hagg. 1994. The accuracy and precision of the "Demirjian system" when used for age determination in Chinese children. *Swed Dent J* 18:113–16.
86. Teivens, A., and H. Mornstad. 2001. A comparison between dental maturity rate in the Swedish and Korean populations using a modified Demirjian method. *J Forensic Odontostomatol* 19:31–35.
87. Eid, R.M., et al. 2002. Assessment of dental maturity of Brazilian children aged 6 to 14 years using Demirjian's method. *Int J Paediatr Dent* 12:423–28.
88. Andersen, E., et al. 2004. The influence of jaw innervation on the dental maturation pattern in the mandible. *Orthod Craniofac Res* 7:211–15.
89. Demirjian, A., and H. Goldstein. 1976. New systems for dental maturity based on seven and four teeth. *Ann Hum Biol* 3:411–21.
90. Nystrom, M., et al. 2000. Dental maturity in Finns and the problem of missing teeth. *Acta Odontol Scand* 58:49–56.
91. Garn, S.M., A.B. Lewis, and J.H. Vicinus. 1963. Third molar polymorphism and its significance to dental genetics. *J Dent Res* 42(Suppl.):1344–63.
92. Johanson, G. 1971. Age determination from human teeth. *Odont Revy* 22(Suppl. 22):1–126.
93. Gustafson, G. 1950. Age determination on teeth. *J Am Dent Assoc* 41:45–54.
94. Smith, B.H. 1984. Patterns of molar wear in hunter-gatherers and agriculturalists. *Am J Phys Anthropol* 63:39–56.
95. Varrela, T.M., et al. 1995. The relation between tooth eruption and alveolar crest height in a human skeletal sample. *Arch Oral Biol* 40:175–80.
96. Iseri, H., and B. Solow. 1996. Continued eruption of maxillary incisors and first molars in girls from 9 to 25 years, studied by the implant method. *Eur J Orthod* 18:245–56.
97. Danenberg, P.J., et al. 1991. Continuous tooth eruption in Australian aboriginal skulls. *Am J Phys Anthropol* 85:305–12.
98. Behrents, R.G. 1985. The biological basis for understanding craniofacial growth during adulthood. *Prog Clin Biol Res* 187:307–19.
99. Holm-Pedersen, P., and H. Loe, ed. 1986. *Geriatric dentistry: A textbook of oral gerontology*. St. Louis, MO: CV Mosby Company.

100. Ten Cate, A.R. 1994. *Oral histology: Development, structure, and function*, ix. 4th ed. St. Louis, MO: Mosby.
101. Kvaal, S.I., et al. 1995. Age estimation of adults from dental radiographs. *Forensic Sci Int* 74:175–85.
102. Prapanpoch, S., S.B. Dove, and J.A. Cottone. 1992. Morphometric analysis of the dental pulp chamber as a method of age determination in humans. *Am J Forensic Med Pathol* 13:50–55.
103. Vandevoort, F.M., et al. 2004. Age calculation using x-ray microfocus computed tomographical scanning of teeth: A pilot study. *J Forensic Sci* 49:787–90.
104. Cameriere, R., L. Ferrante, and M. Cingolani. 2004. Variations in pulp/tooth area ratio as an indicator of age: A preliminary study. *J Forensic Sci* 49:317–19.
105. Berner, C., P.C. Brumit, B.A. Schrader, and D.R. Senn. 2007. *Age estimation from progressive changes in dental pulp chambers*. San Antonio, TX: F41 Odontology Section, American Academy of Forensic Sciences.
106. Adetona, O., J. Ethier, P. Nummikoski, W. Moore, R. Langlais, M. Noujeim, and D. Senn. 2008. *Dental age estimation by calculating the ratio of tooth and pulp volumes using cone beam computed tomography*. Washington, DC: F32 Odontology Section, American Academy of Forensic Sciences.
107. Stein, T.J., and J.F. Corcoran. 1990. Anatomy of the root apex and its histologic changes with age. *Oral Surg Oral Med Oral Pathol* 69:238–42.
108. Solheim, T. 1990. Dental cementum apposition as an indicator of age. *Scand J Dent Res* 98:510–19.
109. Giuliana, G., et al. 1995. Cementum growth in impacted teeth. *Acta Stomatol Belg* 92:7–11.
110. Kato, S., et al. 1992. The thickness of the sound and periodontally diseased human cementum. *Arch Oral Biol* 37:675–76.
111. Woods, M.A., Q.C. Robinson, and E.F. Harris. 1990. Age-progressive changes in pulp widths and root lengths during adulthood: A study of American blacks and whites. *Gerodontology* 9:41–50.
112. Harris, E.F., Q.C. Robinson, and M.A. Woods. 1993. An analysis of causes of apical root resorption in patients not treated orthodontically. *Quintessence Int* 24:417–28.
113. Harris, E.F., B.W. Boggan, and D.A. Wheeler. 2001. Apical root resorption in patients treated with comprehensive orthodontics. *J Tenn Dent Assoc* 81:30–33.
114. Maples, W.R. 1978. An improved technique using dental histology for estimation of adult age. *J Forensic Sci* 23:764–70.
115. Gat, H., et al. 1984. Dental age evaluation. A new six-developmental-stage method. *Clin Prev Dent* 6:18–22.
116. Lucy, D., and A.M. Pollard. 1995. Further comments on the estimation of error associated with the Gustafson dental age estimation method. *J Forensic Sci* 40:222–27.
117. Monzavi, B.F., et al. 2003. Model of age estimation based on dental factors of unknown cadavers among Iranians. *J Forensic Sci* 48:379–81.
118. Bodecker, C.F. 1925. A consideration of some of the changes of the teeth from young to old age. *Dent Cosmos* 67:543–49.
119. Kinney, J.H., et al. 2005. Age-related transparent root dentin: Mineral concentration, crystallite size, and mechanical properties. *Biomaterials* 26:3363–76.
120. de Jonge, T.H. 1950. Das Altern des Gebisses. *Paradontologie* 4:113–16.

121. Lamendin, H., and J.C. Cambray. 1981. Etude de la translucidite et des canalicules dentinaires pour l'apprdciatio de l'age. *J Med Leg Droit Med* 24:489-99.
122. Whittaker, D.K., and M.M. Bakri. 1996. Racial variations in the extent of tooth root translucency in ageing individuals. *Arch Oral Biol* 41:15-19.
123. Vasiliadis, L., A.I. Darling, and B.G. Levers. 1983. The amount and distribution of sclerotic human root dentine. *Arch Oral Biol* 28:645-49.
124. Sognnaes, R.F., B.M. Gratt, and P.J. Papin. 1985. Biomedical image processing for age measurements of intact teeth. *J Forensic Sci* 30:1082-89.
125. Drusini, A., I. Calliari, and A. Volpe. 1991. Root dentine transparency: Age determination of human teeth using computerized densitometric analysis. *Am J Phys Anthropol* 85:25-30.
126. Jensen, B., and J.M. Nielsen. 1968. Age determination in the red fox (*Vulpes vulpes*) from canine tooth sections. *Dan Rev Game Biol* 5:1-15.
127. Fletemeyer, J.R. 1978. Laminae in the teeth of the Cape fur seal used for age determination. *Life Sci* 22:695-98.
128. Stott, G.G., R.F. Sis, and B.M. Levy. 1982. Cemental annulation as an age criterion in forensic dentistry. *J Dent Res* 61:814-17.
129. Lieberman, D.E. 1994. The biological basis for seasonal increments in dental cementum and thier applications to archaeological research. *J Archaeol Sci* 21:525-39.
130. Renz, H., et al. 1997. Incremental lines in root cementum of human teeth: An approach to their ultrastructural nature by microscopy. *Adv Dent Res* 11:472-77.
131. Naylor, J.W., et al. 1985. Cemental annulation enhancement: A technique for age determination in man. *Am J Phys Anthropol* 68:197-200.
132. Kvaal, S.I., T. Solheim, and D. Bjerketvedt. 1996. Evaluation of preparation, staining and microscopic techniques for counting incremental lines in cementum of human teeth. *Biotech Histochem* 71:165-72.
133. Sousa, E.M., G.G. Stott, and J.B. Alves. 1999. Determination of age from cemental incremental lines for forensic dentistry. *Biotech Histochem* 74:185-93.
134. Kagerer, P., and G. Grupe. 2001. On the validity of individual age-at-death diagnosis by incremental line counts in human dental cementum. Technical considerations. *Anthropol Anz* 59:331-42.
135. Charles, D.K., et al. 1986. Cementum annulation and age determination in *Homo sapiens*. I. Tooth variability and observer error. *Am J Phys Anthropol* 71:311-20.
136. Condon, K., et al. 1986. Cementum annulation and age determination in *Homo sapiens*. II. Estimates and accuracy. *Am J Phys Anthropol* 71:321-30.
137. Lipsinic, F.E., et al. 1986. Correlation of age and incremental lines in the cementum of human teeth. *J Forensic Sci* 31:982-89.
138. Wittwer-Backofen, U., J. Gampe, and J.W. Vaupel. 2004. Tooth cementum annulation for age estimation: Results from a large known-age validation study. *Am J Phys Anthropol* 123:119-29.
139. Kagerer, P., and G. Grupe. 2001. Age-at-death diagnosis and determination of life-history parameters by incremental lines in human dental cementum as an identification aid. *Forensic Sci Int* 118:75-82.
140. Leigh, R.W. 1925. Dental pathology of American Indian tribes of varied environmental and food conditions. *Am J Phys Anthropol* 18:179-95.

141. Molnar, S., M.J. Barrett, L. Brian, L. Brace, D. Brose, and J. Dewey. 1972. Tooth wear and culture: A survey of tooth functions among some prehistoric populations. *Curr Anthropol* 13:511–26.
142. Murphy, T. 1959. The changing pattern of dentine exposure in human tooth attrition. *Am J Phys Anthropol* 17:167–78.
143. Scott, E.C. 1979. Increase of tooth size in prehistoric coastal Peru, 10,000 B.P.–1,000 B.P. *Am J Phys Anthropol* 50:251–58.
144. Rose, J.C., and P.S. Ungar. 1998. Gross dental wear and dental microwear in historical perspective. In *Dental Anthropology, fundamentals, limits, and prospects*, ed. K.W. Alt, F.W. Rosing, and M. Teschler-Nicola, 349–86. New York: Springer-Verlag.
145. Ungar, P.S., and M.A. Spencer. 1999. Incisor microwear, diet, and tooth use in three Amerindian populations. *Am J Phys Anthropol* 109:387–96.
146. Litonjua, L.A., et al. 2003. Tooth wear: Attrition, erosion, and abrasion. *Quintessence Int* 34:435–46.
147. Smith, B.G., and J.K. Knight. 1984. An index for measuring the wear of teeth. *Br Dent J* 156:435–38.
148. Dahl, B.L., et al. 1989. The suitability of a new index for the evaluation of dental wear. *Acta Odontol Scand* 47:205–10.
149. Santini, A., M. Land, and G.M. Raab. 1990. The accuracy of simple ordinal scoring of tooth attrition in age assessment. *Forensic Sci Int* 48:175–84.
150. Donachie, M.A., and A.W. Walls. 1995. Assessment of tooth wear in an ageing population. *J Dent* 23:157–64.
151. Pigno, M.A., et al. 2001. Severity, distribution, and correlates of occlusal tooth wear in a sample of Mexican-American and European-American adults. *Int J Prosthodont* 14:65–70.
152. Sarnat, H., et al. 2003. Developmental stages of the third molar in Israeli children. *Pediatr Dent* 25:373–77.
153. Solari, A.C., and K. Abramovitch. 2002. The accuracy and precision of third molar development as an indicator of chronological age in Hispanics. *J Forensic Sci* 47:531–35.
154. Gunst, K., et al. 2003. Third molar root development in relation to chronological age: A large sample sized retrospective study. *Forensic Sci Int* 136:52–57.
155. Arany, S., M. Iino, and N. Yoshioka. 2004. Radiographic survey of third molar development in relation to chronological age among Japanese juveniles. *J Forensic Sci* 49:534–38.
156. Olze, A., et al. 2004. Forensic age estimation in living subjects: The ethnic factor in wisdom tooth mineralization. *Int J Legal Med* 118:170–73.
157. Kasper, K., D. Austin, A.H. Kvanli, T.R. Rios, and D.R. Senn. 2009. Reliability of third molar development for age estimation in a Texas Hispanic population: A comparison study. *J Forensic Sci* 54(3).
158. Thorson, J., and U. Hagg. 1991. The accuracy and precision of the third mandibular molar as an indicator of chronological age. *Swed Dent J* 15:15–22.
159. Schmeling, A., et al. 2004. Forensic age diagnostics of living people undergoing criminal proceedings. *Forensic Sci Int* 144:243–45.
160. Cameriere, R., L. Ferrante, and M. Cingolani. 2004. Precision and reliability of pulp/tooth area ratio (RA) of second molar as indicator of adult age. *J Forensic Sci* 49:1319–23.

161. Chaillet, N., and A. Demirjian. 2004. Dental maturity in South France: A comparison between Demirjian's method and polynomial functions. *J Forensic Sci* 49:1059–66.
162. Carolan, V.A., et al. 1997. Some considerations regarding the use of amino acid racemization in human dentine as an indicator of age at death. *J Forensic Sci* 42:10–16.
163. Bada, J.L. 1987. Paleoanthropological applications of amino acid racemization dating of fossil bones and teeth. *Anthropol Anz* 45:1–8.
164. Masters, P.M., J.L. Bada, and J.S. Zigler Jr. 1977. Aspartic acid racemisation in the human lens during ageing and in cataract formation. *Nature* 268:71–73.
165. Ritz, S., H.W. Schutz, and C. Peper. 1993. Postmortem estimation of age at death based on aspartic acid racemization in dentin: Its applicability for root dentin. *Int J Legal Med* 105:289–93.
166. Helfman, P.M., and J.L. Bada. 1975. Aspartic acid racemization in tooth enamel from living humans. *Proc Natl Acad Sci USA* 72:2891–94.
167. Helfman, P.M., and J.L. Bada. 1976. Aspartic acid racemisation in dentine as a measure of ageing. *Nature* 262:279–81.
168. Mornstad, H., H. Pfeiffer, and A. Teivens. 1994. Estimation of dental age using HPLC-technique to determine the degree of aspartic acid racemization. *J Forensic Sci* 39:1425–31.
169. Waite, E.R., et al. 1999. A review of the methodological aspects of aspartic acid racemization analysis for use in forensic science. *Forensic Sci Int* 103:113–24.
170. Seitz, S., J.R. McCutcheion, P.C. Brumit, B.A. Schrader, and D.R. Senn. 2009. *Age estimation: Aspartic acid racemization utilizing whole teeth*. Denver: Odontology Section, American Academy of Forensic Sciences.
171. Spalding, K.L., et al. 2005. Forensics: Age written in teeth by nuclear tests. *Nature* 437:333–34.
172. Druid, H., K. Spalding, and B. Buchholz. 2009. *Dead victim identification: Age determination by analysis of bomb-pulse radiocarbon in tooth enamel*. Denver: American Academy of Forensic Sciences.
173. Willems, G. 2001. A review of the most commonly used dental age estimation techniques. *J Forensic Odontostomatol* 19:9–17.

Bitemarks

14

DAVID R. SENN
RICHARD R. SOUVIRON

Contents

14.1 Background and History of Bitemarks and Bitemark Cases	306
14.1.1 Chronology	306
14.1.1.1 Before the Twentieth Century	306
14.1.1.2 Twentieth Century	307
14.1.1.3 Twenty-First Century	308
14.1.2 Significant Cases	308
14.1.2.1 *Texas v. Doyle*, 1954	308
14.1.2.2 *Public Prosecutor v. Torgersen* (Norway), 1958	309
14.1.2.3 *Crown v. Hay* (Scotland), 1967	309
14.1.2.4 *Illinois v. Johnson*, 1972	310
14.1.2.5 *California v. Marx*, 1975	311
14.1.2.6 *Illinois v. Milone*, 1976	312
14.1.2.7 *Florida v. Bundy*, 1979 and *Florida v. Bundy*, 1980	313
14.1.2.8 *Florida v. Stewart*, 1979	314
14.1.3 Problem Cases	316
14.1.3.1 Frederik Fasting Torgersen, 1958	316
14.1.3.2 Richard Milone, 1976	320
14.1.3.3 Greg Wilhoit, 1987	322
14.1.3.4 Ray Krone, 1992	323
14.1.3.5 Michael Cristini and Jeffrey Moldowan—Michigan	325
14.1.3.6 Kennedy Brewer—Mississippi	326
14.2 Bitemark Characteristics	332
14.2.1 Bitemark: Definition	332
14.2.2 Bitemarks and Teeth Marks	332
14.2.3 Class Characteristics	334
14.2.4 Individual Characteristics	336
14.2.5 Bitemark Frequency and Distribution, and Biter Demographics	336
14.3 Bitemark Case Management	337
14.3.1 Evidence Collection	337
14.3.1.1 Evidence from the Bitemark or Patterned Injury	337
14.3.1.2 Evidence from Potential or Suspected Biters	342
14.3.2 Analysis of Evidence	344
14.3.2.1 Bitemark Injury Classification Systems	344

	14.3.2.2 Methods of Analysis	346
14.3.3	Comparison of Injury and Dental Evidence	348
	14.3.3.1 Methods of Comparison	348
	14.3.3.2 Reporting Conclusions and Opinions	350
14.4 Scientific Considerations, Bitemark Issues, and Controversies		351
14.4.1	The Uniqueness of the Human Dentition	351
14.4.2	Human Skin as a Medium for Recording Bitemark Patterns	353
14.4.3	Statistical and Mathematical Analyses Relating to Bitemarks	354
14.4.4	ABFO Bitemark Workshop 4	355
14.4.5	The Totalitarian Ego	358
14.5 Forensic Value of Bitemark Evidence		359
14.6 Responsibilities and Consequences of Forensic Odontology Expert Testimony		361
14.7 The Future of Bitemark Analysis		364
References		365

14.1 Background and History of Bitemarks and Bitemark Cases

That the human animal is capable of biting is obvious; that humans often bite each other is surprising to some, and to others a subject of much interest and study. The marks made by human teeth in inanimate objects and in human skin have been reported and recorded in both ancient and modern history. Although scientific information is limited in early recorded history, the anecdotal information is vivid and sometimes astonishing. Reports of the role of bitemarks in legal cases are rare prior to 1950. That role increased rapidly after 1975.

14.1.1 Chronology

14.1.1.1 *Before the Twentieth Century*

Accounts of bitemark cases before the twentieth century can be characterized as ranging from the materially unsubstantiated to the bizarre. Two brief examples illustrate that range.

14.1.1.1.1 1066–1087, William I (the Conqueror) There is no dependable manner to confirm the facts, but folktales persist that claim that William I used his distinctive teeth to bite into and mark the Seal of England in order to verify the authenticity of his correspondence.[1]

14.1.1.1.2 1692, Salem Witch Trials, Reverend George Burroughs Reverend George Burroughs was accused of practicing witchcraft involving, in part, the biting of the subjects he was allegedly inducing into the craft. Although he was in prison at the time of the alleged attacks, the bitemarks were judged to have been made by Burroughs's specter. A specter is defined by Merriam-Webster as a visible disembodied spirit.[2] "Biting was one of the ways which the Witches used for the vexing of the Sufferers, when they cry'd out of G. B. biting them, the print of the Teeth would be seen on the Flesh of the Complainers, and just such a sett of Teeth as G. B's would then appear upon them, which could be distinguished from those of some other mens" (Cotton Mather in Burr[3]). Burroughs's mouth was reportedly pried open in court and his teeth were said to match the bitemarks. Burroughs was convicted, sentenced to death, and hanged.

The above examples notwithstanding, there were other early cases that indicate that bitemark evidence was recognized and utilized in Europe, Asia, and North America, with cases cited in France, Belgium, England, Scotland, Japan, Canada, and the United States. These cases included bitemarks in foodstuffs, other inanimate items, and human skin. Many of the same arguments that are offered in modern cases were argued by both prosecution and defense teams in those cases.[4]

14.1.1.2 Twentieth Century

Presaging a later, more well-known case, a 1906 burglary case in County Cumberland in northern England featured a piece of cheese allegedly bitten by one of two accused burglars. Impressions were made and casts compared to the cheese. The teeth of one of the two accused men were judged to "fit" the bitemark in the cheese, leading to a conviction.[5] Although there were relatively few bitemark cases reported in the first half of the century, the latter half of the twentieth century saw a rapid increase in the number of criminal cases for which the analysis of bitemarks played an important role as an element of the prosecution's case. Some of the most noteworthy twentieth-century cases are listed here in chronological order and will be discussed in more detail in the next sections: *Doyle v. State* (Texas), 1954—burglary; *Public Prosecutor v. Torgersen* (Oslo, Norway), 1958—murder of Rigmor Johnsen; *Scotland v. Hay* (Biggar, Scotland), 1967—murder of Linda Peacock; *People (Illinois) v. Johnson*, 1972—rape and aggravated battery; *People (California) v. Marx*, 1975—murder of Lovey Benovsky; *People (Illinois) v. Milone*, 1976—murder of Sally Kandel; *People (Florida) v. Stewart*, 1979—murder of Margaret Hazlip; *People (Florida) v. Bundy*, 1979—murder of Lisa Levy and Margaret Bowman; *People (Florida) v. Bundy*, 1980—murder of Kimberley Leach; *People (Oklahoma) v. Wilhoit*, 1987—murder of Kathryn Wilhoit; *People (Michigan) v. Moldowan and Cristini*, 1991—kidnapping and rape of Maureen Fournier; *People (Arizona) v. Krone*, 1991 and 1995—murder of Kimberley Ancona; and *People (Mississippi) v. Brewer*, 1995—murder of Christine Jackson.

14.1.1.3 Twenty-First Century

In regard to bitemark cases, forensic odontology in the first decade of the twenty-first century has been beleaguered by the alarming number of court cases involving bitemark testimony that have been challenged and reversed. The legal community, especially individuals and groups that work to prove the innocence of persons who have been wrongly convicted of crimes, has been instrumental in bringing attention and scrutiny to law enforcement practices, prosecutorial behavior and misconduct, and forensic identification sciences, with bitemark analysis being prominent among them. Most of the efforts of the innocence groups have centered around the possibility of the analysis of biological material in or on evidence collected around the time of the incidents that contained or may have contained DNA. The most well-known of those cases that include bitemark analysis as a key part of the investigation, prosecution, and expert testimony are discussed below.

14.1.2 Significant Cases

Lists of "reported" bitemark cases most commonly consist of those that have been appealed and reviewed by state or national appellate or supreme courts. Lists of those cases assembled by Pitluck and others currently include cases in excess of three hundred.[6] The following bitemark cases are considered significant because they were either groundbreaking or controversial, or both.

14.1.2.1 Texas v. Doyle, 1954

The first reported case in the United States involving bitemarks was the appellate case *Doyle v. State of Texas*, 1954. This case involves bitemarks left in cheese at the scene of a burglary. This first U.S. case is significant for its importance when considering the admissibility in a court of law of bitemarks in food. The fact that it was treated as a pattern or tool mark evidence is also significant. The primary testimony was given by a firearms examiner with supporting testimony from a dentist. In addition to this being the first reported American bitemark case, a significant lesson to be learned from this case is the manner in which the evidence from the biter was collected. Mr. Doyle was asked to bite into another piece of cheese, which he did voluntarily. This then was introduced and compared with the cheese from the crime scene to link Mr. Doyle to the burglary. It was challenged on appeal that same year on the grounds that Doyle was not provided his constitutional rights. A court order was not issued for the gathering of incriminating evidence in violation of the Fifth Amendment, the right to protection from self-incrimination and the Fourth Amendment, the protection from illegal search and seizure. The court denied Doyle relief on both issues.[7]

14.1.2.2 Public Prosecutor v. Torgersen *(Norway), 1958*
This case will be discussed in detail in the problem case section to follow.

14.1.2.3 Crown v. Hay *(Scotland), 1967*
The body of fifteen-year-old Linda Peacock was discovered on August 6, 1967, in a cemetery in Biggar, Scotland. She had been strangled and there was a bitemark on her right breast. Gordon Hay, seventeen, had, for some time, been detained at a nearby minimum security school for troubled boys, the Loaningdale Approved School. Drs. Warren Harvey and Keith Simpson made a remarkably detailed examination of many Biggar residents, including the boys at the Loaningdale school, and made dental models on twenty-nine of them judged to be viable suspects. From those 29 the suspect population was reduced to five from whom additional evidence was obtained. Unusual pits in the cusp tips of Hay's right canine teeth were deemed consistent with similar features seen in the bitemark. Hay was tried and found guilty. As a minor he was sentenced to serve an undetermined term characterized as "at Her Majesty's pleasure"[8] (Figures 14.1 to 14.5).

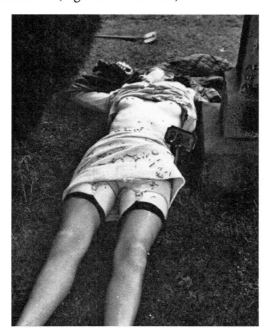

Figure 14.1 Crime scene photograph from 1967 murder of Linda Peacock.

Figure 14.2 Linda Peacock prior to autopsy. Note ligature marks on neck and vertical dried bloodstain anterior to left ear.

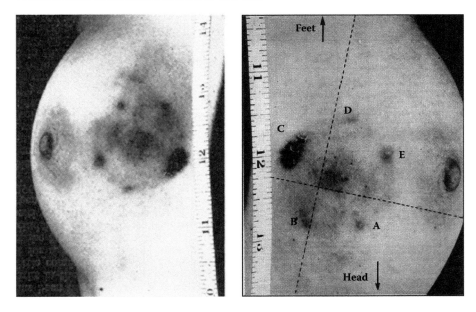

Figures 14.3 and 14.4 Left: Bitemark pattern medial to the areola and nipple of the right breast. Right: Bitemark rotated so that the marks judged to have been made by the maxillary teeth are at the top. Note the circular features within marks A and E.

14.1.2.4 Illinois v. Johnson, 1972

The first U.S. case involving a bitemark in human skin occurred in 1972. In *People (Illinois) v. Johnson*, the defendant was accused of rape and aggravated battery. There was a bitemark on the breast of the victim. An Illinois dentist, Dr. Paul Green, testified that the teeth of Johnson were similar to the bite pattern on the breast of the victim. Johnson was convicted of rape and aggravated battery and his conviction was upheld at the appellate level.[9]

Bitemarks

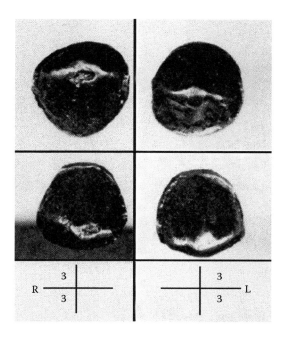

Figure 14.5 Copper models of the canine teeth of Gordon Hay. Note the defects on the cusp tips of the right canines.

14.1.2.5 California v. Marx, 1975

The trial for the first bitemark evidence case in California occurred in 1975. In *People (California) v. Marx*, Walter Marx was charged with the murder of Lovey Benovsky in a case in which the bitemark was the only physical evidence offered by the prosecution. The elderly female victim was sexually assaulted and strangled. In February 1974 Walter Marx was jailed initially for contempt of court for refusing to provide dental casts pursuant to a court order. He later consented to the impressions of his teeth. At autopsy a patterned injury, "an elliptical laceration of the nose," was noted. The pathologist believed it to be a possible bitemark. After the autopsy in February, the body was embalmed and buried in Texas. In March 1974, after Marx finally agreed to the teeth impressions, Benovsky's body was exhumed and a Dallas dentist made photographs and impressions of the nose. The material was sent to Los Angeles where the homicide occurred (Figures 14.6 and 14.7). A team of three forensic dentists performed the analysis on this bitemark. This was the first known case in which a team of forensic odontologists worked together in the examination, testing, evaluation, and comparison of a bitemark on the skin of a victim to the teeth of a suspect. Test bites were performed in this case and a three-dimensional model of the nose was made. Overlays, three-dimensional comparisons, and scanning electron microscopy were also used. None of these techniques had been documented as having been used in previous

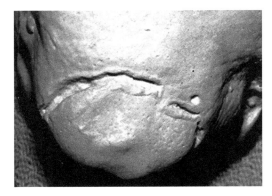

Figure 14.6 Marked three-dimensional bitemark on the nose of Lovey Benovsky.

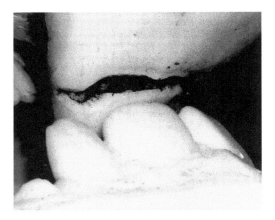

Figure 14.7 The teeth of Richard Marx compared to the nose injury.

cases. The marked three-dimensional nature of the bite in the nose in this case remains an unusual finding, even today. Direct comparisons were also made utilizing the dental casts from the only suspect, Walter Marx, directly to the three-dimensional model of the nose. All three dentists, Drs. Gerald Felando, Reidar Sognnaes, and Gerald Vale, testified at trial that the teeth of Walter Marx made the bitemark in the nose of Lovey Benovsky. The admissibility of the bitemark evidence and the conviction of Walter Marx were upheld on subsequent appeals. Without the bitemark evidence, the prosecution did not have a strong case against Marx. The testimony of a psychiatrist was considered and Marx was convicted of voluntary manslaughter, not murder.[10]

14.1.2.6 Illinois v. Milone, 1976

Within two years of the landmark case in California, an important and controversial case occurred in Illinois. This significant and problematic case will be more fully explored in the next section.

14.1.2.7 Florida v. Bundy, *1979 and* Florida v. Bundy, *1980*

In January 1978, a Sunday night at the Chi Omega House, Florida State University, Tallahassee, two coeds were bludgeoned to death and two others survived their attacks. On the same night at a nearby home another female victim was attacked as she slept. At autopsy, bitemark evidence, in the form of excised skin, was removed from the body of one of the victims. The following Saturday, the tissue was analyzed, photographed, and preserved in formalin. Although the tissue had not been optimally preserved—the tissue was not attached to a retaining ring—it was evident that this was a human bitemark and that there was a pattern suggesting the biter had crooked or broken teeth. Months went by without production of any photographs of the bitemark with a ruler or scale in place. The suspect in the case, Theodore Robert Bundy, a serial killer from the State of Washington who had escaped prison in Colorado and moved to Tallahassee, Florida, was held on suspicions of these two murders and the assault on the three other female victims. Prior to the grand jury indictment of Mr. Bundy, only one photograph out of thousands taken at the scene and during autopsy was produced that included a ruler held near the pattern, meaning that the bitemark could be sized. The state attorney, Larry Simpson, realized the significance of the only physical evidence in this case: the bitemark. In order to obtain impressions and photographs of Mr. Bundy's teeth, it was determined that a search warrant as opposed to a court order would be the path that the prosecution was to take. The warrant documented in thirteen pages the scope of the examination to be undertaken and the history of Mr. Bundy prior to and after the homicides in Tallahassee. No defense attorney was present when the warrant was issued or during the dental examination of Mr. Bundy's teeth. The search warrant and its execution were tightly guarded secrets. Once the material from the suspect had been obtained, the state attorney wanted to affirm that bitemark evidence was accepted in courts throughout the United States. Consequently, independent evaluations of the Bundy material were done by Dr. Lowell Levine in New York and Dr. Norman "Skip" Sperber in California. All agreed independently that the bite pattern left on the victim was of evidentiary value, that it showed not only class but individual characteristics of a double bitemark. They also agreed that the teeth of Mr. Bundy were unusual and distinctive and more likely than not left the bitemark. A weeklong evidentiary hearing was held in Tallahassee, at which time a circuit court judge heard evidence as to bitemark evidence and ruled as to its admissibility in the courts of the State of Florida, that is, a *Frye* hearing. Drs. Souviron, Levine, and Sperber all testified at the evidentiary hearing in Tallahassee. In a change of venue the trial was moved to Miami. The proceedings began in July 1979. At trial the defense called Dr. Duane T. Devore, a board-certified forensic odontologist. His testimony was that Bundy's teeth were "not that unique" and produced preorthodontic treatment dental models of five individuals that had similar lower anterior teeth arrangement. It was a valid idea but a tactical disaster for the defense, as these individuals were eleven to thirteen years old, none of whom were in Tallahassee in January 1978, and none of whom could have bitten the victim. Drs. Richard Souviron, Lowell Levine, and Homer Campbell testified for the prosecution. Mr. Bundy was convicted of the aggravated battery of three of the victims and the murder of the other two Chi Omega sorority sisters, Lisa Levy and Margaret Bowman. He was sentenced to death on both counts of murder and life without parole on the aggravated batteries. Several months later in 1980, Mr. Bundy was again on trial for the murder of twelve-year-old Kimberly Leach of Lake City and was again sentenced to death. The appeals on the Leach murder were exhausted before those of the Florida State University students. At 7:06 a.m. on January 24, 1989, forty-two-year-old Ted Bundy was executed in the electric chair for the murder of Kimberly Leach.[11]

The Theodore Robert Bundy case was significant for forensic odontology. Mr. Bundy was a suspect in approximately forty homicides of young females from the states of Washington, Oregon, Colorado, Utah, and Florida. From a scientific viewpoint, the teeth of Mr. Bundy were very distinctive and the bitemark recorded the pattern clearly and with little distortion. Mr. Bundy acted as his own attorney, taking the deposition of the state's odontologist, Dr. Souviron, a unique occurrence in the history of bitemark cases. The Bundy case brought bitemark evidence to national prominence. As a result, a flood of additional cases in which bitemark evidence was used followed. From 1950 through 1978, the number of "reported" bitemark cases in the United States was fewer than twenty. From 1979 through 2000 the number of cases challenged on appeal was in the hundreds.[6]

14.1.2.8 Florida v. Stewart, 1979

Concurrent with the Bundy trial and at the same courthouse, there was another murder trial involving bitemark evidence. Margaret Hazlip, a seventy-seven-year-old woman, had been sexually assaulted and murdered in February 1979. There were obvious teeth marks on the right hip of Ms. Hazlip. Homicide detectives were of the unusual opinion that Ms. Hazlip had, in effect, bitten herself. There were also pieces of bitten bologna at the crime scene. Dr. Souviron, the prosecution expert, was asked whether the pieces of bologna had been bitten by Ms. Hazlip and if she could have bitten herself on her own hip. Ms. Hazlip had an upper removable partial denture that was found next to the body. If she were lying on her partial it certainly could have made "tooth marks" in the skin of her thigh. The odontologist compared the partial denture to the bologna and in his opinion the bitemarks in the bologna were not made by Ms. Hazlip. The bitemark on her hip was analyzed, and it was determined that the biter profile would indicate that there was a large diastema (space or gap) between the upper two central incisors. Roy Allen Stewart was subsequently arrested and charged with the murder of Margaret Hazlip. Mr. Stewart had a large diastema between his upper central incisors. The defense hired Dr. Lowell Levine to analyze the bite and testify at the subsequent trial of Mr. Stewart. As mentioned previously, the trial took place in the same courthouse and at the same time that jury selection was being conducted in the Bundy trial. Mr. Bundy's defense team (five attorneys) all attended the prosecution's forensic dentist's testimony and made notes to challenge his later testimony at the Bundy trial. The defense expert, Dr. Levine, did not take the stand but provided useful information for the defense to cross-examine Dr. Souviron. Mr. Stewart was subsequently convicted, sentenced to death, and after numerous appeals, executed in the electric chair at Florida state prison in 1994. Because of the bologna used at the Stewart trial to show that Ms. Hazlip could not have bitten the bologna and that Mr. Stewart could, the case was referred to as "the bologna case." Dr. Souviron was referred to as an expert in bologna, implying that the same adjective could be applied to bitemark analysis. This was brought out in a humorous way by Judge Coward in the Bundy trial.[12]

These cases show that bitemark evidence has been used to link a suspect to bitemarks not only in cheese or bologna, but in other objects and in human skin. Bitemark evidence consists of patterned and other features that contain variables, not only in the teeth of suspected biters but in the material bitten, especially if that material is human skin. Three-dimensional information can be critical, not only in making a correlation between the biter and the injury, but also in determining if the injury occurred around the time of death. These cases point out the importance of cooperation and consultation among forensic dentists, the value of applying science to the analysis of bitemark evidence, and underscore the need for caution and the recognition that bitemark evidence can become controversial. The valid question "How can highly qualified experts have different opinions when analyzing the same material?" deserves an answer. The quality of the evidence, the distinctive patterns of a biter's teeth, the abundance or paucity of individual characteristics that are recorded in the bitemark—all go to the value and weight of the evidence. In the 1979 Hazlip case, although the defense expert could not exclude the defendant as being the biter, he was able to provide valuable information to the defense attorney in challenging not only the validity of bitemark evidence, but the credentials and testimony of the expert witness for the state. Each and every bitemark case that has proceeded to trial, and especially those that have been reported, contain valuable information that can help odontologists in obtaining, analyzing, and presenting bitemark evidence in a court of law. Had these cases been reviewed and analyzed by attorneys and odontologists in some of the bitemark cases that followed, many of the issues that made them problem cases may have been and should have been avoided.

In addition to *Illinois v. Milone*, already noted, the following significant and problematic cases will be discussed in the next section, on problem cases: *Oklahoma v. Wilhoit*, 1987; *Michigan v. Cristini and Moldowan*, 1991; *Arizona v. Krone*, 1992 and 1995; and *Mississippi v. Brewer*, 1995.

14.1.3 Problem Cases

In relation to the total number of cases involving bitemarks, the problem cases have been relatively few, but the consequences of those problem cases have been very serious. It can be argued that, if bitemark analysis is properly conducted, there should be no problem cases. What can be learned from these problem cases? Is there a common theme in these cases even though the individuals and circumstances are different? The details of the following few cases will hopefully shed light on the problems and pitfalls of bitemark analysis and help prevent errors in the future.

14.1.3.1 Frederik Fasting Torgersen, 1958

The body of Rigmor Johnsen was found by Oslo, Norway, firemen responding to a fire in the basement of her apartment building on the night of December 7, 1957. She had been sexually assaulted and the cause of death was listed as manual strangulation. There was a bitemark on the left breast. A forensic dentist from the University of Oslo, Professor Ferdinand Strom, collected the bitemark evidence and he and another dentist testified in the original trial in 1958 linking the teeth of Torgersen with the bitemark (Figures 14.8 to 14.11). The only other physical evidence was the presence of nonspecific feces on Torgersen's shoes and some common tree needles in his jacket pockets and the cuffs of his trousers. Torgersen was convicted of murder and sentenced to life in prison. An appeal, also in 1958, affirmed the conviction. Torgersen served sixteen years in prison and was released in 1974. He maintains his innocence and has repeatedly sought a new trial. The court engaged Professor Gisele Bang of Sweden to review the original materials in 1975. In 1976 he determined that Torgersen had likely made the bitemarks. A team of persons believing in Torgersen's innocence succeeded in convincing the Norwegian court to look into the matter again. In 1999 and 2000 Professors Gordon MacDonald and David Whittaker of Scotland and Wales reviewed the material and wrote reports generally supporting the earlier conclusions. In February 2001 Torgersen's defense team brought Dr. David Senn from Texas to Oslo to review the remaining materials in the University of Oslo laboratory of Professor Tore Solheim. Among that evidence was the removed and preserved breast of Rigmor Johnsen. The breast had been placed in Kaiserling's solution and sealed in a plastic container since the autopsy. None

Bitemarks

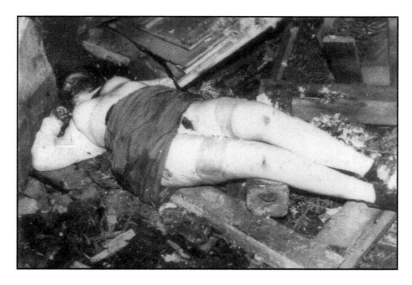

Figure 14.8 Crime scene photograph of the 1957 murder of Rigmor Johnsen.

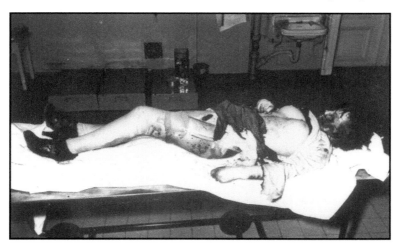

Figure 14.9 Rigmor Johnsen prior to autopsy.

of the investigators after Professor Strom had directly examined the breast. Senn opened the case and examined, photographed, and made impressions of the remarkably well-preserved breast (Figures 14.12 and 14.13). He reviewed all of the material made available and duplicated or photographed the items. Returning to the United States, he examined and tested all of the materials and determined that there were multiple inconsistencies between the tooth marks on the breast and the 1957–1958 models of the teeth of Torgersen. Among the forensic dentists that had previously examined the materials there were three different opinions as to which teeth had made the

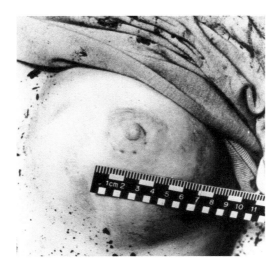

Figure 14.10 The only image of the bitemark with scale from the 1957 examination.

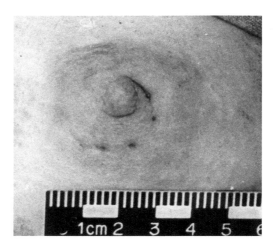

Figure 14.11 Closeup view of the bitemark in 1957.

individual marks. He consulted with three board-certified forensic dentists, Drs. Marden Alder, Paul Stimson, and C. Michael Bowers. He sent them de-identified duplicate materials and solicited their independent opinions. None of the three knew the details of the case or the identity of the victim or suspect. All three independently wrote reports stating that the person represented by the dental models provided could not have made the bitemarks on the breast. The four reports excluding Torgersen were sent to Norway. In 2001 the Supreme Court of Norway held a special hearing to again review the evidence in Torgersen's case. Drs. MacDonald and Whittaker testified

Bitemarks 319

Figure 14.12 Breast removed in 1957 in the just opened case in 2001. (See color insert following page 304.)

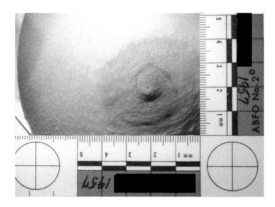

Figure 14.13 Closeup of bitemark on breast, imaged in 2001. (See color insert following page 304.)

for the prosecutor and Dr. Senn testified for the defense. The Supreme Court refused to allow Torgersen a new trial. In most countries that would have been the end of the story, but Torgersen's team succeeded in getting a new hearing before a commission for criminal case review (*Kommisjonen for Gjenopptakelse av Straffesaker*). Dr. Whittaker again testified for the prosecutor (Professor MacDonald had died in the interim) and Dr. Senn testified again for the defense. This commission, to the great surprise of Torgersen's team, also refused to recommend a new trial for Torgersen. Torgersen and his

14.1.3.2 Richard Milone, 1976

A 1972 murder victim, Sally Kandel, had a bitemark on her inner right thigh. It was the opinion of some that the bite had been inflicted after her death (Figures 14.14 and 14.15). Richard Milone was charged with the crime and initially tried in 1976. Three prosecution forensic odontology experts, Drs. Lester Luntz, Harold Perry, and Irvin Sopher, testified that Milone was "without doubt" the biter. Four defense experts, Drs. Lowell Levine, Curtis Mertz, Paul Stimson, and Duane Devore, disagreed.[14,15] This case was the first of several that involved a "battle of the experts" among forensic odontologists. There were appeals, but all of Milone's petitions were denied.[16] An executive clemency petition prompted the governor to ask for a review of the bitemark evidence. Three additional forensic odontologists, none of whom was involved with the initial trial, gave a joint opinion that Richard Milone inflicted the bitemark on the leg of Sally Kandel. Drs. Steven Smith, Raymond Rawson, and Larry Pierce brought the total number of forensic odontologists in this case to ten. They further opined that "there were distinguishing irregularities between the bitemarks found on Macek's victims [more on this below] and Sally Kandel." Although Drs. Smith and Rawson were asked on many occasions to disclose the bases for their opinions, the information has never been presented publicly or to forensic odontology groups.[15] The appeals court stated, "When coupled with the positive identification of the defendant as the perpetrator of the bite on the victim's thigh, we conclude and hold that the defendant's guilt was proved beyond a reasonable doubt" and "we therefore affirm the judgment."[16] The defense had attempted to offer the testimony of Dr. Homer Campbell to bolster the claim of their client's innocence to no avail. This brought the total number of forensic odontologists in this case to eleven, six stating that Milone made the bitemark and five opining that he did not.

The case had been further complicated by the introduction of evidence that a convicted serial killer, Richard Macek, may have also killed Sally Kandel. Many of the features of the case were similar, including the slitting of eyelids and the presence of a bitemark, but Macek had, unfortunately, had his teeth removed before he became a suspect. In the presence of a psychiatrist and the chief jailer Macek made a written confession to the murder of Sally Kandel, giving details that, according to the psychiatrist who later wrote a book containing the information, he could not have known if he were not present when Kandel was murdered. These details included what Sally Kandel wore the day of the murder, the position in which the body was

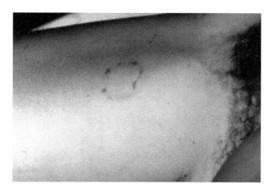

Figure 14.14 Bitemark on right thigh of Sally Kandel. (See color insert following page 304.)

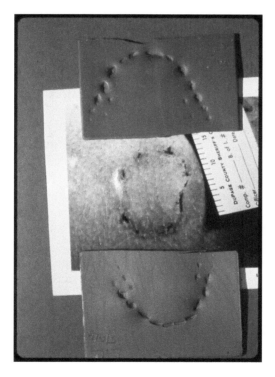

Figure 14.15 Kandel bitemark with Milone wax exemplars.

found, and this chilling statement, "Her right leg was raised higher than the other. And I bit her on the right thigh."[17] Macek reportedly later recanted the confession. He committed suicide in his jail cell in 1987.

Milone was released on bail in 1994. Milone appealed the earlier denial of a *writ of habeas corpus*.[16] The Court of Appeals, in *Milone v. Camp*, held that the potential prejudice did not so greatly outweigh the probative value of the evidence comparing a bitemark on a murder victim's thigh with the dental impressions taken from the defendant as to deny the defendant a fundamentally fair trial in violation of the Sixth Amendment, even though, at the time of the murder prosecution, the science of forensic odontology was still in its infancy. The evidence had some probative value, and the defense cross-examined the state's witnesses and presented witnesses that testified that he could not have made the bitemark. Nevertheless, the jury convicted Milone.[18]

The Milone case is noteworthy for the large number of dental experts involved, the extreme range of opinions by those experts resulting in a so-called battle of experts, and the confusion added by the confession of another convicted killer.

14.1.3.3 Greg Wilhoit, 1987

In 1987 Greg Wilhoit was accused of the murder of his ex-wife. A bitemark was the key piece of physical evidence at his trial. Drs. R.T. Glass and R.K. Montgomery testified that Mr. Wilhoit's teeth matched the bitemark, and in addition, bacteria from the bitemark were specifically unique to Mr. Wilhoit and rarely found in the general population. At Wilhoit's trial, no dental expert was called by the defense even though Dr. Thomas Krauss, a board-certified forensic odontologist, had been hired by the Wilhoit family. Dr. Krauss refuted both the bacterial evidence and the matching of the teeth of Wilhoit to the bitemark. The defense decided not to use Dr. Krauss at the trial, relying only on cross-examination of the prosecution's experts. Wilhoit was convicted and sentenced to death. Following his conviction Dr. Krauss worked with the attorneys handling the appeal and sent the bitemark evidence to eleven other board-certified forensic odontologists. All eleven odontologists independently excluded Wilhoit as the person who inflicted the bite. All the while Mr. Wilhoit was serving time on death row. As in all death sentence cases, the new defense attorneys appealed the conviction in part based on Dr. Krauss's work. In 1991, an evidentiary hearing was granted by the court of appeals. Both of the odontologists for the State of Oklahoma repeated their earlier testimony linking the teeth and the bacteria found in the bitemark to Mr. Wilhoit. The defense presented as experts Dr. Krauss and Dr. Souviron, both of whom stated that the bitemark had not been made by Mr. Wilhoit and the bacterial evidence was flawed, as more than 50% of the general population would be expected to have the same types of bacteria reportedly found on the victim's bitemark. As a result of the lengthy hearing, Mr. Wilhoit was granted a new trial. The state decided not to pursue the case and Mr. Wilhoit was finally a free man after spending over five years in prison.[19] There are many lessons to be learned from this case. Importantly, had the prosecution's dentists sought independent second opinions (or indeed eleven independent second opinions) and been willing to accept the possibly that their earlier interpretations could be wrong, Mr. Wilhoit would very likely never have been tried or convicted.

14.1.3.4 Ray Krone, 1992

Late on December 28 or early in the morning of December 29, 1991, Kimberley Ancona, a Phoenix bartender, was attacked and killed in the men's restroom of the CBS Lounge in Phoenix, Arizona. Dr. John Piakis, a Phoenix dentist, was called to the scene by the police. He examined the victim and noted a probable bitemark on the victim's left breast. The police investigation led them to a postal worker who frequented the CBS Lounge, Ray Krone. Dr. Piakis collected dental information from Krone and told them that his teeth were consistent with the bitemark. He was not an experienced forensic dentist, so he consulted his mentor, a well-known San Diego, California, forensic dentist, Dr. Norman Sperber.[20] Drs. Piakis and Sperber disagree on what took place after the consultation, but Dr. Sperber states that he told Dr. Piakis that Krone's teeth were not consistent with the bitemark, "not even close."[21] The police and prosecutors then sought the opinion of Dr. Raymond Rawson of Las Vegas. Dr. Rawson had lectured to the Arizona Homicide Investigators Association and was known to them as experienced in bitemark analysis. Dr. Rawson did a comprehensive analysis and developed a videotaped presentation of his analysis and experiments. At trial Dr. Piakis testified that Krone was the probable biter, and Dr. Rawson that Krone was the biter with reasonable medical certainty (Figures 14.16 to 14.18). In part of his testimony, Dr. Rawson reportedly stated, "The question should not be are bitemarks as good as fingerprints but are fingerprints as good as bitemarks" (transcript of original trial in *State v. Krone*[22]). No defense expert odontologist testified at Krone's first trial. Krone was found guilty and sentenced to death. In 1995 the Supreme Court of Arizona reversed the decision on procedural grounds and remanded the case for a new trial. Four board-certified forensic odontologists, Drs. Vale, Campbell, Sperber, and Souviron, were contacted by Krone's defense team. They independently excluded Ray Krone as the person who caused the bitemark. When one of those experts confronted Dr. Rawson, the state's expert, prior to the second trial and asked him to reconsider his opinion, Dr. Rawson reportedly stated, "I am in too deep."[12] Ray Krone was convicted a second time, again based largely on the same forensic dental expert's bitemark testimony. This second trial resulted in a sentence of life in prison.[22] In 2002 DNA evidence from the clothing of the victim was finally analyzed. Not only did the DNA profile exclude Ray Krone, but when compared to a state DNA database, it pointed to another Arizona prison inmate, Kenneth Phillips, who at the time of the Ancona murder lived near and frequented the bar where the murder occurred. When confronted, he obliquely confessed to the crime, reportedly stating that he only remembered struggling with the victim then awakening

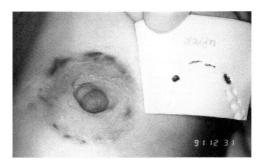

Figure 14.16 Side-by-side comparison of Ancona bitemark to Krone test bite in expanded polystyrene. (See color insert following page 304.)

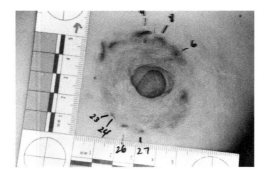

Figure 14.17 Ancona bitemark: Labels indicate one odontologist's opinion of marks made by specific teeth. (See color insert following page 304.)

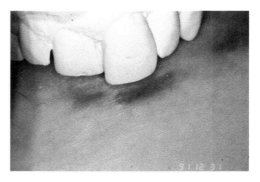

Figure 14.18 Direct comparison of Krone maxillary dental model to bitemark injury on breast of Ancona. (See color insert following page 304.)

the next morning with blood on his shoes. Dr. Piakis subsequently had the opportunity of compare Phillips's dentition to the bitemark and stated that Phillips's teeth were more consistent with the bitemark than Krone's.[23] Ray Krone was released and exonerated after spending thirteen years in prison, part of that time on death row, for a murder someone else committed.

The case of Ray Krone is a tragic indictment of law enforcement and legal prosecution practices and of the faulty application of bitemark analysis. This activity included overstating and overdramatizing the results of tests and experiments and failure to follow accepted guidelines by not seeking second opinions and disregarding or discounting the unsolicited opinions received. The homicide detectives failed to thoroughly investigate and follow all leads, and the prosecutors exhibited tunnel vision and willingness to shop for expert opinions that supported their theory of the crime. During an interview by a prosecutor before the retrial, one defense odontologist remarked, "I hope you have other important evidence ... the bitemark evidence is bad" and was bluntly told, "Doctor, this is a bitemark case and has always been a bitemark case."[24] The investigation, prosecution, and bitemark analysis and testimony combined to produce real tragedy. This triumvirate committed errors that compounded to produce a gross miscarriage of justice. This case is described in detail in a book authored by Jim Rix, Ray Krone's cousin and the sponsor of his defense.[25]

From a forensic odontology perspective, the initial opinion in any bitemark case should be referred to as "an investigative or preliminary opinion." The final opinion or evidentiary opinion should be formulated only after consultation with others and a thorough review and repeated review of all evidence, including scene photographs, video tapes, autopsy evidence, biological evidence, all bitemark photographs, impressions, bite prints, the tissue, and any other possibly useful information. One or more second opinions from other competent forensic odontologists should be sought and considered.

14.1.3.5 *Michael Cristini and Jeffrey Moldowan—Michigan*

Until the uniqueness of the human dentition is established, the use of mathematical degrees of certainty in associating bitemarks to individuals is not science. The Supreme Court of Michigan ruled that that type of testimony was inadmissible after several cases in that state in which bitemarks were associated to a suspect with statements of mathematical degrees of certainty. The Court made its ruling based on appeals of several of these cases. The 1991 case of the kidnapping, assault, and rape of Maureen Fournier featured the victim's eyewitness identification of the five men who participated in the attack and the two who allegedly bit her. Both Michael Cristini and Jeffrey Moldowan were convicted based on the victim's identifications and two forensic odontologists' testimony that the bitemark associations were positive. One of the dentists, Dr. Allan Warnick, testified that one of the marks was made by Moldowan and the odds that someone else made the mark were 3 million to one.[26]

This was not the only time that Dr. Warnick testified as to mathematical probability in a bitemark case. In another case he testified that "the chances of someone else having made the mark would be 4.1 billion to one."[27] On appeal in 1997, Drs. Homer Campbell and Richard Souviron independently reviewed the evidence and reported that, in their opinion, Moldowan and Cristini could be excluded.[12] In 2002 a Michigan Court of Appeals jury acquitted Moldowan. In 2004 Mr. Cristini was also granted a new trial. The court ruled that no testimony regarding mathematical degrees of certainty for bitemarks would be heard. This likely stemmed from other cases in which Dr. Warnick had given testimony. In one he said "that out of the 3.5 million people residing in the Detroit metropolitan area, the defendant was the only one whose dentition could match the individual who left the possible bite mark on the victim's cheek."[28]

In the Cristini retrial the prosecution expert, Dr. G. Berman, testified that Cristini made the bitemark with a high degree of certainty, and the defense expert, Dr. Souviron, testified that Cristini could be excluded. In an unusual twist in this trial, one of the original odontologists for the prosecution in the first trial in 1991, Dr. Pamela W. Hammel, took the stand for the defense and testified that she had erred in the original trial. She stated that she had been lied to and misled by the other prosecution expert. She stated further that she originally had doubts about the orientation of the bitemark, and after gaining more experience and reviewing the evidence, she realized her error. It took a great deal of courage for her to admit the error, but it was absolutely the right thing to do. The jury acquitted Mr. Cristini. He had spent thirteen years in prison. In December 2008 it was reported in the media that Mr. Cristini had been arrested and charged with eight counts of first-degree criminal sexual conduct allegedly involving a five-year-old child.[29]

What can be learned from this case? First, that eyewitness testimony may or may not be accurate—here the victim may have been wrong about the identity of the biters. She accused others that were later proven to be elsewhere at the time of the crime. Second, there is no scientific basis for mathematical degree of certainty with bitemark evidence on skin. Third, unlike in other cases, one of the experts had the courage to take the stand and admit an earlier error.

In the above detailed problem cases there was agreement among both the defense and the prosecution experts that these were indeed human bitemarks. The disagreements were related to features and orientation of the bitemarks and to who could have or who could not have inflicted the bites. The problems were compounded in some cases by the use of mathematical degrees of certainty or overreaching statements of the value and certainty of bitemark evidence.

14.1.3.6 Kennedy Brewer—Mississippi

The most flagrant departures from best practices in forensic odontology have occurred when a patterned injury not caused by human teeth has been misdiagnosed as a human bitemark. The most recent and highly publicized of these cases is that of Kennedy Brewer in Mississippi. Brewer was convicted in 1995 of the murder and sexual assault of Christine Jackson. The body of the three-year-old victim had been found in a nearby creek on a Tuesday morning, the third day after her Saturday night disappearance. The prosecution's dental expert, Dr. Michael West, examined Christine Jackson on May 9, 1992, and wrote in his May 14, 1992, report that nineteen human bitemarks were found on the body, and that "the bitemarks found on the body of Christina [sic] Jackson are peri-mortem in nature." "The bitemarks found on the body of Christina [sic] Jackson were indeed and without doubt inflected [sic] by Kennedy Brewer."[30]

Five days is a very short time period for examining, analyzing, and comparing nineteen patterned injuries. Dr. West later testified that "indeed and without doubt" and that "to a reasonable degree of medical certainty" the teeth of Mr. Brewer made five of those marks, and that it was "highly consistent and probable that the other fourteen bite mark patterns were also inflicted by Brewer" (West in original trial transcript in *Brewer v. State*[31]) (Figures 14.19 to 14.22). The defense expert, Dr. R. Souviron, testified that the patterned injuries on the body were not human bites at all but were patterns that were made by other means. "There could be insect activity there. There could be fish activity or turtle activity or who—God knows what" (Souviron in original trial transcript in *Brewer v. State*[31]). The jury convicted Mr. Brewer and he was sentenced to death. Appeals were denied until a 2007 analysis of the victim's vaginal swabs containing spermatozoa indicated the DNA profiles of two men. Neither profile included Brewer but did point to another man, Justin Albert Johnson, who, ironically, had also been an early suspect in Jackson's murder. Johnson later confessed to killing Christine Jackson and another young girl who had been similarly sexually assaulted and murdered. In that earlier case, Levon Brooks had also been wrongly convicted based, in part, on Dr. West's bitemark testimony conclusion. He testified that "it could be no one else but Levon Brooks that bit this girl's arm."[32] If Johnson, the confessed killer of both, had been arrested and convicted of that earlier murder, Christine Jackson could be alive today.

How can this have happened? How can an "expert" ignore the circumstances and disregard the crime scene information? How can patterns with no class or individual characteristics of human teeth in patterned injuries found on a body that had been in water for more than two days be judged to be human bitemarks? To then associate those patterns to a suspect with *any* level of certainty seems unthinkable. Perhaps, an understanding of alternative explanations to human teeth causing the marks should have been considered more seriously, especially in a case in which human bitemarks seemed unlikely. In fact, Dr. Souviron provided viable and testable theories for possible alternatives—the marks may have come from activity by insects, fish, turtles, or other sources not readily apparent. There is ample evidence

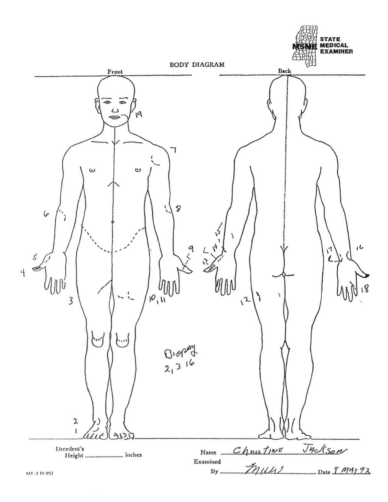

Figure 14.19 Location of nineteen patterned injuries labeled by MHW as bitemarks.

available to suggest that any number of aquatic organisms, even crayfish, could have made the marks on the victim. Preliminary laboratory evidence generated at Millersville University by an American Board of Forensic Entomology (ABFE)–certified forensic entomologist, Dr. John R. Wallace, suggested that crayfish, which were very abundant in the stream where the victim was found, were likely suspects and could have left such marks on the victim's body through normal feeding activity. The alternate theories for the genesis of these marks should have been explored.

Incredibly, the odontologist in this case associated only the upper incisor teeth to all of the "bitemarks"; there were no lower teeth marks identified. Neither the forensic pathologist nor Dr. West performed a simple test to determine if the patterns on the skin were in fact bitemarks of the type

MICHAEL H. WEST
Deputy Medical Examiner Investigator
FORREST COUNTY MISSISSIPPI

P. O. Box 15846
Hattiesburg, MS 39402

Business: (601) 264-2474
Residence: (601) 264-1422

```
Diane Brooks                    May 14, 1992
DMEI Noxubee Co.
Rt. 4 Box 115-6
Macon, MS 39340

RE: Christina Jackson

Dear Diane,

        On 5 May 92, at the request of Dr. Steve Hayne, I did travel
to the Rankin Co. morgue. I examined the remains of one Christina
Jackson,  a  black  female DOB 5-21-88. She  had  been  raped  and
sodomized. It was decided to hold her body until the weekend  and
examine it then.

        On  8 May 92 Noxubee Co. Sherif Dep. Bud Permenter brought
four  suspects  to my office for Dental  impressions.  They  were
Kennedy  Brewer  21y/o  black  male, Gloria  Jackson  24y/o  black
female, Dwayne Graham 17y/o black male and Leshone Williams 15y/o
black male.

        On 9 May 92, I returned to the Rankin Co. Morgue to  examine
and  compare the Dental study models to the remains of  Christina
Jackson.  Nineteen (19) human bitemarks were found  and  compared
to the dental study models.

OPINION

        The  bitemarks  found on the body of Christina  Jackson  are
peri-mortem in nature.
        The  bitemarks found on the body of Christina Jackson  were
indeed and without doubt inflected by Kennedy Brewer.

        If I can be of any further assistance in this  matter, please
feel free to contact me at home or office. A statement for my fee
is enclosed, please remit upon receipt.

Sincerely,

Michael H. West BS DDS ABFO DMEI SCSA

cc: SO
    DA
```

Figure 14.20 MHW report regarding nineteen "bitemarks" found on Christine Jackson. The report speaks for itself.

seen in human biting activity. An incision through a mark will reveal if there is the subepidermal hemorrhage often associated with human bitemarks (Figures 14.23 and 14.24). Alternatively, either could have harvested tissue from one or more of the patterned injuries. Dr. West had a history of similarly outrageous findings in other cases (Keko, Harrison, Maxwell). He had identified shoe marks on human skin and knife handle rivets on the hand of a murder suspect. He had made dramatic, overreaching statements in court, including conclusions to absolute certainty, "indeed and without doubt," and incredible estimates of his own error rates, "something less than my savior, Jesus Christ."[33] He had been suspended from the American Board of Forensic

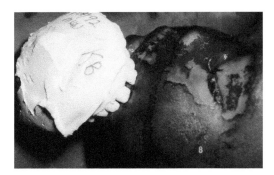

Figure 14.21 Direct comparison of Brewer's maxillary model to Jackson's face (pattern 19 on the MHW body chart). (See color insert following page 304.)

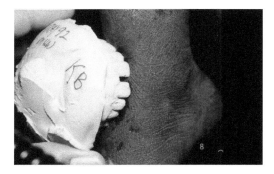

Figure 14.22 Direct comparison of Brewer's maxillary model to Jackson's foot (pattern 1 or 2 on MHW body chart). (See color insert following page 304.)

Odontology (ABFO). He had also resigned from the American Academy of Forensic Sciences (AAFS) and the International Association for Identification (IAI) seemingly to avoid suspension or expulsion from those groups for ethics violations. He subsequently was reinstated by the ABFO but later resigned to avoid facing additional ethics violation charges. Dr. West widely distributed by email his June 2006 resignation letter. The letter contained his general indictment of the ABFO replete with Latin quotes, plus support for his own work and this curious statement, "This action, these latest 'ethics' complaints, in my opinion, is only a ruse to affect the upcoming retrial of Kennedy Brewer in Mississippi, nothing more, nothing less.... *Age quod agis!*"[34]

These problem cases afford forensic odontologists the opportunity to learn valuable lessons. That the liberty, and indeed the life, of a human being is often in question and may depend upon the quality of experts' opinions is of paramount importance and cannot be overemphasized. We cannot be too cautious, too conservative, or too diligent when analyzing the potential association of a suspect to a bitemark. We should also recognize that eyewitnesses may be wrong or may have reason to lie. Prosecutors and investigative

Bitemarks

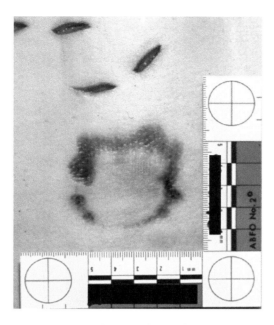

Figure 14.23 Bitemark and stab wounds made through clothing. (See color insert following page 304.)

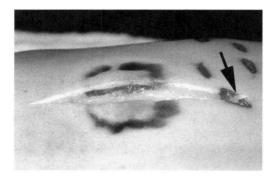

Figure 14.24 Incision through bitemark and stab wound. Note subepithelial hemorrhage in bitemark, none in stab wound (arrow). (See color insert following page 304.)

authorities may develop a theory of a crime that targets the wrong suspect and may give forensic dentists incorrect or misleading information about that chief suspect.

The ABFO has published guidelines and standards for bitemark analysis. These have been updated and changed. Some of those changes were made, in part, as a result of problem cases.

Education programs in forensic odontology were relatively rare before 1980. Since that time more programs and more comprehensive programs have become available. These include programs at the Armed Forces Institute of Pathology, the University of Texas Health Science Center at San Antonio, McGill University in Montreal, the Miami-Dade County Medical Examiner's Office, the University of Detroit–Mercy School of Dentistry, and others. Annual programs at the American Academy of Forensic Sciences and the American Society of Forensic Odontology highlight ongoing research and casework in forensic dentistry. The American Board of Forensic Odontology sponsors workshops in dental identification, bitemark analysis, and expert witness testimony. Research, education, and due diligence are required if forensic odontology is to progress and, in the process, minimize the probability of the occurrence of future problem cases.

14.2 Bitemark Characteristics

14.2.1 Bitemark: Definition

A bitemark is a patterned injury in skin or a pattern in an object caused by the biting surfaces of human or animal teeth. The ABFO manual defines a bitemark as (1) a physical alteration in a medium caused by the contact of teeth and (2) a representative pattern left in an object or tissue by the dental structures of an animal or human. The manual then describes a bitemark as "a circular or oval patterned injury consisting of two opposing symmetrical, U-shaped arches separated at their bases by open spaces. Following the periphery of the arches are a series of individual abrasions, contusions, and/or lacerations reflecting the size, shape, arrangement, and distribution of the class characteristics of the contacting surfaces of the human dentition"[35] (Figures 14.25 and 14.26). There has been much discussion about whether *bitemark* should be written as one word or two (*bite mark*) or hyphenated (*bite-mark*), for any of the forms the meaning is understandable and grammatically correct. Some will say one word is preferable as *fingerprint* is one word. Others will prefer two words, as in *tool mark*. The authors contend that this is a pointless argument and choose to use the single word form except in quoted material that uses other forms.

14.2.2 Bitemarks and Teeth Marks

Some distinguish between the terms *bitemark* and *tooth mark* or *teeth marks*, pointing out that marks can be left by teeth without the action of biting. This may occur when skin or other objects contact the teeth instead of a biter intentionally closing his jaws, and hence his teeth, into skin or an object. This is a valid distinction. A fist striking the teeth in an affray is a common and dangerous example often referred to as a clenched fist injury or "fight bite."

Bitemarks

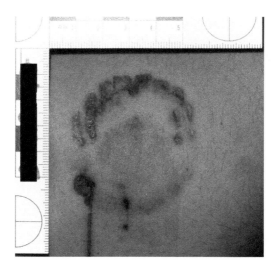

Figure 14.25 Bitemark with high forensic/evidentiary value. (See color insert following page 304.)

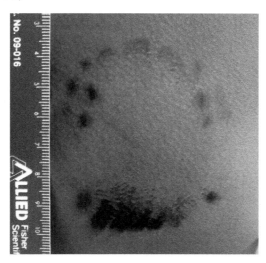

Figure 14.26 Another bitemark with high forensic/evidentiary value. (See color insert following page 304.)

A bitemark may be caused by a human biting another human, by an animal biting a human, or by either biting an object. Teeth marks are passive, as they involve no active, intentional or reflexive jaw movement, whereas in bitemarks the jaw muscles are active, causing the jaws and thence the teeth to move into the bitten substrate. Teeth mark examples include marks left by teeth on steering wheels, dashboards, or other objects during motor vehicle accidents. If struck in the mouth, the teeth of the victim may leave imprints on the fist or other object. Teeth marks may be found on the inner aspect of a victim's upper and lower lips after an attack. Pressure applied to the lips, which are consequently pressed against the teeth, leaves the teeth marks, patterned injuries that may indicate asphyxiation by force. Teeth mark injuries have not been widely discussed in the literature separate from bitemarks. The teeth marks left in different locations during motor vehicular accidents may be used to help distinguish passengers from drivers. Teeth marks have been found on exterior parts of vehicles during hit-and-run accidents. Teeth marks left on fists or hands as when striking a victim in the mouth can be crucial evidence placing a subject at a scene and showing that there was a violent interaction. These may be particularly helpful when a victim's body may have been severely affected by trauma or fire. Teeth marks on the hand or arm of a police officer or other law enforcement official may be used to support or disprove conflicting testimony in a police custody scenario, for example, a policeman's claim that the detainee bit him versus the prisoner stating, "He had me in a strangle hold and his forearm was forced into my mouth." It is important to note the nature of the patterned injury and the location of the teeth marks to determine which scenario is more likely. Teeth marks, unlike a bitemark, are more often single arch in nature. Bitemarks, on the other hand, more often show multiple teeth from both arches.

14.2.3 Class Characteristics

The definition above includes descriptions of both class and individual characteristics of bitemarks. The ABFO manual describes a class characteristic as "a feature, trait, or pattern that distinguishes a bitemark from other patterned injuries. Thus, a bitemark class characteristic identifies the group from which it originates: human, animal, fish, or other species."[35] Class characteristics include the characteristic arch forms of the biter and allow distinguishing between the upper and lower arch forms in a bitemark. Confirming the presence of class characteristics should be the first step in evaluating a bitemark. If unable to distinguish upper from lower arches, the evidence should not be considered to be of sufficient quality for *comparison* analysis. Obviously, every curved pattern injury was not caused by teeth. Misinterpretations of injuries that look similar to dental arches have led to errors in evaluation. It must be emphasized that, if unable to distinguish between upper and lower dental arches in a patterned injury, or to be able to identify which marks were made by specific individual teeth, forensic dentists should not attempt to compare the injury to suspect information unless that pattern contains one or more unique individual characteristics that may also be seen in a suspect biter's tooth or teeth. In some cases it may be possible to exclude individuals based on lower quality bite pattern information (Figures 14.27 to 14.29). An example would be a contusion pattern that exhibited a gap or gaps and a suspect that

Bitemarks

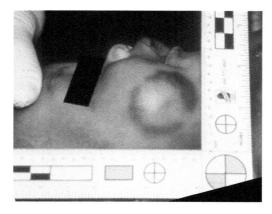

Figure 14.27 Bitemark with useful but limited forensic/evidentiary value. (See color insert following page 304.)

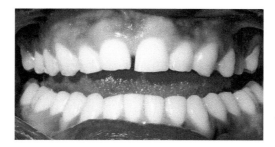

Figure 14.28 Suspect for bitemark seen in Figure 14.27. (This suspect could not be excluded as a possible biter for this bitemark.)

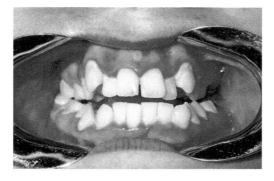

Figure 14.29 Suspect for bitemark seen in Figure 14.27. (This suspect was excluded as a possible biter for this bitemark.)

had no missing or malposed teeth. This information would come, not from comparison, but from the profile generated from the analysis of the mark.

14.2.4 Individual Characteristics

The ABFO manual defines an individual characteristic as "a feature, trait, or pattern that represents an individual variation rather than an expected finding within a defined group."[35] Individual characteristics are specific features found within the class characteristics. The ABFO manual distinguishes between individual characteristics as either arch or dental characteristics. Arch characteristics that qualify as individual distinguish one person's arch from another's and may include "a combination of rotated teeth, [teeth in] buccal or lingual version, mesio-distal [*sic*, mesial or distal] drifting [of teeth], and [variations in] horizontal alignment [of teeth that] contribute to differentiation between individuals." A dental characteristic within a bitemark is one that is specific to an individual tooth and is "a feature or trait within a bitemark that represents an individual tooth variation." Examples include the recording in a bitemark of the contour of the lingual surface of an upper anterior tooth with a depression indicating access for a root canal. A feature that makes one tooth different from all others is a dental characteristic and may include such things as fractures, wear patterns, developmental defects, and restorations. The ABFO manual further states, "The number, specificity, and accurate reproduction of these dental characteristics in combination with the *arch characteristics* contribute to the overall assessment in determining the degree of confidence that a particular suspect made the bitemark (e.g., unusual wear pattern, notching, angulations, fracture)."[35] When seen in bitemarks, individual characteristics allow the forensic dentist to eliminate, limit, or identify suspects depending on their degree of distinctiveness (Bernstein in Dorion[36]).

The majority of patterned injury cases analyzed by forensic odontologists involve one or more suspected human bite patterns in human skin. The quality of the evidence, including the amount of information in each mark and the distinctiveness of the pattern, dictate whether or not an association with the biter can be established. While much information is available in the literature on methodology, a standardized system of analyzing marks and evaluating the evidentiary value of those marks has not yet been developed, tested, and accepted by the forensic odontology community.

14.2.5 Bitemark Frequency and Distribution, and Biter Demographics

Several forensic dentists, including Harvey (1976), Vale and Noguchi (1983), Pretty and Sweet (2000), and Freeman, Senn, and Arendt (2005), have published studies of the frequency and distribution of bitemarks.[5,37–39] Freeman

Bitemarks

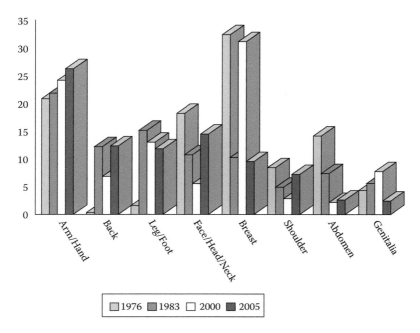

Chart 14.1 Comparison of bitemark locations in four studies by percentage of bitemarks reported.

et al. reported that females were more likely to be bitten than males, with adult females the most often bitten group. The areas of the body most often bitten included arms, backs, legs, faces and breasts overall, with some sex-related differences, women's breasts, for example, were more often bitten than men's. The four studies were in general agreement, although Harvey in 1976 and Pretty and Sweet in 2000 reported higher rates of bitemarks on breasts. In 1976 Harvey reported no bitemarks on the back and very few on legs and feet, but higher rates of abdomen bites (Chart 14.1). These reports indicate that almost any area of the body can be bitten, and that in violent exchanges the attacker, the attacked, and sometimes both may bite. This is useful information and emphasizes the propriety of taking dental impressions on deceased victims with bitemarks and interviewing living victims about the possibility that they also bit their attacker.

14.3 Bitemark Case Management

14.3.1 Evidence Collection

14.3.1.1 Evidence from the Bitemark or Patterned Injury

In this chapter we will not deal with the detailed technical aspects of evidence collection. This subject is adequately covered in the ABFO bitemark guidelines[35] and in *Bitemark Evidence*.[36] Instead, this chapter will focus on practical, legal, and philosophical aspects. In an ideal world the forensic odontologist would be involved with a case involving suspected bitemarks from the beginning, at the scene, during evidence collection, evidence analysis, and continuing to the courtroom. We do not live in an ideal world. In the real world the odontologist may not be contacted until much later and may be influenced, positively or negatively, by many factors, including his or her lack of experience, ego, and enthusiasm, to be part of the investigative team in a criminal case involving bitemarks. Case management begins with the first contact from the agency or person soliciting information from the forensic odontologist. Case management is critical during evidence collection and analysis, and continues through the trial and the final judicial appeal. The odontologist may be asked to review photographic evidence of a pattern injury taken by a crime scene investigator untrained in bitemark evidence collection. The only record of the injury pattern may have been photographed with non-state-of-the-art equipment. Images may have been collected with a Polaroid or other snapshot-type camera, perhaps without a scale or ruler, and at a distance and angle that precludes proper analysis. Lighting or flash equipment may be substandard or lacking. Alternately, the injury may have been photographed by emergency room or hospital personnel. Medical intervention may have distorted or even obliterated portions of the patterned injury. The odontologist must obtain as much information as possible to improve his or her chance of arriving at a conclusion that has scientific validity. This includes but is not limited to all scene photographs, even those in which the patterned injury is not visible. If the victim survives the attack, photographs of the injury should be taken as soon as possible, with follow-up photography during the healing process. If the victim does not survive, photographs of the patterned injury should be made at the scene, at the hospital, and at the morgue prior to the embalming and internment. If the body is later exhumed, procedures that preceded burial must be considered. In summary, the odontologist must review *all available evidence,* especially all photographic evidence. The likelihood of making an error in interpretation is inversely proportional to the abundance of the available evidence. Forensic odontologists may be asked during their careers to evaluate potential bitemark evidence on both living and deceased persons, including embalmed and exhumed bodies. If the bitten person is living, he or she may also be an eyewitness, capable of identifying the biter. As in all eyewitness accounts, this information should be considered with skepticism. In police custody cases in which the officer has been bitten, he or she may be able to describe how the pattern injury occurred and exactly who bit them. The odontologist may be called upon to confirm or challenge the officer's account. On the other hand, living bitemark victims may have reasons to falsely testify that a pattern injury is a bitemark when indeed it may be a pattern that was caused by another means that may mimic a human bitemark. Forensic odontologists will most often become involved with the investigation of cases that involve homicide, sexual assault, and abuse. Abuse or assault victims may or may not be hospitalized with serious injuries. Surviving victims may not be seen by an odontologist immediately. Several hours or sometimes even days may pass before the forensic odontologist is called. Occasionally emergency room personnel take photographs of the patterned injuries during early treatment. In sexual assault cases the victim may be seen at a "rape treatment center," and the personnel at the center usually photograph and swab patterned injury sites. In some jurisdictions it is rare for DNA or salivary samples to be taken on the living victim. Swabbing for salivary traces is a practice that must become standard in all cases involving suspected bitemarks. Occasionally a victim may also have inflicted bites on the attacker. A thorough and detailed history should be taken from the victim as soon as medically possible. The odontologist should ascertain that DNA or salivary samples have been taken from the bite wound. Often the medical examiner or emergency room personnel will be trained in the value of biologic evidence from a bite wound and will have taken the necessary samples. Dental impressions from the victim can be taken after the victim has sufficiently recovered. This is always done on a voluntary basis and there have been no known cases in which victims who have bitten their attacker refused the dental impressions. Photographing the bite wound days or even weeks subsequent to the attack may provide additional information. This is problematic since victims often will not return to have photographs taken of their injuries. In cases of abuse in which the victim is hospitalized the odontologist usually does not see the victims while they are in initial stages of care but only after they have been stabilized. This may be days after the incident occurred. Interviews with family members that are present may provide additional information to help with the analysis. It should be remembered that voluntary information given by family members may not be admissible in court. The eyewitness testimony of the victim may or may not be accurate, as illustrated in the case of a sexual assault victim in a Michigan attack (People [Michigan] v. Moldowan and Cristini). Although the reasons are unclear, subsequent DNA evidence showed that the victim seems to have given false or mistaken testimony as to the identity of her attackers.

Protocols are somewhat altered when examining deceased victims. When called to the medical examiner's office to examine patterned injuries that may or may not be bitemarks, the odontologist may be unable initially to clearly distinguish class and individual characteristics. This requires judgment and experience. If the patterned injury is judged to have sufficient class and individual characteristics of a bitemark, the injury should be photographed and biologic samples (DNA) collected prior to any manipulation of the tissue. Photography is the single most important method of preserving evidence of a bitemark. Various photographic techniques should be considered, including, but not limited to, color photography, black-and-white photography, ultraviolet, infrared, fluorescent, and other appropriate alternate light modalities. This subject is discussed in detail in Chapter 11. In some real-world cases, the odontologist may not have the opportunity to see the bitemark much less supervise or contribute to the case management. Often specialists in photography (forensic imaging departments) in major metropolitan medical examiners' offices will provide the photographic expertise. Nevertheless, the odontologist should be prepared to accomplish the necessary photography, be familiar with the appropriate techniques, and have the necessary equipment. Following the collection of the biologic evidence and photography, the next step in case management of a bitemark is evaluating the necessity and techniques for documenting the patterned injury in the third dimension. If tooth indentations or other three-dimensional features are present, dusting with fingerprint powder will allow the odontologist to view a pseudo-three-dimensional depiction of the mark and provide additional enhancement for photographic documentation. Fingerprinting of the bite, or "bite print," is best done with carbon or magnetic as opposed to fluorescent powders. Once the bite has been lightly dusted with the powder it is photographed, and then it can be lifted with the use of a gel lifter or standard fingerprint tape. This not only preserves the evidence for later analysis and presentation in court, but it will document the third dimension if present. The three-dimensional indentations in skin from the bite may be helpful in approximating the time of the bitemark in relation to the time of death of the victim. Bites on decedents do not always leave indentations. However, if present, they may be valuable in determining that the bite was inflicted after death or very close to the time of death. The lifted print, along with the subsequent impression of the bite, is rarely used for comparison to a suspect, but is far more valuable to document the third dimension. Subsequent to lifting the bitemark print with the gel lifter, the odontologist may then proceed to making impressions of the bitemark. If impressions are indicated, they should be made utilizing standard impression techniques and the most accurate impression materials; modern vinyl polysiloxane (VPS) impression materials work well. Note that if the body has been refrigerated it will take additional time for the impression material to set. There are faster setting materials available, and these should be considered. An impression from certain areas of the body may be made more difficult with gravity causing lighter-bodied materials to flow. The use of a retaining ring around the impression material may help prevent this problem. Before the impression material sets, mechanical retention devices for dental stone or plaster backing may be placed. There are several acceptable techniques for this. Gauze, paper clips, or other items can be partially inserted into the impression material, with the protruding portions acting as retention for the backing or base. Once the material is fully set, and the orientation recorded, the impression is removed and a negative of the bitemark is recorded. The orientation of the impression and case number should be placed on the backing and photographed *in situ*. Specifically, the placement of anatomical orientation markers as well as the case number, date, and the initials of the odontologist should be recorded on the stone backing. Photographic documentation of the entire procedure from start to finish is advisable. Being the most important type of evidence for recording a bitemark, too many photographs cannot be taken. Subsequent to the impression and with the permission of the medical examiner, the skin containing the bitemark injury may be removed in total (excisional) or may be biopsied (incisional) for microscopic analysis. If the injury is not on the face of the individual and if the medical examiner approves, the bite can be removed and preserved for later analysis. The procedure for removing and preserving the bitemark is well described in the ABFO guidelines[35] and in *Bitemark Evidence*.[36] In summary, the forensic odontologist has three main roles in evidence collection: (1) to physically document and collect evidence related to the patterned injury; (2) to perform an analysis of the injury and, if appropriate, determine from class and individual characteristics a dental profile; and (3) to obtain as much scene and other information from the investigative authorities as possible. This information should include scene photographs, body position, and presence of clothing or other objects that may have altered the biting mechanism or themselves produced patterned injuries on the body that may mimic a bitemark. If a bitemark is found the odontologist is responsible for examining that mark, and the remainder of the body if the mark is on human skin, to determine whether or not there are additional bitemarks. Confusion may arise if there are artifactual injuries that mimic bitemarks. The odontologist may be called upon to document these other injuries. He or she may be asked to analyze the information and at a later date be called to testify as to what possibly caused them. The initial investigative procedures, if properly performed, can be of great help in preventing or minimizing errors in interpretation. The appropriate handling of these three responsibilities of case management in the beginning of a bitemark case may well prove valuable as the case progresses through the criminal justice system.

The dental profile determined at the time the bite is analyzed could turn out to be among the most important articles of evidence and certainly may help to eliminate suspects. The dental profile has been used to support "probable cause" for the authorities to secure a court order or search warrant.

As discussed previously, in an ideal world, the odontologist will collect and analyze the evidence from a patterned injury and will have the opportunity to conduct his or her investigation from crime scene through court presentation. In the real world this rarely happens. Often the odontologist, especially if he or she is asked to review a case by the defense, will have only photographs and evidence collected by others with which to work. Rarely will he or she be provided scene photographs, and often only the images of the patterned injury from police or morgue photographs. It bears repeating that it is imperative that the forensic odontologist strongly request scene photographs and transcripts of the statements from the victim, medical care providers, pathologists, medical examiners, and eyewitnesses whenever possible. He or she should know the history and the circumstances of the event in order to be able to complete the most accurate analysis of the evidence and formulate the most accurate opinion. Analysis of a patterned injury without all of the information greatly increases the possibility of errors. Bitemarks are not fingerprints; bitemark analysis is not DNA analysis. Bitemarks are unique items of physical evidence that should be handled differently from other forensic evidence. Bitemarks are items of physical evidence that have unique forensic value with the potential to include and exclude. This is a powerful tool in the pursuit of truth and justice.

14.3.1.2 *Evidence from Potential or Suspected Biters*
The evidence required from a potential biter may include dental history and examination, photographs, dental impression, bite records, and biological evidence, including saliva samples. Again, the ABFO guidelines[35] and *Bitemark Evidence*[36] give complete and detailed instructions on the protocols and procedures for collecting evidence.

14.3.1.2.1 Legal Considerations There are legal issues in obtaining dental evidence from a suspect. Even though there may be voluntary consent, a written, signed consent by the subject is strongly advised. The subject may be a victim who bit a perpetrator of a crime or a suspected biter in a crime, usually a rape, assault, or homicide. They may also be potential biters of other material left or found at a crime scene. It is extremely important that the investigators, police, prosecutors, or crime scene technicians know in advance of the potential of a person changing his or her teeth if made aware that there is a possibility that teeth can be associated to a bitemark. Consequently, the subject (suspected biter) should be "kept in the dark" until such time as steps to legally collect dental information have been satisfied. A suspect in a violent crime, especially in a homicide, will not be inclined to give dental evidence voluntarily. The legal process begins with probable cause, and here the dental profile of the biter from the bitemark or bite wound may assist police, prosecutors, and judges to satisfy the legal requirement to obtain a warrant or court order to obtain the dental evidence.

A search warrant is most often used for the search of the physical property of a suspect, such as his house, car, or computer, but can also be used to obtain dental evidence in a bitemark case. A search warrant has certain advantages over the court order to obtain dental evidence. With a search warrant the suspect and his or her defense attorneys are not necessarily made aware of the evidence collection in advance. Also, during examination based upon a search warrant of the body of the suspect, the defense counsel will not be present. The police and prosecutor will draw up the warrant and present it to the judge for his signature. The warrant will outline the details of the crime and the areas of probable cause, circumstantial or otherwise, that may link the suspect to the crime. The dental profile from a bitemark may be helpful in establishing probable cause. The execution of the warrant usually takes place at a location that has a medical or dental facility, but may be done at any location the authorities deem appropriate. It is most often done in the medical or dental clinic of a jail. The suspect may refuse to cooperate by refusing to answer questions. He may refuse to open his mouth, object to the taking of biological samples, and resist being photographed. This can be a real problem if the warrant is not drawn up properly. It must contain language that will allow the taking of the required evidence with the use of "reasonable force." In the case of Ted Bundy, who refused to cooperate, he was advised that force would be used, including total anesthesia if necessary, and that he could be injured during the process. After he saw the determination and size of the guards he decided to cooperate. Despite the effectiveness of search warrants, the majority of bitemark evidence is obtained following the issuance of a court order. The court order stipulates the collection of the same evidence as the search warrant. Probable cause is necessary for the prosecutor to obtain a court order from the judge. The defense attorney, may, if he chooses, be in attendance when the records are obtained and advise his client not to answer any questions other than exactly what is spelled out in the order. The order should contain the name of the defendant and the date, time, and place that the evidence will be collected in general terms. It should name the forensic odontologist who is to obtain the evidence and spell out in general terms the items to be taken. These should include the dental history and oral examination, photographs, impressions, bite registrations, and where applicable, any other biological evidence the state may request. Saliva, blood, hair, and urine are taken by a lab technician or physician's assistant. It is important that the document includes the statement that "reasonable force" can be used to obtain the necessary evidence. If omitted from the warrant or court order, the defendant can refuse and cannot be forced to submit. His punishment for refusal may be contempt of court and jail. If he is already in jail the order will usually be reissued with the appropriate language.

 A recommended protocol when obtaining records pursuant to a court order or search warrant is to document the identity of all persons present, photograph the defendant with orientation images that picture him from head to toe, and include images that show his full face as well as his teeth. A log should be kept of the evidence and retained with the other evidence in your possession throughout the analysis process and eventual court presentation. The documentation of custody and recording of what, when, and who had possession of the evidence is referred to as the chain of custody. It is important to keep all evidence within your custody until time of trial and document any transfers or loans of that material. If you are required to release the material, you must document the release and return with appropriate chain-of-custody documentation.

Protocols, when followed, help to prevent mistakes. Plan ahead and be prepared. There is usually no second chance to gather evidence. Have a prepared list of what you need to take to the facility and bring an assistant, preferably another odontologist, when possible. The specific and detailed steps for dental evidence collection from suspects are discussed in the ABFO Bitemark Analysis Guidelines[35] and *Bitemark Evidence*.[36]

14.3.2 Analysis of Evidence

To properly analyze evidence after it is collected the evidence must be understandable and understood. It is dangerous to continue analysis of substandard evidence. Consequently, a standardized method of characterizing and classifying bitemarks must be developed.

14.3.2.1 Bitemark Injury Classification Systems

14.3.2.1.1 The ABFO Bitemark Scoring Guide In 1981 the Bitemark Standards Committee of the ABFO developed a system of analyzing and scoring bitemarks. Beginning in 1982, the scoring guide was published in the ABFO manual. In 1986, Rawson, Vale, Sperber, Herschaft, and Yfantis published a statistical analysis of an experimental examination of cases using the ABFO scoring guide for bite mark analysis.[40] The paper was complex and controversial. In 1988 the same authors, minus the statistician, published a letter in the *Journal of Forensic Sciences* essentially withdrawing their recommendation of the scoring system and stating "the authors believe that further research is needed regarding the quantification of bite mark evidence before precise point counts can be relied upon in court proceedings."[41] The scoring system has languished with no further development since that time. It remained a part of the ABFO manual through 2004 but was deleted from the 2005 manual and has not reappeared.

The authors believe that a standardized system of classifying and scoring or rating bitemarks should be developed. The system should classify bitemarks by type and by forensic or evidentiary value. Pretty proposed in 2006 and later published the proposal for a severity and significance scale that combined both[42] (Chart 14.2). This or a similar scale in concert with the ABFO scoring system or a derivative would be an excellent starting point for developing a truly useful method for assisting forensic odontologists to determine whether a bitemark contains sufficient evidentiary value to warrant further analysis. There would still be disagreements among examiners,

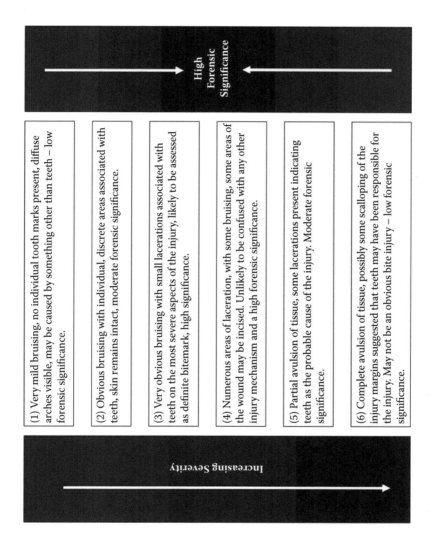

Chart 14.2 Pretty Bitemark Severity and Significance Scale.

but the disagreements would be based upon the same determining factors, known to all, allowing subsequent examiners to judge for themselves. It would also assist legal officials and others to assess the probative value of specific bitemarks and of specific bitemark opinions. After development, the system should be used to assess all bitemarks for forensic value. Those that are scored as substandard should not be used for comparison to suspects. As earlier stated, these lower quality injuries may be helpful for developing biter profiles that lead to exclusions but are not suitable for comparison to suspected biters' dental information. Those marks that have substantial forensic value allow proceeding to the next phase, analysis of the bitemark injury or injuries. Those analyses should be completed before comparisons to any suspect material are initiated.

14.3.2.2 Methods of Analysis

Current methods include pattern, metric, microscopic, three-dimensional, and computer assisted. All are mentioned or discussed in detail in the *ABFO Diplomates Reference Manual*[35] and *Bitemark Evidence*.[36] They are also discussed in other works, including a chapter by Souviron, Sweet, Golden, and Bowers in the Spitz and Fisher's forensic pathology text[43] and a chapter by Souviron in the Dolinak et al. forensic pathology text.[44]

Methods that utilize three-dimensional analysis are currently underdeveloped and rarely used. They and computer-assisted technologies must be explored, developed, and implemented. Forensic odontologists must commit to continually search for new methods to improve and validate the methods of analysis.

14.3.2.2.1 Variables and Bitemark Analysis

The bitemark is a pattern left by teeth that was actively produced by a biter. Teeth marks can be produced passively. Technically, bitemark patterns can be made by humans or animals, land and aquatic, by birds, and by some insects. Forensic odontologists must know and understand the differences and be able to explain them to investigators, litigators, and triers of fact. In previous sections we discussed that the initial investigator of a patterned injury must consider whether the pattern is a human bitemark or an injury that mimics a human bitemark. The odontologist must be able to locate and identify, if present, class, individual, and specific dental characteristics. When considering the dynamic actions possible when one human bites another, the variables are legion. The volume of material bitten is significant: Was a large amount of tissue taken into the mouth or a small nibble? These two simple possible scenarios can change the appearance of the pattern, even when made by the same teeth. Unfortunately, in violent exchanges between humans the possible scenarios are never only two and they are never simple. Consider that there are two humans involved, the biter and the person bitten. Both can and will move, twist, flex, and bend. The force, duration, volume of tissue bitten, area of the body bitten, and the strength of the biter will all have an effect on the nature of the bitemark. Considering just one of these dynamics, the area of the body bitten, there are many possibilities as to how the bitemark will appear. A bite on the buttocks or shoulder may be less distorted than one on an arm, breast, or stomach. There will always be some level of distortion in any bitemark left on skin. The responsible odontologist will accept this and explore the nature, type, and degree of distortion.

Bitemarks on living victims change over time. The vital reaction of the body to a compression wound will vary greatly depending upon numerous factors, including but not limited to the features of the teeth; the force, duration, and direction; the area bitten; the volume of tissue; the type of tissue; movement by the person bitten; and the age, skin type, and health status of the victim. Other possible mitigating factors: Was the bite through clothing? Did the victim struggle? Did the wound bleed? Did the bitten person live? If the person bitten died, was the bitemark made prior to death, around the time of death, or after death? Was the victim dark or light skinned? The biter and the person who is bitten each present variables that will affect the nature of the actual bite wound. These variables present challenges to bitemark interpretation, but they also offer opportunities to aid in the investigation of a crime.

If no saliva is found on or around a bite wound it may be because the bite was through clothing. If the clothing was preserved and the area of the bite located, the clothing may contain the saliva and hence the DNA of the biter. Thus, the bitemark did produce the biologic evidence for a biological (DNA) match. The distorted bite with very limited individual characteristics, no specific dental characteristics, and only general class characteristics may still be of value to the overall prosecution of a case. If the indentations (a third dimension) are present in the bitemark, the victim may have been bitten at or near the time of death, or after death, but certainly not hours before. This specific type of evidence is not available from fingerprints or DNA. Clinical testing has shown that bitemarks on living volunteers will produce indentations initially, but vital reactions (swelling, wheal formation, and subepidermal hemorrhage) will eliminate the indentations in a very short period of time. Biting is violence and bitemarks are painful. This can be important information for investigators and at trial. Forensic odontologists should be aware of the potential value of the evidence that bitemarks present beyond associating the teeth of a biter to a bitemark.

With the omnipresent variables in bitemark analysis, law enforcement and judicial officials should not expect bitemark evidence to develop to the same level of scientific certainty as DNA or even fingerprints. DNA and fingerprints have the advantage of both individuality and permanence. This is not true of teeth. Teeth may move, break, wear, and be lost. Teeth may be filled, crowned, or decorated. The questions of the uniqueness of the human dentition and human skin's ability to record the features of teeth will be discussed in Sections 14.4.1 and 14.4.2. The challenge for forensic odontologists is to consider all of the known and possible variables with bitemark evidence when forming their initial investigative opinions, during their analyses, and when forming their final opinions.

14.3.3 Comparison of Injury and Dental Evidence

The methodology for comparing information from bitemarks to the teeth or facsimiles of the teeth that may have caused them has developed over time into a robust, comprehensive, and detailed process. The bitemark analysis guidelines developed by the ABFO are described and discussed in the *ABFO Diplomates Reference Manual*[35] and *Bitemark Evidence*.[36] In this section discussion will center not on methodology, but on issues related to that methodology.

14.3.3.1 Methods of Comparison

Comparisons between bitemarks and the teeth that may have made them are accomplished by comparing features of the teeth of suspected biters to features in the bitemark. The methods used may be direct and indirect. Direct methods do not include directly placing the actual teeth of suspected biters against skin or images, but signify that exemplars or models of the teeth are employed for comparisons.

14.3.3.1.1 Direct Comparisons on Skin Direct comparison of dental models to bitemarks on skin is an acceptable screening technique if properly conducted. Opaque stone dental models impede the view of the underlying marks, making meaningful, in-depth feature comparison impossible. Some useful initial screening information may be gleaned from such comparisons. However, the placement of stone dental models directly on skin and moving them to facilitate visualization or mimic possible bite mechanisms can cause serious problems, including a real possibility of creating iatrogenic artifacts. Such artifacts could then be confused with actual injury pattern information. Videotapes of direct comparisons in actual casework have demonstrated this exact scenario in some of the problem cases.

14.3.3.1.2 Dental Exemplars Dental exemplars are facsimiles of the biting surfaces of teeth that facilitate comparisons to injury patterns. Exemplars of the solid- and hollow-volume type have been created by various means, ranging from freehand tracing onto transparent acetate sheets to radiographs of opaque materials placed into teeth indentations into wax or other media to computer-generated methods. In 1996 Dr. Heidi Christensen developed a method of scanning dental models on a flatbed scanner, then generating solid- and hollow-volume overlays using Adobe Photoshop.[45] The method was quickly adopted and adapted by others. In 1998 Sweet and Bowers compared five methods in use at the time and stated that the computer-generated overlay method was superior to other modalities, citing improved accuracy and objectivity.[46] Sweet et al. published revised methodology later that same year.[47] In the 2000 book and the 2003 second edition CD version, Johansen and Bowers provided detailed instructions for the computer-generated method.[48] The computer-generated method is now considered by many to be the gold standard for comparing the teeth of putative biters to life-sized photographs or exemplars of bitemarks. The problem with this method and the earlier methods is that they virtually disregard the three-dimensional features of teeth. The computer-generated method depends upon scans of the biting surfaces of teeth using flatbed optical scanners followed by use of various Photoshop tools to select or illustrate the biting surfaces based upon that scanned image. Flatbed optical scanners are not laser scanners and record no three-dimensional information. They record light reflected from the dental models to "highlight" the biting surfaces. Light reflects similarly from the biting surfaces of almost all teeth independently of their proximity to the bed of the scanner. Although some features, such as rotations and variations in labial or lingual position, are fairly accurately recorded, a tooth that may be several millimeters "shorter" or "longer," that is, farther from or closer to the incisal plane, will reflect light very similarly to one that is on plane. Hollow-volume overlays created by this method often similarly depict the outlines of teeth that, because they are millimeters less prominent, could not have participated in the bite with the same force as the surrounding teeth. A fractured incisor with part or all of the incisal portion missing may reflect light with no discernable differences when viewed two-dimensionally. Experienced forensic odontologists will carefully adjust the hollow-volume overlays by visually comparing the possibilities to the three-dimensional models. They will eschew the "magic wand" and utilize the "wise eraser" and "experienced pencil." This, of course, greatly reduces the objectivity but not the accuracy of the method.

The perfect method to produce exemplars or overlays has not yet been developed. Dailey in 2002 reported a method of considering the third dimension.[49] Dailey embedded dental models in a contrasting color of dental stone, then scanned or photographed the models. By selectively grinding away parts of the biting edges and reimaging in a stepwise manner, he created a series of images that mimic the slices of computed tomography. The method requires training, time, and skill to produce useful results and destroys the model.

Foy et al. demonstrated a cone beam computed tomography (CBCT) method that creates serial hollow-volume overlays at investigator-controlled levels. The method works similarly whether imaging the putative biter or his or her dental models. Automated software that facilitates comparing the serial hollow-volume overlays at various operator-controlled angulations and levels to compatible images of the injury could greatly enhance the process. Indirect three-dimensional methods comparing three-dimensional depictions of the teeth and merged two- and three-dimensional exemplars of the bitemark will be even better. Thali, de las Heras, and Dalle-Grave et al. have taken steps to develop three-dimensional methods.[51,52] Computer-generated two-dimensional hollow-volume overlays' days as the so-called gold standard may be rightfully numbered.

14.3.3.2 Reporting Conclusions and Opinions

The ABFO Bitemark Analysis Guidelines were revised in 2006. The revision included changes in recommended terminology for both indicating degree of confidence that an injury is a bitemark and relating or associating bitemark and suspected biters. Gone are the confusing and numerous terms for confidence that a mark is a bitemark that once included "more likely than not" and the now famous "less likely than not." They have all been replaced by simple and straightforward choices: "bitemark," "suggestive" [of a bitemark], and "not a bitemark."[35] These changes are to be applauded. The changes for terms relating bitemarks to suspected biters are equally simple and straightforward. In the order listed in the manual they are "reasonable dental/medical certainty," "probable," "exclusion," and "inconclusive." They are listed in the order of the highest positive correlation between biter and bitee to the lowest positive correlation. Had they been listed in the order of degree of certainty, "exclusion" would be listed first, followed by "reasonable dental/medical certainty," "probable," and "inconclusive," reflecting the maxim that in bitemark analysis exclusion is both more possible and more certain. In practical casework these guidelines present a problem. Some cases fall into that chasm between probable and exclusion, causing forensic odontologists to lose sleep trying to find features that would either raise or lower the determination. In the end, some cases will go neither up nor down and end up in an indeterminate state. Many odontologists have adopted DNA terminology and have chosen "cannot exclude" to characterize those cases. As these are guidelines, not standards, every forensic odontologist will have to determine how to handle those cases until the ABFO considers revising the guidelines.

The ABFO Standards for Bitemark Terminology were modified in 2005.

1. Terms ensuring unconditional identification of a perpetrator, or without doubt, are not sanctioned as a final conclusion.
2. Terms used in a different manner from the recommended guidelines should be explained in the body of a report or in testimony.
3. All boarded forensic odontologists are responsible for being familiar with the standards set forth in this document.[35]

The ABFO Standards for Bitemark Analytical Methods were modified in 2006.

1. All diplomates of the American Board of Forensic Odontology are responsible for being familiar with the most common analytical methods and should utilize appropriate analytical methods.
2. A list of all the evidence analyzed and the specific analytical procedures should be included in the body of the final report. All available evidence associated with the bitemark must be reviewed prior to rendering an expert opinion.
3. Any new analytical methods not listed in the previously described list of analytical methods should be explained in the body of the report. New analytical methods should be scientifically sound and verifiable by other forensic experts. New analytical methods should, if possible, be substantiated with the use of one or more of the accepted techniques listed in these guidelines.[35]

14.4 Scientific Considerations, Bitemark Issues, and Controversies

14.4.1 The Uniqueness of the Human Dentition

The concept that every person's teeth are unique is widely accepted by dentists. This is based on clinical experience and has not been conclusively scientifically established. Bitemark analysis is based on the concept that the biting surfaces of the anterior dentition, usually the six or eight most anterior maxillary and mandibular teeth, are sufficiently distinctive that well-trained forensic odontologists can distinguish the incisal portion of one person's anterior teeth from another's. This has not been sufficiently scientifically tested and confirmed. There is little doubt that if the features of human teeth are scrutinized with a fine enough measuring device, and if that device takes into account all three spatial axes, that is, three-dimensional analysis, then everyone's teeth are in fact unique. Sognnaes, Rawson, Gratt, and Nguyen in 1982[53] and Rawson, Ommen, Kinard, Johnson, and Yfantis in 1984[54] attempted in their research to establish the concept of uniqueness. The 1982 article examined the teeth of five pairs of monozygotic twins by making test bites then comparing the patterns. The researchers found that there were significant variations between the twin pairs, concluding that "identical twins are not dentally identical."[53] In the 1984 article "Statistical Evidence for the Individuality of the Human Dentition," the authors proposed "to establish the scientific base for the statistical analysis of the uniqueness of the human dentition" and stated as a conclusion, "This mathematical evaluation of a general population sample demonstrates the uniqueness of the human dentition beyond any reasonable doubt, thus placing the odontologist's statements about individuality beyond the realm of theory and into the realm of supported fact."[54] Had these two groups of researchers reviewed an earlier paper published on the subject, by MacFarlane et al. in 1974? If so, they did not include the paper in their list of references. The study involved the evaluation of dental characteristics seen in two hundred study casts of adult dental patients. MacFarlane et al. concluded that many of the features they examined and reported were interrelated. Features in the same arch and in the opposing arch had an effect on the positions of other teeth. MacFarlane et al. concluded that they could not combine and multiply features that were not independent, and that they had not determined that the human dentition was unique.[55] In 2007, Kieser et al. published a paper utilizing a novel geometric morphometric and procrustes superimposition technique to compare the biting surfaces of the anterior teeth of fifty orthodontic dental casts. In this study the researchers confronted MacFarlane head-on and stated that because the earlier study had "relied on highly subjective examinations of the casts by multiple examiners and failed to publish a table of their results," the conclusions in MacFarlane on the independence of the features were flawed. They further discussed the independence of the variables and state that their results "suggest a low, non-significant level of correlation between dental size/shape and arch shape," leading them to conclude that "the product rule can be applied to the assessment of these data," and "it appears that the incisal surfaces of the anterior dentition are in fact unique." They are careful to state that they did no investigation into whether or not the unique features would be transferred to skin.[56]

In counterpoint, critics of these conclusions concerning the statistical uniqueness of the human dentition, as well as bitemark analysis in general, abound. The criticisms come from within as well as outside forensic odontology groups. Most of the critical articles center on the lack of scientific vigor in the methods and conclusions of forensic dentists in bitemark cases. Saks and Koehler concluded in a 2005 article: "Simply put, we envision a paradigm shift in the traditional forensic identification sciences in which untested assumptions and semi-informed guesswork are replaced by a sound scientific foundation and justifiable protocols."[57] The foundations and protocols they envisioned include the implementation of research models that challenge the core assumptions of forensic fields, reveal the strengths and weaknesses of earlier methods, and apply new knowledge on a case-by-case basis. Scientists must select the methods that most closely apply to each case. They encouraged the collection of data on the frequency at which attributes and variations occur and the development of objective, computer-aided, or other programs to perform the actions that may currently be performed more subjectively. Finally, they recommended increased consultation, true blind examinations, and the institution of proficiency testing. These changes and others are needed to bring bitemark analysis (and some other identification sciences) toward the realm of true sciences.

The chief bitemark evidence and analysis critic within forensic odontology is Dr. C. Michael Bowers, a California forensic odontologist and attorney. He has authored books, written articles, given affidavits, and published his opinions on his website. Dr. Bowers appears to feel that bitemark analysis should only be used to exclude or to associate an individual as a "possible biter." Stronger or more discriminatory positive associations seem to be, in his view, *not* scientifically feasible. Interestingly, the ABFO deleted the "possible" designation when it revised the bitemark analysis guidelines in 2006. Only "reasonable medical/dental certainty," "probable," "exclusion," and "inconclusive" remain as recommended conclusions.

Dr. Bowers coauthored a textbook, now in its second edition, teaching methods for extensive bitemark detail analysis, metric analysis, exemplar creation, and feature comparison.[48] The techniques described in the book are very detailed and seem to be in significant contrast with the level of discrimination that Dr. Bowers advocates elsewhere as appropriate in bitemark analysis.

In spite of past or recent claims to the contrary, it may not be possible to mathematically or statistically prove the uniqueness of the anterior human dentition related to the information found in bitemarks. Consequently, a path similar to that recommended by Saks and Koehler seems the most sensible: Continue research into uniqueness, but collect data and build databases on the frequency with which those features and patterns of the anterior dentition appear, especially those features that may also be discernable in bitemark patterns. This work has already begun. Dr. L. Thomas Johnson and a Marquette University team reported development of a computerized method of collecting data on dental characteristics (oral presentations in Johnson et al.[58,59]). The method may be the first step toward the creation of a database of the frequency at which dental characteristics and combinations of characteristics occur in a population. Dr. Roger Metcalf reported on an alternate method at the same 2008 meeting (oral presentation in Metcalf et al.[60]). That method is currently being investigated at the University of Texas Health Science Center in San Antonio.[52] One of these methods or derivatives, or perhaps other totally new methods, is likely the appropriate step for moving forensic odontology and bitemark analysis along the correct path toward scientific vigor.

14.4.2 Human Skin as a Medium for Recording Bitemark Patterns

In a 2001 article, Pretty and Sweet analyzed a 1984 uniqueness study and stated, "Rawson has proven what his article claims, although perhaps not to the mathematical or statistical certainty expressed."[61] They added that the uniqueness of the human dentition is not the important question, rather "it is the rendition of these asserted unique features on human skin that is the unknown quantity."[61] Dentists understand the need for accurately depicting complex features and routinely work with impression materials that are capable of very accurately recording the features of the biting surfaces of teeth. There is almost universal agreement among forensic dentists that human skin is a very poor material for faithfully and accurately recording those features. Most early bitemark-related papers that discuss skin concentrate on the distribution of bitemarks, the classification of bitemarks, and the analysis of distortion in bitemarks or in preserved skin with bitemarks.[5,37–39,62–65] Although the earlier papers contained useful information, until recently there has been a paucity of research by forensic odontologists on the properties and behavior of human skin when bitten.[66,67] In 2008 at the AAFS meeting, a team of researchers from the State University of New York, Buffalo (SUNY-Buffalo) presented four papers examining biomechanical, macroscopic, and microscopic features of human skin and the features seen in skin when the bitemarks are made through clothing. These studies are a significant step toward a better understanding of bitten skin.[68–71] Finding answers to questions regarding the features and behavior of skin when bitten is crucial to establishing a scientific basis for bitemark analysis. The SUNY-Buffalo team has since published the first article concerning the biomechanical factors.[72] It must be demonstrated that teeth can or cannot dependably transfer discernable features to skin during biting, and that those features can be successfully imaged or recorded. Those features must then be scientifically analyzable, the distortion accounted for, and a statistical or mathematical basis for comparisons established.

14.4.3 Statistical and Mathematical Analyses Relating to Bitemarks

As discussed in the section on the uniqueness of the human dentition, the validity of any statistical and mathematical analysis of bitemarks is suspect until the uniqueness of the human dentition is confirmed. To date, only three significant papers have discussed or attempted this task.[53–55] In each of these studies, the statistical analysis problems center around the use of the product rule. Also known as Leibniz's law and the probability rule for independent variables, the rule is most commonly applied to problems in differential calculus. For forensic research applications the following definition may be most useful: "The probability rule for independent variables, or product rule, states that the probability of the simultaneous occurrences of two independent events equals the product of the probabilities of each event."[73] For three or more independent events the multiplication of probabilities continues by multiplying the product of the first two by the probability of the third and so on, as many times as there are independent variables. The same concept is used in DNA analysis and fingerprint analysis. In the most often cited work on the subject of the uniqueness of teeth, the use of the product rule is essential to the conclusion.[54] As discussed earlier in the section on uniqueness, there are divergent views on the independence of those variables (events), that is, the positions and characteristics of the biting surfaces of the anterior teeth. Until significant research shows that the dental features are indeed independent, mathematical or statistical certainty cannot be assigned to either the features of the biting surfaces of the anterior dentition or to the marks that those teeth make in skin.

As an alternative to those mathematical or statistical methods, research currently under way on collecting and recording data on the frequency of dental variation features is encouraging. This same type of analysis is used in associating persons based on mitochondrial DNA (mtDNA) and in analyzing patterns of present, missing, unrestored, and restored teeth for dental identification[74,75] (see also Chapter 7). To date there is insufficient scientifically confirmed information to support the association of bitemark patterns on human skin and sets of teeth with statistical or mathematical degrees of certainty.

14.4.4 ABFO Bitemark Workshop 4

One of the most contentious issues in forensic dentistry began as a well-intentioned educational exercise. This 1998–1999 exercise was the fourth in a series of American Board of Forensic Odontology (ABFO)–sponsored workshops started in 1984 and intended to develop guidelines and best practices for the scientific analysis of bitemarks. Thirty-two ABFO diplomates participated in the 1998–1999 workshop and examined four cases. The same seven sets of dental models served as the potential biters in all four cases. In only one of the cases, a bitemark in cheese, was the identity of the true biter known. The other cases came from casework selected by the workshop leaders. In only one of the three cases of bitemarks on skin was the true biter known. This knowledge was based, at least in part, on the victim's identification of the biter and the biter's subsequent confession.[76] Participants were asked to analyze each case and answer questions relating to the evidence. Results were tabulated and two nondiplomates were authorized to perform a statistical analysis of the results and produce a paper for submission to a refereed journal. The first journal to which the paper was submitted, the *Journal of Forensic Sciences*, rejected the paper, citing the inappropriate design of the workshop for statistical analysis. Perhaps unfortunately, considering the later misinterpretations, the article was submitted to other journals and was ultimately published in 2001 in the journal *Forensic Science International*. In the paper the authors stated that the primary objective of the study was "to determine the accuracy of examiners in distinguishing the correct dentition that make a bitemark," and the secondary objective was "to determine whether examiner experience, bitemark certainty, or forensic value had an effect on accuracy." They discussed the weaknesses in the study design, including the uncertain nature of using unverified casework due to the "possibility that the original examining dentist was wrong."[77]

As casework, the bitemarks on skin depicted patterned injuries for which the true biter was unknown. Consequently, mathematical or statistical analysis of the opinions of workshop participants was not possible without assuming that a true cause-effect relationship existed in each of the cases. As stated in the ABFO "Position Paper on Bitemark Workshop 4," "It was designed as an educational exercise whose primary purpose was designed to survey the degree of agreement (or disagreement) between diplomates confronted with cases of varying amounts and quality of bitemark evidence."[78] The ABFO further stated in the paper, "In conclusion, the ABFO states that due to the limitations imposed by the case material used in Bitemark Workshop #4, one cannot draw accurate statistical conclusions applicable to actual casework. Bitemark Workshop #4 was neither designed as, nor can it be used as, a proficiency test for forensic odontology. Tests of consistency and validity (necessary in a proficiency examination) were neither accomplished nor attempted; and, as subsequent reviewers of the data correctly pointed out, the construction of the examination and the workshop was not designed to produce an examination that had statistical validity and statistical consistency. The ABFO reiterates that it remains the responsibility of each individual examiner to determine if sufficient evidence exists to go forward with a meaningful bitemark analysis."[78]

Problems appeared when others implied that the Bitemark Workshop #4 was a bitemark proficiency examination of ABFO diplomates and mischaracterized the results. Although it is unclear whether his actions were intentional mischaracterization or well-intentioned error, the most enigmatic generator of misinformation about this workshop is an ABFO diplomate, Dr. C. Michael Bowers of California. In 2002 the Supreme Court of Mississippi was considering a petition for postconviction relief in a death penalty case involving alleged bitemarks. Included in the petition was this material described by the court:

> In support of this claim, Brewer presents the affidavit of Charles Michael Bowers, D.D.S., along with the draft of an unpublished article outlining the shortcomings of bite-mark evidence and the error rate as determined by the American Board of Forensic Odontology. He urges that this Court should not tolerate a science that, as Brewer claims, is more likely than not to identify the wrong suspect."[79] In that 2001 affidavit, Dr. Bowers wrote, "Recently, a study completed by the ABFO of participating members, regarding the reliability of bite mark identification evidence produced data on the accuracy of results in bite mark identification forensic casework. These results counter balance the years of assured self-confidence shown by the dentists testifying on bite mark

evidence. This study, called the ABFO Bitemark Workshop No. 4, underscores the gap between odontologists' opinion-based testimony and their ability to *correctly* identify a defendant as the creator of a bite mark. The median overall error rate is 12.5%—out of a maximum possible error rate of 27%. Thus, they were wrong nearly half the time they tried to identify the source of a bitemark. More specifically, it is their false positive error rate—the tendency to conclude that an innocent person's dentition matches the bitemark—that accounts for the bulk of that overall error rate. On average, 63.5% of the examiners committed *false positive errors* across the test cases. If this reflects their performance in actual cases, then inculpatory opinions by forensic dentists are more likely to be wrong than right." (Affidavit of Dr. Bowers in *Brewer v. State*[79])

Considering the ABFO position paper, and Arheart and Pretty's cautions about the study design, Bitemark Workshop 4 does not appear to have been, as Dr. Bowers claimed, a "study regarding the reliability of bite mark identification evidence," nor did it "produce data on the accuracy of results in bite mark identification forensic casework." The allegations that "they were wrong nearly half of the time" and "63.5% of the examiners committed *false positive errors* across the test cases" seem to directly misinterpret and contradict the information reported by Arheart and Pretty in the published results. In fact, in their paper Arheart and Pretty reported, "The ROC area calculated by the non-parametric trapezoidal method is 0.86, a fairly high accuracy, indicating that the examiners are able to correctly identify the dentition belonging to a particular bitemark. Bootstrap 95% confidence intervals are 0.82–0.91."[77] They also wrote, "This survey indicates that analysis of bitemark evidence is a relatively accurate procedure among experienced forensic odontologists when the results are examined in combination."[77] They concluded, "The results of the present survey indicate that bitemark examination is an accurate forensic technique, at least with cases such as used in this study."[77] These are very different conclusions from those reported by Dr. Bowers and those critics of bitemark analysis who later quoted him.

That the results of the ABFO Bitemark Workshop 4 have not been accurately reported by some is undeniable, and the effect of those inaccurate reports has been decidedly negative. The authors of the *Forensic Science International* paper correctly stated in closing, "This study, despite its limitations, has opened the debate into evidence-based forensic dentistry. Forensic odontologists must ensure that the techniques they employ are backed by sound scientific evidence and that the decisions they present in Court serve to promote justice and to strengthen the discipline. Committed to these high professional standards, the ABFO is proactive in the continuing education of odontologists."[77]

Ultimately, the publication of this information may have a more positive effect. Because the number of cases in the study was too few, the reasons can be neither properly analyzed nor clearly understood from these data. However, the results show an interesting distribution. A review of Arheart and Pretty's ABFO Bitemark Workshop, Table 3, p. 111, reveals that five of the thirty-two participants achieved diagnostic accuracy scores (AUC) of 1.0 on each case, that is, 100% overall, and another four achieved AUC scores of 0.99 or 99% overall. In ROC analysis, the area under the ROC curve defines the diagnostic accuracy. Sixteen, half of the participants, scored 92% or better overall, and twelve had no single score lower than 91%. It seems clear from these data, in spite of the shortcomings of the study's design, that some participants were able to accurately analyze the material and were clearly more skilled in analysis of that material than others. This information supports the shared opinion of this chapter's authors that bitemark analysis, when performed by *some* experienced forensic odontologists, following appropriate guidelines, can be a very accurate discipline. It also illustrates, conversely, that some forensic dentists should not be independently or individually responsible for bitemark analysis cases until their skills are sufficiently developed and demonstrated. These data further support the recommended requirements for seeking second opinions, the need for true proficiency testing for forensic odontologists in bitemark analysis, and mandatory remedial education for those not performing well on those proficiency tests. It may also indicate that the qualifications required to apply for board certification should be modified to include an increase in the level of bitemark analysis experience required and the mandated oversight of the bitemark-related activities of new diplomates.

14.4.5 The Totalitarian Ego

Some have said that the egocentricity of some forensic odontologists is a major factor in the generation of the problem cases in bitemark analysis. There are theoretical sociological and psychological bases for these opinions. Greenwald discussed the relationships between the ego and cognitive biases in his 1980 article.[80] The discussion centers on the effect that cognitive biases, especially egocentricity, beneffectance (accepting personal responsibility for desired but not undesired outcomes), and cognitive conservatism (resistance to cognitive change), have on the ego and human behavior. He theorizes that those biases combine to negative effect, especially in individuals involved in "higher level organizations of knowledge, perhaps best exemplified by theoretical paradigms in science."[80] In a later work Tavris paraphrases Greenwald, "The ego is a self-justifying historian, which seeks only that information that agrees with it, rewrites history when it needs to, and does not even see the evidence that threatens it."[81] That quote seems to accurately apply to the actions and statements of some forensic odontologists involved with problem cases. Greenwald, Tavris, and others assert that an integral part of cognitive conservatism, resistance to certain kinds of change, is the tendency toward susceptibility to confirmation bias, a persistent problem in the identification sciences. The Madrid error in fingerprint analysis seems a prime example.[82] That the ego and egocentricity of some forensic odontologists have contributed to errors in bitemark analysis problem cases is undeniable. Most, but not all, of those who bear the responsibility for the errors made in these problem cases steadfastly insist that *others do not or cannot see what they see*. They appear to have that cognitive disconnect consistent with the theories of Greenwald, a totalitarian ego.

Forensic odontologists must learn to deal with these effects before the consequences appear in their work. They must recognize personal signs of ego-related hazards, and take draconian steps to reduce expectation or confirmation bias. This can be greatly facilitated by continuing the development and positive modification of guidelines and protocols designed to minimize those effects. Additionally, periodically testing the proficiency of forensic odontologists in bitemark analysis must become a requirement.

14.5 Forensic Value of Bitemark Evidence

Bitemark analysis, like all of the forensic sciences, has played and should continue to play an important role in the criminal justice system. There will be new developments, new techniques, and new challenges. Although errors occur in all endeavors, forensic scientists have a greater responsibility to minimize errors because of the consequences to people's lives. Bitemark analysis has not been exempt from the challenges accompanying the advances in scientific techniques and the learning curves associated with them. In the majority of the bitemark cases reported over the last fifty plus years in the United States, this evidence has been used primarily to link a suspected biter to a specific victim. Bitemarks, unlike fingerprints and DNA, lack the specificity and durability that these two modalities employ; human teeth change over time and their uniqueness remains unproven. Bitemark evidence, however, has other advantages that are useful to the criminal justice system apart from the specifics of linking a specific individual to the crime or victim. The use of bitemark analysis in an abuse case is a practical example. Many abuse cases involve suspects that are in frequent contact with the victim. In those cases fingerprints and DNA are of almost no value, as they would be expected to be present in abundance. A bitemark, on the other hand, is not accidentally or casually inflicted and is an indication of intimate and violent interaction. If bitemark analysis can lead to the inclusion or exclusion of suspects, that is very powerful evidence indeed. For a bitemark to be useful for analysis it must contain abundant information and the teeth that made the mark must be very distinctive. The specificity of evidence is inversely proportional to its variables. Because of the high number of variables involved in bitemark evidence, the specificity of associating a bitemark to a single suspect is low.

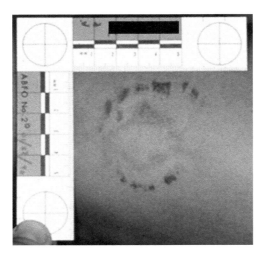

Figure 14.30 Bitemark with high forensic/evidentiary value showing class and individual characteristics. (Profile suggested that biter had a missing lower incisor and broken or malposed upper right incisors.) (See color insert following page 304.)

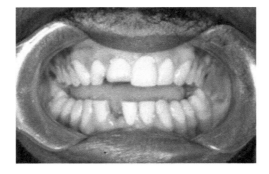

Figure 14.31 Suspect for bitemark seen in Figure 14.30. (This suspect was located, arrested, and confessed to the attack.)

Those variables include, but are not necessarily limited to, the elastic and other properties of human skin, the potential, constantly changing nature of human teeth, and the complex dynamics of the act of biting. This does not mean that bitemark evidence should not be used but that it should be used only in those cases for which it has merit, just as fingerprints and DNA should be used in those cases where their special advantages are useful. For bitemarks these useful areas most often involve showing evidence of violence. Bitemarks indicate violence, pain, a struggle, with the possibility of both offensive and defensive bites. Additionally and importantly, bitemarks in skin *may* produce a profile of the biter. If the individual and specific characteristics of the teeth are recorded in skin, forensic odontologists may be able to determine tooth position, spacing between teeth, broken or missing teeth, and other individualizing features (Figures 14.30 and 14.31).

With the use of technology developments and improvements, especially three-dimensional or pseudo-three-dimensional techniques such as laser scanning (LS), cone beam computed tomography (CBCT), or scanning electron microscopy (SEM), forensic odontologists may be able to discern even more detail in the bitemarks and in the individual teeth of biters. With bites in other substrates, be they foodstuffs such as cheese, cookies, bologna, or chewing gum, or nonfood items, like expanded polystyrene cups, pencils, or golf tees, with almost any item that can go into the mouth, forensic dentists may be able to isolate specific individual tooth characteristics that are found in the teeth of the suspected biter. Although bites in material other than skin do not indicate violence, they may have the ability to show greater detail of both class and specific individual dental characteristics.

The characterization of bites in human skin in relation to the time of death of victims is another area in which bitemark evidence may prove to be valuable. As in the 1975 Marx case with the bitemark on the nose, or in other cases where the three-dimensional nature of the marks, especially the retention of indentations from the teeth, played a prominent role, a forensic odontologist may be able to offer an opinion related to when the wound was inflicted in relation to the time of death. Forensic odontologists must be very careful to not overstate the significance or accuracy of this finding, and limit the opinion, if and when indicated, to "around the time of death." To state that the bitemark was inflicted at any specific interval of time prior to or after death is not currently scientifically supportable.

Advances in science when applied to bitemark analysis and interpretation are very likely to provide greater assistance to the examining and testifying expert and to enhance the value of evidence in specific cases.

14.6 Responsibilities and Consequences of Forensic Odontology Expert Testimony

As expert witnesses in criminal trials, forensic odontologists are under oath; they swear to tell "the truth, the whole truth and nothing but the truth." In fact, they are required to answer truthfully only the questions asked. It is the responsibility of attorneys to ask the appropriate questions to get the whole truth from expert witnesses. Experts for the prosecution in a criminal bitemark case may be pressured to provide a "positive link" between the putative biter and the bitemark. With the best evidence and a limited or closed population odontologists may be able to associate the two with reasonable medical or dental certainty, but *never* with absolute certainty. Terms stating or connoting absolute certainty ("Mr. X bit Ms. Y," "only one person in X million or billion could make such a mark," or "indeed and without doubt") are inappropriate and scientifically unsupportable. The meanings of reasonable scientific, medical, or dental certainty are consistent with the legal phrase "beyond reasonable doubt." The word *reasonable* in both terms implies that the person considering the question did so "to the best of his or her ability and after due consideration." A reasonable doubt must not be an imaginary or frivolous doubt, nor should it be based upon bias, sympathy, or prejudice. Rather, it should be based on logic, reason, and common sense. Determinations with reasonable dental/medical/scientific certainty, like determinations beyond reasonable doubt, should be logically derived from the evidence or absence of evidence. The burden on the prosecution expert in a homicide case is a grave one indeed. As we have seen from the tragic errors of the past, a misidentification of a bitemark can lead to the incarceration of an innocent person and, consequently, freedom for the guilty person, who can continue to kill. The consequences to the expert who misidentifies a bitemark or gives improper testimony can be terrible, legally, financially, emotionally, personally, and publicly. They may be barred from giving bitemark testimony in a court of law in the future. The expert may be subject to legal action brought by the person wrongly accused/convicted. He or she may be exposed in the media. Accounts may be written in newspapers, weblogs, magazines, and books, generating public hostility. As a result, bitemark evidence may be referred to as a "junk science" or worse. All this is in addition to how the expert must feel knowing he was even partly responsible for the incarceration of an innocent man. These are grave consequences for erroneous analysis and testimony. The defense expert has just as much responsibility to be truthful and objective as the expert for the prosecution, but errors by defense experts do not carry the same legal, financial, or public burden as the errors by a prosecution expert. Defense expert errors may contribute to a guilty suspect being freed, but the defense expert will rarely be publicly humiliated or sued. Defense expert errors are perceived to be less serious than those made by prosecution experts, partly because of the belief that it is better for many guilty persons to go free than to convict one innocent man. In 1970 the U.S. Supreme Court ruled that the highest standard of proof is grounded on "a fundamental value determination of our society that it is far worse to convict an innocent man than to let a guilty man go free."[83]

Expert testimony in civil bitemark cases differs from that in criminal cases. The burden of proof is different, requiring only "preponderance of evidence," not "beyond reasonable doubt." Expert testimony for a plaintiff if in error will never cause an innocent person to lose his life or liberty, only his money and maybe his reputation. The expert witness will rarely be the subject of legal action after a trial unless it can be proven he or she knew facts and lied about them. Bitemark testimony in tort cases is rare,

Bitemarks

and it usually involves domestic cases or civil action after a criminal case has been adjudicated. For example, childcare givers or facilities may be sued by parents whose child has been bitten. The question will likely be not who made the bite, but only the age of the biter: Was the biter a child or an adult? The plaintiff sues if their child was bitten by a caregiver. The defense will likely contend that the bite was from another child. The case may proceed or be dismissed hinging upon odontologists' expert opinions. ABFO guidelines recommend that forensic dentists consult other forensic dentists in bitemark cases. This practice is advisable but places responsibility on the dentist who is consulted, which may also have consequences. If consulted but not called to testify, and the testimony given by the primary forensic dentist results in posttrial lawsuits, the consulted dentist(s) may also be drawn into a difficult situation. The situation can be ameliorated by writing clear reports and keeping excellent records.

14.7 The Future of Bitemark Analysis

In April 2007, the National Academy of Science's Committee on Identifying the Needs of the Forensic Sciences Community sponsored an inquiry into forensic identification sciences, including bitemark analysis.[84] The committee interviewed a forensic odontologist representing organized forensic odontology education and the ABFO, asking questions about the current status and the future of bitemark analysis. The questions included: What is the state of the art? Where is research conducted? What is the scientific basis that informs the interpretation of the evidence? What are the major problems in the scientific foundation, methods, and practice? The responses to those questions and the conclusions relayed to the National Academy relate directly to the future of bitemark analysis. The standards for bitemark analysis must be closer to the state of the art. This means that forensic odontologists must be capable of using all known evidence collection and comparison modalities and select those modalities appropriate for the case in question. They should employ blinding techniques to inhibit bias and observer effects in all appropriate phases of their work. Competent forensic odontologists will seek second or multiple second opinions from other independent, blinded, competent forensic odontologists. They will engage in continuous study and research to improve themselves and forensic odontology and recognize and abide by appropriate codes of ethics and conduct. They understand the scientific method and use the method in tests and procedures to the greatest extent possible.

In February 2009, the National Academy of Sciences Committee released their report titled *Strengthening Forensic Science in the United States: A Path Forward*.[85] The 254 page pre-publication version of the report, henceforth referred to as the NAS report, contains a fairly comprehensive analysis of the current state of forensic science as it is practiced in the U.S. Chapter 5, Descriptions of Some Forensic Science Disciplines, details their surveys of specific areas of forensic science. Included in that list and beginning on page 5–35 of that chapter is a section titled Forensic Odontology. The only discussion of any aspect of forensic odontology other than bitemark analysis appears in the first paragraph, "Although the identification of humans remains by their dental characteristics is well established in the forensic science disciplines, there is continuing dispute over the value and scientific validity of comparing and identifying bite marks." (5–35) Many of the problems with the scientific foundation of bitemark analysis noted earlier in this chapter (Sections 14.4 and 14.6) are repeated in the NAS report. They summarize by stating, "Although the majority of forensic odontologists are satisfied that bite marks can demonstrate sufficient detail for positive identification, no scientific studies support this assessment, and no large population studies have been conducted." They also state, "In numerous instances, experts diverge widely in their evaluations of the same bite mark evidence. The committee received no evidence of an existing scientific basis for identifying an individual to the exclusion of all others." (5–37) They closed the section on Forensic Odontology by stating, "Some research is warranted in order to identify the circumstances within which the methods of forensic odontology can provide probative value."

The authors think that the assessment by the committee's report was a fair and reasonable evaluation in whole but included questionable conclusions drawn from the flawed information from the various reports and analyses of data from the now infamous Bitemark Workshop #4. This in no way absolves forensic dentists from the responsibility to perform research and establish scientific bases for bitemark analysis. The final NAS report conclusion seen at the end of the previous paragraph is mildly encouraging. The authors think that considerable research supported by funding is warranted and needed.

Bitemark analysis is too valuable to the investigation and adjudication of certain crimes to be abandoned, discounted, or overlooked. The use of bitemark analysis to exclude suspects is powerful and important. The scientific basis for associating unknown biters to tooth marks or bitemarks must be established. Currently, the association of one individual, *in an open population*, to a bite pattern on human skin to the highest level of association currently recommended by the ABFO, that is, to a reasonable dental, medical, or scientific certainty, and based on *pattern analysis alone*, cannot be scientifically supported. In closed or limited population cases, it *may* be possible to associate a biter and a bitemark with reasonable dental, medical, or scientific certainty *for that limited population.*

Forensic odontology certifying bodies must begin to properly test and periodically retest their certified members for proficiency in bitemark analysis. The requirements for board certification in North America, as they relate to bitemark analysis, are inadequate. Receiving board certification after being the principle investigator on one bitemark and co-investigator on another cannot be justified as sufficient experience. A remarkably modern list of recommendations was written as "Suggested Procedure for Future Cases" by Dr. Warren Harvey et al. following their analysis of the bitemarks in the 1967 murder of Linda Peacock. Included at the end of the list was this perceptive and prudent advisory statement, "Perhaps after the 5th or 6th case a forensic odontologist might have acquired the skill, knowledge and experience necessary properly to assess skin abrasions in bite-marks; lesser mortals will not lose face but will gain in wisdom by humbly sitting at the feet of a forensic pathologist who may have spent a lifetime specialising in this subject."[8] The language may be considered archaic but the sagacity is "spot on," and the "sitting at the feet" could be considered a metaphor for demonstrating an effort to control the "totalitarian ego."

The future of bitemark analysis depends upon those forensic odontologists who have impeccable ethics combined with protocols, procedures, and opinions firmly rooted in science, and their egos in check. They must have committed to continuously study, experiment, and learn, and if called upon to do so, they must have the vision, energy, and courage to make necessary changes.

References

1. Lerner, K.L.B. 2006. *Odontology, historical cases.* World of Forensic Science. http://www.enotes.com/forensic-science/odontology-historical-cases.
2. Merriam-Webster, Inc. 2005. *The Merriam-Webster dictionary*, xviii. Springfield, MA.: Merriam-Webster.
3. Burr, G.W. 1914. *Narratives of the witchcraft cases 1648–1706*, 216–23. New York: Charles Scribner's Sons.
4. Pierce, L. 1991. Early history of bitemarks. In *Manual of forensic odontology*, ed. D. Averill. 2nd ed. Colorado Springs, CO: American Society of Forensic Odontology.
5. Harvey, W. 1976. *Dental identification and forensic odontology*, xii. London: Kimpton.
6. Pitluck, H.M. 2000. *Bitemark case citations, bitemark management and legal update.* Las Vegas: S.W. 2d.
7. *Doyle v. State.* 1954. 159 Tex. C.R. 310, 263 S.W.2d 779.
8. Harvey, W., et al. 1968. The Biggar murder. Dental, medical, police and legal aspects of a case in some ways unique, difficult and puzzling. *J Forensic Sci Soc* 8:157–219.
9. *People v. Johnson.* 1972. 8 Ill. App.3d 457, 289 N.E.2d 772.
10. *People v. Marx.* 1975. 54 Cal. App.3d 100, 126 Cal. Rptr. 350.

11. *Bundy v. State.* 1984. 455 So.2d 330 (Fla. 1984), 349.
12. Souviron, R. 2008. Personal communication to D.R. Senn, editor.
13. Eskeland, S. 2008. Personal communication to D.R. Senn, editor.
14. Levine, L.J. 2008. Personal communication to D.R. Senn, editor.
15. Stimson, P.G. 2008. Personal communication to D.R. Senn, editor.
16. *People v. Milone.* 1976. 43 Ill. App.3d 385, 356 N.E.2d 531.
17. Morrison, H. 2004. *My life among the serial killers.* New York: Harper Collins.
18. *Milone v. Camp.* 1994. 22 F.3d 693 (7th Cir. 1994), 698.
19. *Wilhoit v. State.* 1991. 809 P.2d 1322 (Ct. of Crim. App. of Okla.).
20. Piakis, J. 2008. Personal communication.
21. Sperber, N.D. 2008. Personal communication to D.R. Senn, editor.
22. *State v. Krone.* 1995. 182 Ariz. 319, 897 P.2d 621 (Ariz. Sup. Ct.).
23. Piakis, J. 2008. Personal communication to D.R. Senn, editor.
24. Confidential personal communication to D.R. Senn, editor. 2008.
25. Rix, J. 2007. *Jingle jangle: The perfect crime turned inside out.* 1st ed. Zephyr Cove, NV: Broken Bench Press.
26. *People v. Moldowan.* 2002. 466 Mich. 862, 643 N.W.2d 570.
27. *Ege v. Yukins.* 2005. 380 F.Supp.2d 852, 878 (E.D. Mich. 2005).
28. *People v. Ege.* 1996. No. 173448, 1996 WL 33359075, at *1 n.1 (Mich. Ct. App. Sept. 17, 1996).
29. Stolarz, C. 2008. Warren man once cleared of rape now charged with molesting 5-year-old. In *Detroit News*, December 5.
30. West, M.H. 1992. Forensic report: Examination of Cristine Jackson. May 14.
31. *Brewer v. State.* 2002. 819 So.2d 1169 (Miss. 2002).
32. *Brooks v. State.* 1999. 748 So.2d 736 (Miss. 1999).
33. *State v. Van Winkle.* 1995. 658 So.2d 198 (La., 1995).
34. West, M.H. 2006. Letter of resignation from ABFO, June, 2006, M. H. West, D.D.S. Broadcast email.
35. ABFO. *ABFO Diplomates Reference Manual.* www.abfo.org.
36. Dorion, R.B.J. 2005. *Bitemark evidence.* New York: Marcel Dekker.
37. Vale, G.L., and Noguchi, T.T. 1983. Anatomical distribution of human bite marks in a series of 67 cases. *J Forensic Sci* 28:61–69.
38. Pretty, I.A., and Sweet, D. 2000. Anatomical location of bitemarks and associated findings in 101 cases from the United States. *J Forensic Sci* 45:812–14.
39. Freeman, A.J., Senn, D.R., and Arendt, D.M. 2005. Seven hundred seventy eight bite marks: Analysis by anatomic location, victim and biter demographics, type of crime, and legal disposition. *J Forensic Sci* 50:1436–43.
40. Rawson, R.D., Vale, G.L., Sperber, N.D., Herschaft, E.E., and Yfantis, A. 1986. Reliability of the scoring system of the American Board of Forensic Odontology for human bite marks. *J Forens Sci* 31:1235–60.
41. Vale, G.L., et al. 1988. Discussion of "Reliability of the scoring system of the American Board of Forensic Odontology for human bite marks." *J Forensic Sci* 33:20.
42. Pretty, I.A. 2007. Development and validation of a human bitemark severity and significance scale. *J Forensic Sci* 52:687–91.
43. Spitz, W.U., Spitz, D.J., and Fisher, R.S. 2006. *Spitz and Fisher's medicolegal investigation of death: Guidelines for the application of pathology to crime investigation,* xxx. 4th ed. Springfield, IL: Charles C. Thomas.

44. Dolinak, D., Matshes, E.W., and Lew, E.O. 2005. *Forensic pathology: Principles and practice*, xxii, 690. Amsterdam: Elsevier/Academic Press.
45. Christensen, H.L., and Alder, M.E. 1996. *Generating transparent bitemark overlays using a scanner, microcomputer, and laser printer*. Nashville, TN: F17, Odontology Section, American Academy of Forensic Sciences.
46. Sweet, D., and Bowers, C.M. 1998. Accuracy of bite mark overlays: A comparison of five common methods to produce exemplars from a suspect's dentition. *J Forensic Sci* 43:362–67.
47. Sweet, D., Parhar, M., and Wood, R.E. 1998. Computer-based production of bite mark comparison overlays. *J Forensic Sci* 43:1050–55.
48. Johansen, R., and Bowers, M.C. 2000. *Digital analysis of bite mark evidence: Using Adode Photoshop*. Santa Barbara, CA: Foresnic Imaging Institute.
49. Dailey, J.C. 2002. *The topographic mapping of teeth for overlay production in bite mark analysis*. Atlanta, GA: Odontology Section, American Academy of Forensic Sciences.
50. Foy, C.B., Ethier, J., and Senn, D.R. 2008. *Exemplar creation in bitemark analysis using cone beam computed tomography*. Washington, DC: F33, Odontology Section, American Academy of Forensic Sciences.
51. Thali, M.J., et al. 2003. Bite mark documentation and analysis: The forensic 3D/CAD supported photogrammetry approach. *Forensic Sci Int* 135:115–21.
52. Dalle Grave, C.M., dos Santos, A., Brumit, P.C., Schrader, B.A., and Senn, D.R. 2009. *Three dimensional analysis of human anterior teeth*. Denver: Odontology Section, American Academy of Forensic Sciences.
53. Sognnaes, R.F., et al. 1982. Computer comparison of bitemark patterns in identical twins. *J Am Dent Assoc* 105:449–51.
54. Rawson, R.D., et al. 1984. Statistical evidence for the individuality of the human dentition. *J Forensic Sci* 29:245–53.
55. MacFarlane, T.W., MacDonald, D.G., and Sutherland, D.A. 1974. Statistical problems in dental identification. *J Forensic Sci Soc* 14:247–52.
56. Kieser, J.A., et al. 2007. The uniqueness of the human anterior dentition: A geometric morphometric analysis. *J Forensic Sci* 52:671–77.
57. Saks, M.J., and Koehler, J.J. 2005. The coming paradigm shift in forensic identification science. *Science* 309:892–95.
58. Johnson, L.T., Radmer, T.W., and vanScotter-Asbach, P.J. 2006. *Quantification of individual characteristics of the human dentition*. Seattle: Odontology Section, American Academy of Forensic Sciences.
59. Johnson, L.T., Radmer, T.W., and Wirtz, T.S. 2008. *The verdict is in: Can dental characteristics be quantified?* Washington, DC: Odontology Section, American Academy of Forensic Sciences.
60. Metcalf, R.D., Brumit, P.C., Schrader, B.A., and Senn, D.R. 2008. *On the uniqueness of the human dentition*. Washington, DC: Odontology Section, American Academy of Forensic Sciences.
61. Pretty, I.A., and Sweet, D. 2001. The scientific basis for human bitemark analyses—A critical review. *Sci Justice* 41:85–92.
62. Rothwell, B.R. 1995. Bite marks in forensic dentistry: A review of legal, scientific issues. *J Am Dent Assoc* 126:223–32.
63. Barbenel, J.C., and Evans, J.H. 1974. Bite marks in skin—Mechanical factors. *J Forensic Sci Soc* 14:235–38.

64. Sheasby, D.R., and MacDonald, D.G. 2001. A forensic classification of distortion in human bite marks. *Forensic Sci Int* 122:75–78.
65. DeVore, D.T. 1971. Bite marks for identification? A preliminary report. *Med Sci Law* 11:144–45.
66. Millington, P.F. 1974. Histological studies of skin carrying bite marks. *J Forensic Sci Soc* 14:239–40.
67. West, M.H., et al. 1990. The use of human skin in the fabrication of a bite mark template: Two case reports. *J Forensic Sci* 35:1477–85.
68. Bush, M.A., Miller, R.G., Dorion, R., and Bush, P.J. 2008. *The role of the skin in bite marks: Biomechanical factors and distortion.* Washington, DC: F6, Odontology Section, American Academy of Forensic Sciences.
69. Miller, R.G., Bush, P.J., Dorion, R., and Bush, M.A. 2008. *The role of the skin in bite marks: Macroscopic analysis.* Washington, DC: F7, Odontology Section, American Academy of Forensic Sciences.
70. Bush, P.J., Miller, R.G., Dorion, R., and Bush, M.A. 2008. *The role of the skin in bite marks: Microscopic analysis.* Washington, DC: F8, Odontology Section, American Academy of Forensic Sciences.
71. Phillips, B.G., Bush, P.J., Miller, R.G., Dorion, R., and Bush, M.A. 2008. *The role of the skin in bite marks: Clothing weave transfer.* Washington, DC: F9, Odontology Section, American Academy of Forensic Sciences.
72. Bush, M.A., et al. 2009. Biomechanical factors in human dermal bitemarks in a cadaver model. *J Forensic Sci* 54:167–76.
73. ThinkQuest Education Foundation. Independent variables. Probability Central. http://library.thinkquest.org/11506/prules.html/.
74. Adams, B.J. 2003. Establishing personal identification based on specific patterns of missing, filled, and unrestored teeth. *J Forensic Sci* 48:487–96.
75. Steadman, D.W., Adams, B.J., and Konigsberg, L.W. 2006. Statistical basis for positive identification in forensic anthropology. *Am J Phys Anthropol* 131:15–26.
76. Souviron, R. 2008. Personal communication.
77. Arheart, K.L., and Pretty, I.A. 2001. Results of the 4th ABFO Bitemark Workshop—1999. *Forensic Sci Int* 124:104–11.
78. ABFO. 2003. Position paper on Bitemark Workshop 4. http://www.abfo.org/WorkSH42.htm.
79. *Brewer v. State.* 2002. MS 424 (MS, 2002).
80. Greenwald, A.G. 1980. The totalitarian ego: Fabrication and revision of personal history. *Am Psychol* 35:603–18.
81. Tavris, C., and Aronson, E. 2007. *Mistakes were made (but not by me)*, x. 1st ed. Orlando, FL: Harcourt.
82. Templeton, H. 2008. A statistical modeling approach to fingerprint identification …: The Madrid error. http://www.henrytempleman.com/madrid_error.
83. In the matter of Samuel Winship, appellant. 1970. 397 U.S. 358, 90 S.Ct. 1068, 25 L.Ed.2d 368.
84. Committee on Science, Technology, and Law. 2007. Identifying the needs of the forensic sciences community. http://www8.nationalacademies.org/cp/meeting-view.aspx?MeetingID=1846&MeetingNo=2.
85. National Research Council (U.S.) 2009. *Strengthening forensic science in the United States: A path forward.* Washington, DC: National Academies Press.

Abuse

The Role of Forensic Dentists

15

JOHN D. MCDOWELL

Contents

15.1 Introduction and Background	369
15.2 Child Maltreatment	370
15.3 Abuse during Pregnancy	371
15.4 Physical Injuries during Dating	372
15.5 Spouse Abuse (Intimate Partner Violence)	372
15.6 Elder Abuse	374
15.7 Conclusion	376
References	378

15.1 Introduction and Background

It has often been stated that intrafamily and intimate partner violence (IPV) is epidemic in the United States. Over the previous decades, best estimates indicate that somewhere between 20 and 50% of U.S. households have experienced some form of violence. Many of the injuries associated with inflicted (also termed nonaccidental or intentional) trauma are seen in the maxillofacial complex. These inflicted injuries are often treated on an immediate or delayed basis by general dentists, specialists within the dental profession, physicians, or ancillary members of the oral health care team. This chapter will provide useful information when determining if the facial/dental injuries are accidental in nature or if the trauma is more likely to be the result of nonaccidental trauma. The key features that are helpful in differentiating accidental injuries from inflicted injuries are:

1. Injuries at variance with history given
2. Injuries at various stages of healing
3. Interpersonal difficulties
4. Delayed presentation for care

When presented with injuries such that the history and physical examination suggest the trauma might not be accidental in origin, the oral health care provider should always include *inflicted trauma* in the differential diagnosis. Health care providers must remember that all forms of familial and intimate partner violence are interrelated. Since the sequelae of violence affects all age groups—especially those living together—one form of violent behavior cannot be separated from another.

Domestic violence and violence within the family—child abuse, spouse abuse, abuse/neglect of the disabled, and elder abuse—are common in Western society. Violent assaults can result in injuries to the developing fetus, children, adolescents, adults, and the elderly; in fact, no age group is free of the potential for violent acts. The most recent estimates from the Centers for Disease Control and Prevention indicate that approximately 25% of women have been raped or physically assaulted by an intimate partner.[1] Although most researchers in the field of domestic violence report that men and women are nearly equally involved in assaults, women experience more chronic injuries and assaults from men than men do from their female intimate partner. More than 40% of women who experience partner rapes and physical assault suffer some form of physical injury.[1] These soft and hard tissue physical injuries can take the form of bruises, lacerations, contusions, gunshot wounds, avulsion tissue injuries, broken bones (including the alveolar bone, jaws, and other bones of the maxillofacial complex), bitemarks, and fractured, subluxated, or avulsed teeth.

Intimate partner violence can take many forms, to include psychological/emotional, physical, and sexual abuse. Intimate partner violence can begin in the dating relationship, while cohabitating, while married, when separated or divorced, and can continue into the later years of life. Unfortunately, the violence can often result in death, serious physical injury, disfigurement, and emotional injuries.

15.2 Child Maltreatment

Most legal definitions of child abuse/maltreatment and neglect generally utilize the same or very similar criteria for classifying physical or emotionally injurious actions directed against children. One of the most useful definitions comes from the Child Abuse Prevention and Treatment Act (CAPTA), Public Law 104-235. CAPTA defines *child maltreatment* (abuse) as any act or failure to act resulting in imminent risk of death, serious physical or emotional harm, sexual abuse, or exploitation of a child by a parent or caretaker who is responsible for the child's welfare.[2] CAPTA more specifically defines *child abuse* as any physical act (i.e., burns, broken bones, etc.), sexual act (i.e., touching, fondling, sexual assault, or incest), or emotional insult (i.e., isolation, belittling, or calling names).[2] CAPTA defines *neglect* as happening

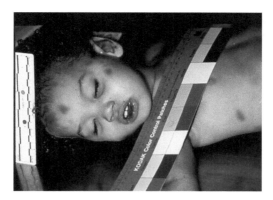

Figure 15.1 Example of abuse resulting in fatal injuries to four-year-old child.

when a parent or responsible caretaker fails to provide adequate supervision, food, clothing, shelter, or other basics for a child.[2] All forms of abuse and neglect may result in serious harm to a child (Figure 15.1).

In the United States, child protective services best estimates indicate that there are approximately eight hundred thousand annual cases of child maltreatment. Childhood exposure to abuse, neglect, and parental violence has been associated with risky behaviors, smoking, using illicit drugs, and overeating.[1] Depression, suicidal behavior, perpetrating youth and intimate partner violence, and negative health conditions, such as heart disease and cancer, have also been associated with child abuse and neglect.[1]

15.3 Abuse during Pregnancy

Although child abuse/neglect is one of the most common forms of violence, injuries can result from behaviors that begin prior to birth (during pregnancy). The effects of alcohol, drugs, poor nutrition, and physical trauma are well documented in the professional and lay literature. Assault of the pregnant women can result in a wide range of emotional and physical trauma, including the potential for injuries that result in death (homicide), perinatal death, low-birth-weight live births, and preterm delivery.[3] In two separate studies of nearly two thousand women aged eighteen to sixty-five, it was reported that approximately 15% of women reported intimate partner violence or abuse during a pregnancy.[3,4] Greater frequency of abuse was associated with increased health risks to the mother and her developing fetus.[3] Koening et al.[4] reported that women more commonly experienced violence during than after their pregnancy (61% of those women reporting abuse during or after pregnancy were abused only during their pregnancy, 21.7% were repeatedly abused, and 16.7% were abused only after delivery).[4] The American College of

Obstetricians and Gynecologists (ACOG) has produced an excellent publication that describes abuse during pregnancy.[5] The ACOG reports that during pregnancy, the abuser is more likely to direct blows at the pregnant woman's breasts and belly, sometimes resulting in maternal homicide or miscarriage.[5] However, while some of the inflicted trauma is directed at the breast and belly, the abusive acts can take many forms, including pushing, hitting, slapping, kicking, choking, beating, or attacking with a weapon.[5] Although these figures are alarming, not every pregnant women is the victim of abuse—in fact, most women are not abused during pregnancy. In some cases of women reporting being in an abusive relationship, abuse might actually decrease during pregnancy. Some women report that they actually feel safe only when carrying a child.[5] Since some of the trauma may be in areas not commonly seen by oral health care providers, dentists, hygienists, and ancillary personnel should be aware that there is a need for medical (or law enforcement) evaluation of pregnant women with suspicious maxillofacial injuries.

15.4 Physical Injuries during Dating

Violence in intimate (to include dating) relationships can begin at an early age. Dating violence (also termed physical dating violence [PDV]) has been defined as physical, sexual, or psychological violence within a dating relationship. A study of dating violence among students in grades 7 to 12 found that physical and psychological violence was 12 and 20%, respectively.[6] These self-reports of violence behaviors indicated that there was hitting, slapping, or some other form of physical harm during the dating period. Students with poorer grades (mostly Ds and Fs), blacks, non-Hispanics, and students from the northeastern United States were at greatest risk for dating violence and victimization.[6] In addition to the risk for physical injury and death associated with physical violence during dating, other forms of secondary risk associated with dating violence included sexual intercourse (protected and unprotected), attempted suicide, substance abuse, episodic heavy drinking, and physical fighting.[7,8] (Figure 15.2).

15.5 Spouse Abuse (Intimate Partner Violence)

Spouse abuse (intimate partner violence) is a major cause of morbidity and mortality in the United States. These data are supported by the Federal Bureau of Investigation (FBI) Supplementary Homicide Reports, 1976–2004.[9]

Abuse

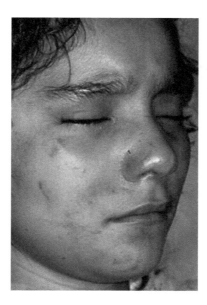

Figure 15.2 Physical injuries associated with a sexual assault on a fifteen-year-old girl. Note facial injuries including a patterned injury on right cheek (by history, a bitemark inflicted during the assault).

The report indicates that:

- Approximately one-third of female murder victims were killed by an intimate.
- Approximately 3% of male murder victims were killed by an intimate.
- Of all female murder victims, the proportion killed by an intimate declined slightly until 1995, when the proportion began increasing (most reports indicate that the rate has been stabilizing in recent years).
- Of male murder victims, the proportion killed by an intimate has dropped during the reporting period.
- Annually, at least fifteen hundred women were murder victims from domestic violence.
- Approximately one-third of injuries presenting to the emergency department were nonaccidental—the result of deliberate, intentional acts of violence.
- Approximately one-third of women over the age of fifteen who are homicide victims were killed by their husbands, ex-husbands, or boyfriends.

Homicides may also result in the murderer taking his or her own life following the violent act that resulted in the death of the intimate partner. It has been reported that 74% of all murder-suicides involved an intimate partner. Of these reported cases, 96% were females killed by their intimate partner, with 75% of these cases occurring within the home.[10]

Notwithstanding the physical injuries suffered during violent intimate partner relationships, there is also a significant financial burden placed upon families. Estimates indicate that the annual medical expense associated with domestic violence is at least $3 billion to $5 billion. Also, businesses are reported to lose another $100 million in lost wages, sick leave, absenteeism, and loss of productivity.[11]

Crandall et al.[12] have reported that women who suffered intentional blunt trauma exhibited very different injury patterns than those hospitalized for motor vehicle accidents and falls. The risk for facial injury was much higher among the domestic violence victims than was seen in other mechanisms of injury. Head injuries were also more common in women victims of intimate partner violence.[12]

The author's published thesis[13] also found that compared to women who were victims of motor vehicle accidents, women who were intentional trauma victims were more likely to:

- Present for care on a delayed basis (not presenting immediately after the incident causing the injuries associated with the chief complaint)
- Have had a previous facial/dental injury
- Have had a previous emergency department visit for injuries associated with intimate partner violence

Multiple injuries have also been reported to suggest intimate partner violence. One published study indicates that 85% of intimate partner violence victims were found to have injuries on more than one area of the body. The most common sites for injury were the eye, side of the face, throat and neck, upper and lower arms, upper and lower legs, mouth, outside of the hand, back, and scalp.[14]

Injuries to the shoulder and back were less common in intimate partner violence cases than were injuries to the shoulder and back found in those cases known to be caused accidentally. Of importance to dentists, 79% of the injuries were in areas clearly visible (injuries to the head and hands).[14] Very similar patterns were found in a study by Sheridan and Nash[15] and a study by Petridou et al.[16]

Intimate partner violence often overlaps in families and those in intimate relationships. An abused child often has an abused mother. An abused mother often has an abused child. Abuse knows no age group limitations and is seen in persons of all ages from the very young through the very old (Figures 15.3 to 15.5).

15.6 Elder Abuse

It is difficult to know the exact number of individuals over the age of sixty-five that are abused and neglected in the United States every year. However,

Abuse

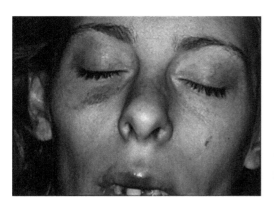

Figures 15.3 to Figure 15.5 Facial and oral injuries in victims of domestic violence. Note chronic inflicted facial trauma resulting in fractured and discolored teeth.

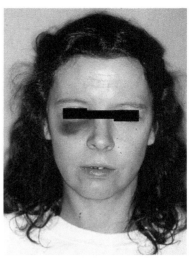

Figure 15.4

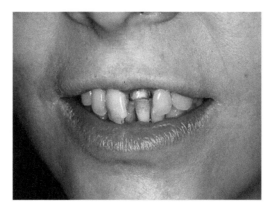

Figure 15.5

the best available estimates from the National Center on Elder Abuse (2005 report) indicate that:

- Between 1 and 2 million Americans age sixty-five and older have been injured, exploited, or otherwise mistreated by someone upon whom they depended for care.
- The frequency of elder abuse ranges from 2 to 10% based on various surveys.
- Approximately one in fourteen incidents (excluding incidents of self-neglect) of abuse or neglect come to the attention of authorities.
- For every case of elder abuse, neglect, exploitation, or self-neglect reported to authorities, about five go unreported.

- In the year 1996, nearly one-half million adults age sixty and over were abused or neglected in a domestic setting.

Like other forms of abuse, physical abuse of an elderly individual can appear in many different patterns. Some of the common signs can present in the head and neck complex. Other signs are traumatic hair and tooth loss, rope or strap marks indicating physical restraint, multicolored bruises indicating injuries at various stages of healing, and injuries suggesting healing "by secondary intention" (possibly indicating inappropriate or delayed presentation for care). The National Committee for the Prevention of Elder Abuse also reports that some of the indicators of elder abuse can include:

- Injuries that are unexplained or are implausible
- Family members providing different explanations of how injuries were sustained
- A history of similar injuries or numerous hospitalizations, or both
- Victims brought to different medical facilities for treatment to prevent medical practitioners from observing a pattern of abuse
- Delay between onset of injury and seeking medical care

Many of these indicators are very similar to those signs and symptoms of abuse/neglect seen in younger populations.

Injuries to the head and neck area are not uncommon in elder abuse. Zeitler reported that approximately 30% of known elder abuse cases presented with neck and facial injuries.[17] Injuries to the oral soft tissues, jaw fractures, and fractured or avulsed teeth have been reported to be indicators of elder abuse.[18] Oral health care providers must be aware that signs of intentional trauma are often seen in the orofacial structures (Figures 15.6 to 15.8).

15.7 Conclusion

Violence is a widespread problem in the United States. Many of the injuries associated with inflicted trauma are seen in the maxillofacial complex. Because of the injury locations, oral health care providers may be the first to have the opportunity to diagnose and treat the victims of nonaccidental (inflicted) trauma. This is valid for victims of all ages. Plans should be in place in each oral health care facility whereby intervention can appropriately begin on behalf of the suspected victim of violent behavior. Without intervention, the assaults may increase, leading to serious sequelae, including death by homicide.[19]

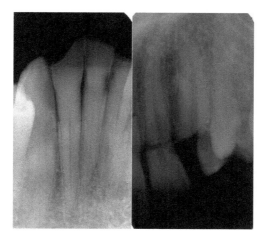

Figure 15.6 Periapical radiographs of fractured anterior teeth resulting from trauma in abuse.

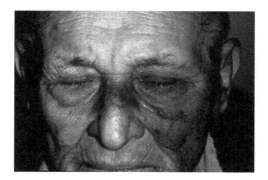

Figure 15.7 Ninety-three-year-old male with zygomaticomaxillary complex fracture. (See color insert following page 304.)

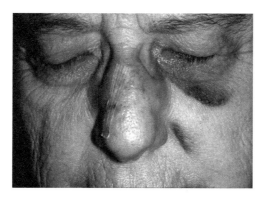

Figure 15.8 Eighty-two-year-old female with fractured nasal bones resulting from inflicted trauma. (See color insert following page 304.)

References

1. Centers for Disease Control and Prevention. 2007. *Preventing intimate partner violence, sexual violence and child maltreatment*, 1–7. cdc.gov/ncipc/pub-res/research_agenda/07_violence.htm.
2. Child Abuse Prevention and Treatment Act (CAPTA). Available through the National Clearinghouse on Child Abuse Information, 330 C Street, SW, Washington, DC 20447.
3. Coker AL, Sanderson M, and Dong B. 2004. Partner violence during pregnancy and adverse pregnancy outcomes. *Paediatric Perinatal Epidemiol* 18:260–69.
4. Koening LJ, Whitaker DJ, Royce RA, Wilson TE, Ethier K, and Fernandez MI. 2006. Physical and sexual violence during pregnancy and after delivery: A prospective multivariate study of women with or at risk for HIV infection. *Am J Public Health* 96:1052–59.
5. *Women's health: Domestic violence*. Available through ACOG's Copyright Clearing Center, 222 Rosewood Drive, Danvers, MA 01923.
6. Black MC, Noonan R, and Legg M. 2006. Physical dating violence among high school students—United States 2003. *MMWR* 19:532–35.
7. Grunbaum JA, Kinchen S, Ross J, Hawkins J, Lowry R, Harris WA, McManus T, Chyen D, and Collins J. 2003. Youth risk behavior surveillance—United States. *MMWR* 53:1.
8. Halpern CT, Oslak SG, Young MI, Martin SI, and Kupper LL. 2001. Partner violence among adolescents in opposite-sex romatic relationships: Findings from the National Longitudinal Study of Adolescent Health. *Am J Public Health* 91:1679–85.
9. FBI. 2006. Supplementary homicide reports, 1976–2004. http://www.ojp.usdoj.gov/bjs/homicide/intimates.htm.2004.
10. Violence Public Policy Center. 2006. *American roulette: Murder-suicide in the United States*. Washington, DC: Violence Public Policy Center.
11. National Center for Injury Prevention and Control. 2003. *Costs of intimate partner violence against women in the United States*. Atlanta (GA): Centers for Disease Control and Prevention.
12. Crandall ML, Nathens AB, and Rivara FP. 2004. Injury patterns among female trauma patients. Recognizing intentional injury. *J Trauma Injury Infect Crit Care* 57:42–45.
13. McDowell JD. 1993. A comparison of facial fractures in victims of motor vehicle accidents and battered women. MS thesis, University of Texas Graduate School of Biomedical Science, San Antonio.
14. Reijnders UJ, van der Leden ME, and de Bruin KH. 2006. Injuries due to domestic violence against women: Sites on the body, types of injury and the methods of infliction. *Nederlands Tijdschrift voor Geneeskunder* 150:429–35.
15. Sheridan DJ and Nash KR. 2007. Acute injury patterns of intimate partner violence victims. *Trauma Violence Abuse* 8:281–89.
16. Petridou E, Browne A, Lichter E, Dedoukou X, Alexe D, and Dessypris N. 2002. What distinguishes unintentional injuries from injuries due to intimate partner violence: A study in Greek ambulatory settings. *Injury Prev* 8:197–201.
17. Zeitler DL. 2005. Domestic violence. *J Oral Maxillofacial Surg* 63:20–21.
18. Wiseman M. 2008. The role of the dentist in recognizing elder abuse. *JCDA* 74:715–20.
19. Jackson A, Veneziano C, and Ice W. 2005. Violence and trauma: The past 20 and the next 10 years. *J Interpers Violence* 20:470–78.

Jurisprudence and Legal Issues

16

ROBERT E. BARSLEY
THOMAS J. DAVID
HASKELL M. PITLUCK

Contents

16.1 Legal Overview	379
16.1.1 Introduction	379
16.1.2 The American Legal System	380
16.1.2.1 Criminal Litigation	380
16.1.2.2 Civil Litigation	381
16.1.3 Court Systems	381
16.1.3.1 State Courts	381
16.1.3.2 Federal Courts	382
16.1.3.3 The U.S. Constitution	383
16.1.4 Court Decision Reporting Systems	383
16.2 Forensic Dentists and Criminal Litigation	384
16.2.1 Expert Witnesses	384
16.3 Forensic Dentists and Civil Litigation	385
16.3.1 The Civil Litigation Process	386
16.3.1.1 Dentists as Defendants	388
16.3.1.2 Dentists as Fact Witnesses	388
16.3.1.3 Dentists as Expert Witnesses	388
16.3.2 Standard of Care Litigation (Malpractice)	388
16.3.3 Personal Injury Litigation	389
16.3.4 Dentists as Expert Witness Defendants	390
16.4 Case Law	390
16.5 Conclusion	392
References	393

16.1 Legal Overview

16.1.1 Introduction

An understanding of the law and the functioning of the legal system is essential for anyone who practices forensic dentistry. Inevitably, the opinion expressed in a forensic dental case is likely to lead to testimony. This chapter will guide the reader through the legal and court systems of the American justice system, explain the need for, and the use of, the expert witness in legal proceedings, and conclude with a review of several case law examples in which forensic dental testimony played an important role.

16.1.2 The American Legal System

The American legal system has at its foundation several important principles. First and foremost, it is an adversarial system in which opposing parties enjoy representation by attorneys who advocate for their client's position. This occurs in criminal cases, civil cases, and administrative law cases. The heart of the advocate's representation is his or her duty to investigate fully the circumstances and events surrounding the legal action coupled with the ability to subject the opposing side's witnesses to a vigorous cross-examination, allowing the testimony of the witness to be thoroughly tested in front of the trier of fact—judge—or jury. A second principle is known as *stare decisis*, Latin for "to stand by that which is decided," or the principle to adhere to prior precedential court decisions furthering the maxim "rule of law, not by man." A third and bedrock principle is that the U.S. Constitution establishes the rights of citizens, which neither courts nor legislators can reduce or intrude upon.

The American legal system can be studied from diverse perspectives. One can consider the division between criminal cases and civil cases, focus on the differences between federal courts and state courts, or study the processes of the courts and how a legal case proceeds through the legal system. A brief discussion of each may be helpful.

16.1.2.1 Criminal Litigation

A forensic dentist will most commonly be involved in criminal cases such as the prosecution or defense of a homicide. In a criminal case the government (commonly referred to as the State or the People) brings an action to impose sanctions for violations of the criminal code. The prosecution team is usually led by an elected individual, often known as the district or state's attorney, who is empowered through the office to pursue charges against individuals who have been arrested or cited for alleged violations of the law. The prosecutor's office and those who represent him or her in court are not duty bound to secure convictions but rather to present the state's case fairly and demonstrate before the judge and jury that the accused has, by presentation of evidence, performed the elements of the crime. For instance, in a homicide, often defined as the killing of one human being by another, the prosecutor may require testimony from many sources to prove who was killed, how that individual was killed, when the killing occurred, that the accused had the opportunity to kill, and that the accused did in fact wield the instrumentality that resulted in death. The prosecution must prove each of the elements to the trier of fact, in most cases a jury, beyond a reasonable doubt. In the U.S., a verdict of not guilty cannot be appealed, nor can the accused be retried for the same crime or for any other one based on the same elements of the crime.

16.1.2.2 Civil Litigation

Civil cases include divorce and custody, sales and contracts, slander and libel, as well as negligence or tort cases, which include malpractice cases. These are actions brought by a private party against another private party. While any dentist might be sued by a patient with an allegation of malpractice, forensic dentists are often involved in these types of cases as an expert witness for either the defense or plaintiff. In order to prevail in a malpractice case, the plaintiff must demonstrate by expert testimony that the defendant dentist performed substandard treatment, that is, rendered care that failed to equal or exceed the standard of care. The defendant dentist's attorney must counter the allegations by providing the testimony of a dentist stating that the care was at least equal to or exceeded the prevailing standard. A mere preponderance of the evidence, 50% plus a little bit, is all that is required for the prevailing side. Either side can appeal the verdict—a successful plaintiff can complain that the damage award is insufficient, or an unsuccessful defendant dentist can argue the finding of liability itself or complain to the appeals court that the damage award is too high.

16.1.3 Court Systems

16.1.3.1 State Courts

It is important to realize that each state has its own system of sovereign courts. Most dental expert testimony will be given in cases that arise through state courts. A case arising and decided in one state system has no legal impact on how a similar case may be decided in any other state; however, there are similarities. Most states have a three-tiered system of courts. The district or trial court (local or regional) that enjoys original jurisdiction over civil and criminal cases that arise within a defined geographical boundary is the court in which the case is first heard, with evidence and testimony being presented. Cases in these venues are usually argued in front of a jury, although a defendant or the parties can sometimes agree to forgo a jury and allow the judge to act as the trier of fact and rule on the law. In cases such as divorce and child custody, a jury trial is not allowed in most jurisdictions. Sitting in review of the decisions rendered at this first level are the intermediate appellate courts, oftentimes known as the circuit court of appeal, although other names may be used. These courts, comprised of multiple judges, hear the initial appeal sought by parties dissatisfied with decisions from the district courts. No testimony is heard in these courts; rather, an appointed panel of some of the appellate judges receives written arguments from each party citing the errors and mistakes that are alleged to have occurred in the original trial. This panel may request oral arguments from the attorneys, but testimony from witnesses is not heard. The judges meet to discuss the case and render a written decision, which may uphold the original verdict, partially uphold the verdict, reverse the verdict, or order a new lower court trial in the case along with instructions to the district court on how or from whom testimony may be given. A party to the case at this level may, if desired, request that a larger panel of appeals judges rehear the case (often referred to as an *en banc* hearing), or may appeal the case to the final state court level—often known as the state supreme court. The supreme court may refuse to hear the case at all (cert. denied), or it may decide to take the case and hear arguments. Similar to the original appeal, no testimony is heard, the justices receive only the written briefs (arguments) of the attorneys, and once again an oral argument may be scheduled. After discussion, the justices issue a written opinion on the case that constitutes the final state action. The opinion may uphold the original verdict, partially uphold the verdict, reverse the verdict, or order a new trial in the case along with instructions to the district court on how or from whom testimony may be given. A party, as a final appeal, can request a rehearing by the state supreme court, an occurrence rarely granted. In some criminal cases, after a defendant has exhausted his or her appeals in the state courts, he or she may appeal certain matters to the federal court system. These federal appeals can take years to conclude, as evidenced by some capital or death penalty cases.

16.1.3.2 Federal Courts

A similar three-tiered arrangement is found in the federal court system. The U.S. District Courts retain original jurisdiction for questions of federal law that arise within their respective district, of which there are currently ninety-four, including locations within the District of Columbia, Puerto Rico, and the territories of Guam, the Virgin Islands, and the Northern Mariana Islands. A district may encompass an entire state or merely a portion of a state. A jury may or may not be empanelled to hear the witness testimony. Appeals from the verdicts in these courts proceed to the U.S. District Courts of Appeal, of which there are twelve, each usually encompassing all of the district courts in several states. Written arguments are made, oral arguments may be heard, but witness testimony is not heard at the appellate level. As in the state system, a group of judges is empanelled to hear and discuss each case, and a written decision is handed down. Once again, that decision can uphold, reverse, remand, or supplement the district court verdict.

A rehearing or an *en banc* hearing can be requested, or an appeal to the U.S. Supreme Court can be made. The U.S. Supreme Court accepts fewer than two hundred appeals annually from across the nation.

16.1.3.3 The U.S. Constitution

As stated above, the U.S. Constitution embodies the supreme law of this country. Each state has at the heart of its law a state constitution that can grant further rights to its citizens beyond those established in the U.S. Constitution but cannot reduce the rights granted by the U.S. Constitution. Who then is to decide questions of law within a state, between citizens of differing states, and any conflicts that might require a balancing of constitutionally guaranteed rights between the United States and a state constitution? In essence, matters involving the constitution, statutes, and regulations from a state or a political subdivision of that state affecting a citizen or citizens of that state (including crimes committed by anyone within that state) are resolved in the state court system. Similarly, cases involving federal criminal or other statutes and regulations (federal income tax, for example) are heard entirely within the federal court system. Civil matters involving citizens of different states may be heard entirely in the federal system, as well as matters that involve two or more state governments, or cases that involve disputes with federal agencies. Additionally, federal courts may hear cases alleged to involve a "federal question," usually an allegation that a right guaranteed by the U.S. Constitution has been abridged. Although the federal courts can remand cases that are brought into their system back to state courts for definitive action, cases usually remain in federal court. Oftentimes the final disposition of a state criminal case is decided by the actions of a federal court relative to whether or not the accused received a fair trial in accordance with the U.S. Constitution and the Bill of Rights.

16.1.4 Court Decision Reporting Systems

Over the years, the legal system has devised a useful shorthand method to reference the decisions of the various appeals courts. Prior to the widespread use of electronic documents, each appellate jurisdiction or each state judiciary would publish printed volumes containing the verbatim written decisions handed down. Over the years, the number of individual volumes was reduced as courts and even states banded together with commercial publishing firms to complete this important task. Today, most appellate decisions are posted on the various courts' own websites on the day of decision. However, the decisions are also "published" electronically in various media for distribution to law libraries, attorneys, and commercial publishers. An ability to interpret the shorthand will not only allow an individual access to the decision, but also impart knowledge as to which court decided the case and, in some cases, whether or not further appeal is likely. The most important piece of information is the *Reporter*—the title of the volume in which the case can be found. These are commonly abbreviated; for example, the *Southern Reporter*, Third Edition, is written as So.3d and includes state cases from several southern states. Federal appellate cases are found in the *Federal Reporter*, abbreviated as F, with F.3d representing the Third Edition. The reporters U.S. and SCT refer to cases from only the U.S. Supreme Court (one published by the court itself and the other by a commercial publisher). Listed before the volume or reporter listing is a number representing the exact volume, and listed after it is a number that is the page in the volume on which the opinion begins. Occasionally a second number follows (after a comma) which points the reader to the exact page upon which the legal point in question is discussed; for example, 509 U.S. 579, 585 (from the 1993 *Daubert v. Merrell Dow* U.S. Supreme Court case) alerts the reader to search for the material on page 585 rather than reading from the beginning (page 579). Next, within parentheses are the date of the decision, and if the case is from an intermediate-level appellate court, the name of that court will be noted. Finally, the abbreviation *cert. denied* may be listed indicating that an appeal has been refused. The advent of electronic databases and reporters such as WestLaw, Lexis-Nexis, and FindLaw coupled with the Internet may allow one to access a decision with as little information as one of the party names and the date or court.

16.2 Forensic Dentists and Criminal Litigation

16.2.1 Expert Witnesses

Because neither the judge nor the jury may fully understand the complexities in matters before the bar, the law allows for the use of the expert witnesses. Unlike other (fact) witnesses, an expert witness is allowed to testify or present his or her opinion. That opinion is based upon the expert's training, education, and experience. An expert may conduct tests or other activities that assist him or her in reaching that opinion. However, the expert's testimony and opinion must be grounded in accepted theory and practice. Although each state and the federal system have individual rules that dictate how expert testimony can be presented and used, most have a similar basis. The original case cited for acceptance of an expert arose in 1923 in *Frye v. United States* in the federal district court for the District of Columbia and involved a precursor to what is popularly referred to as the lie detector.[1] Because the theory underlying the device had never gained general acceptance in the scientific community, the testimony of the expert was not allowed to be presented. The use of this test—general acceptance—spread throughout the country with the practical effect that each judge would decide, often based on the statements of the expert himself or herself (or those from the opposing side), who and what could or could not be heard. But without published guidelines as a basis, the decision to hear or not hear an expert rested more in the mind of the judge than in the veracity of the science. Over the years, rising dissatisfaction with this standard culminated in the 1993 case *Daubert v. Merrell Dow Pharmaceuticals*.[2] This U.S. Supreme Court decision established a four-part test for expert testimony: (1) that the theory is testable (has it been tested?), (2) that the theory has been peer reviewed (peer reviewing usually reduces the chances of error in the theory), (3) the reliability and error rate (100% reliability and zero error are not required, but the rates must be reported), and (4) the extent of general acceptance by the scientific community. This case enunciated the guidelines that formally establish the trial judge as the "gatekeeper" to determine the admissibility of scientific evidence. Over the next several years two other cases refined the test for acceptance of an expert's testimony. The first, in 1997, *General Electric v. Joiner*, established the principle that absent manifest error, the decision of the trial judge in his role as gatekeeper to admit or not admit expert scientific testimony would not be disturbed on appeal.[3] A later 1999 case established that all expert testimony is subject to the *Daubert* rules.[4] The Federal Rules of Evidence, which apply in all federal courts, state in Rule 702: "If scientific, technical or other specialized knowledge will assist the trier of fact to understand the evidence or determine a fact in issue, a witness qualified as an expert by knowledge, skill, experience, training, or education, may testify thereto in the form of opinion or otherwise if (1) the testimony is based on sufficient facts or data, (2) the testimony is the product of reliable principles and methods, and (3) the witness has applied the principles and methods reliably to the facts of the case."[5] The majority of states now incorporate similar rules regarding expert testimony, although some cling to *Frye* or variations somewhere between the two.

16.3 Forensic Dentists and Civil Litigation

Much of the civil litigation in America is based on a legal concept called tort law. There are other branches of civil law encompassing contracts, property, wills and successions, trusts, divorce, and custody. A tort is defined as a legal or civil wrong. Ordinarily, a legal wrong is handled as a civil matter only. However, occasionally an intentional tort, such as an assault or battery, can also become a criminal matter. In addition to intentional torts, there are also negligent torts and strict liability torts. An accidental injury could be a negligent tort. However, product liability cases involve strict liability tort law. Unlike the criminal justice system, where charges are brought based on violations of the penal code, in the civil justice system, tort claims are made on the basis of injury or harm. Based on this allegation of injury or harm, a claim is made for damages. The remedy for these damages is monetary, unlike the criminal justice system, where life or liberty is at stake. Only certain types of damages are recoverable in the civil litigation system. Recoverable damages include loss of earnings, reasonable medical expenses, pain and suffering, and in some cases punitive damages.

Dentists may become involved in almost any type of civil litigation. However, as health care professionals, dentists are more likely to become involved in two types of civil litigation: standard of care (malpractice) and personal injury. Dentists may become involved in these types of litigation in many ways—often as a practicing dentist against whom malpractice is alleged or as an expert witness testifying on behalf of the plaintiff or the defendant dentist. A dentist may also be called to testify as a subsequent treating dentist (one who cared for the patient after the alleged malpractice occurred) or even as a dentist who cared for the patient prior to the incident in question. In these situations the dentist plays a role similar to an expert, from whom opinion testimony may be sought.

16.3.1 The Civil Litigation Process

As a defendant in a civil litigation matter, the process invariably begins with a summons. This serves the defendant with notification that a claim (or suit) has been filed. A legal answer to the summons or claim must be filed with the court in a timely fashion or a default judgment may be issued in favor of the complainant. Therefore, a dentist should immediately notify his or her insurance carrier whenever a claim or summons is received. Typically, the summons will be answered in writing and the response will trigger one of three actions by opposing counsel: (1) dismissal of the suit, (2) settlement of the suit, or (3) initiation of the discovery process. If either of the first two responses occurs, the process is essentially concluded.

However, if the suit is not dismissed or settled, the evidence gathering stage of the suit begins, and this process is called discovery. Plaintiff experts often become involved prior to filing suit, in order to determine whether a claim should be filed. However, defense experts usually do not become involved until after a case is filed and discovery begins. The discovery process allows evidence relevant to the suit to be gathered by both parties. This process is controlled by the court and is meant to eliminate surprises in court. By sharing information with both parties, settlement of the suit is encouraged before trial. The discovery of evidence takes place by means of subpoenas, interrogatories, and depositions. An expert witness must prepare for deposition and related matters with the same degree of attention to detail as if preparing for courtroom testimony.

Interrogatories are a formal set of written questions that are asked of one party by the other. Both parties are allowed to pose questions to each other. This clarifies matters of evidence and helps to determine what will be presented at trial. Depositions are a formal process for obtaining potential evidence under oath. This information is recorded for later use in court. There are two types of depositions for obtaining information: discovery depositions and evidentiary depositions. A discovery deposition is intended to "discover" evidence that may be used at trial. An evidentiary deposition, on the other hand, is meant to gather evidence that will be used as testimony at trial. It is often done if a witness is unavailable for appearance at trial and is often videotaped for presentation to the jury.

A subpoena is a written command issued by a court of proper jurisdiction that compels a person to appear at a certain time and place. In addition, there are also subpoenas that compel someone to bring certain items with him or her when he or she appears to answer a subpoena. This type of subpoena is called a subpoena *duces tecum*. The items to be produced must be specified in the subpoena, and they must also be in the possession of the individual receiving the subpoena *duces tecum*.

If settlement is not reached by the time the discovery process is concluded, then a trial takes place to determine the outcome of the case. The burden of proof is placed on the plaintiff in a civil suit. Unlike criminal court, where the burden of proof is "beyond a reasonable doubt," the burden of proof required by a civil court is "a preponderance of the evidence." Generally speaking, this is interpreted to mean more likely than not or slightly greater than 50%. The verdict rendered by a jury in a civil matter is either for or against the plaintiff. If the verdict is against the plaintiff, there are no damages and the matter is concluded. However, if the verdict is for the plaintiff, then the jury must decide how much to award the plaintiff in damages.

The decision of how much in damages to award to a plaintiff in a successful civil suit is based on several factors. In states using the concept of contributory negligence, the jury must decide whether the plaintiff did anything to contribute to his own injury. If the answer is no, then the plaintiff may be awarded 100% of the calculated damages. If the answer is yes, then the jury must then determine how much the plaintiff contributed to his or her own injuries. This determination is made on a percentage basis and the net award to the plaintiff is calculated by subtracting the amount of contributory negligence from 100%. Damages are awarded by the jury based on three different factors: (1) loss of earnings, (2) reasonable medical expenses, and (3) pain and suffering. Of these three factors, loss of earnings and reasonable medical expenses are generally straightforward determinations. However, determination of something as intangible as "pain and suffering" is often an emotional decision on the part of the jury and tends to vary a great deal among individuals. This is the reason that many states now "cap" or limit the amount that a plaintiff may receive for pain and suffering. The statutes in some states allow an additional damage award known as punitive damages to be levied in those instances that exceed the required threshold—such as wanton, willful, intentional misbehavior, or reckless disregard on the part of the defendant.

16.3.1.1 Dentists as Defendants

If a dentist has been named as a defendant in a civil suit brought by a patient alleging a violation of some standard of care related to diagnosis or treatment of their professional dental needs, the defendant dentist is served with a summons—a notification from the court that a suit has been filed against him or her and that a legal answer is required within a deadline. The summons is usually served in person; however, it may be served at the dentist's residence or last known address, and may in some cases be served by U.S. mail. If a legal answer to the complaint is not received by the deadline, the court may issue a default judgment granting the relief sought by the plaintiff.

16.3.1.2 Dentists as Fact Witnesses

A dental patient's attorney may request information on the patient's behalf as to the circumstances of alleged dental injuries. This may consist of only an informal discussion with the patient's attorney, providing copies of treatment records, or signing a notarized statement. This request may also be formalized by the issuance of a subpoena requiring testimony, in the form of either a deposition or actual trial testimony. In the case of deposition testimony, reasonable compensation for time away from the office should be agreed upon. Whenever sworn testimony is required, the matter of compensation should be broached with the party that issued the subpoena and whether or not testimony is sought as an expert or merely as a fact witness.

16.3.1.3 Dentists as Expert Witnesses

The dentist is retained (hired) by either plaintiff or defense counsel to determine if there are violations of standard of care (SOC) or dental injuries that have been caused by the negligence of others. After reviewing the evidence provided, the dentist is asked to render an opinion in the form of a verbal or written report. The dentist may then be asked to provide sworn testimony, in the form of either a deposition or testimony at trial. In order to testify as an expert, the dentist must qualify by means of special training, education, or experience. A well-qualified expert can expect to be compensated accordingly.

16.3.2 Standard of Care Litigation (Malpractice)

One of the most important concepts to understand in standard of care litigation is that a bad treatment outcome does not entitle the patient to compensation in the form of damages. In order for a plaintiff to prevail in a standard of care case, there are essential elements of professional negligence that must be proven by a "preponderance of the evidence." These essential elements are as follows:

Association: A professional relationship must exist between the plaintiff and the dentist.
Breach: The dentist must have violated one or more appropriate standards of care (SOC).
Causation: The violation(s) of SOC must be the cause of a negative outcome for the plaintiff.
Damage: There must be actual damages associated with a negative outcome.

If all of these elements cannot be proven to the satisfaction of the jury, the defendant will almost always prevail. Some plaintiff experts focus too much on breach of the standard of care (SOC) and overlook the other essential elements, not understanding the legal necessity to prove all four parts of the case. Usually, the most difficult parts of the case to establish are causation and damages. For this reason, it is often more prudent to examine the evidence for causation and damages and work backward from there. An expert in a SOC case should not act as an advocate for one side or the other, but rather determine whether there is evidence to prove the essential legal requirements of the case. The opinions of a dental expert must address what the legal system requires, not what the dental community desires.

16.3.3 Personal Injury Litigation

The mere fact that someone sustains an injury does not automatically make it someone else's fault. This concept is the essential basis of personal injury litigation. As is true with standard of care (SOC) litigation, there are essential elements that must be proven in court in order to prevail. The essential elements that must be established are:

Negligence of another: There must be negligence on the part of someone else.
Relevant negligence: The negligence of another must be related to the injury that is claimed (often chronologically).
Causation negligence: The injury in question would not have occurred except for the negligence of another.
Damages: There must be actual damages associated with the injury in question.

All four of these elements must be established in order to prevail in court. However, in today's litigious society, some individuals file meritless claims against corporations in the belief that these companies will settle for an amount less than the cost of going to trial. In many cases, this philosophy has been successful. The unfortunate consequence of this shortsighted strategy on the part of corporations is that their willingness to settle based on the amount of the claim has led to a dramatic increase in the number of personal injury claims filed. There is probably more fraud in personal injury claims than any other part of the civil litigation process.

16.3.4 Dentists as Expert Witness Defendants

An expert witness dentist can be named as a defendant in a civil suit brought by either of the parties in a case in which he or she provided expert witness opinions or testimony. The claim may be based on a deviation from accepted methodology in the field of expertise, rather than on a technically incorrect opinion itself. An expert witness has an obligation to conduct himself or herself within certain professional guidelines. Actions outside these guidelines tend to invite civil litigation. Professional guidelines for expert witnesses are often not as well recognized as those relating to the clinical practice of dentistry. One who acts as an expert witness must be aware of what is required in this regard. Based on the possibility of civil litigation, prudence would dictate sufficiently broad professional liability insurance coverage for these activities. In some cases, liability insurance covering clinical practice may extend to these activities also. However, do not make such an assumption unless a written clause in the policy or a policy rider states that forensic consulting is covered. Other possibilities for liability coverage for forensic consulting include government agency coverage, homeowner's umbrella coverage, or a separate policy for forensic consulting only. Dentists providing forensic consulting services for a government agency may have coverage as a government agent or be afforded qualified immunity in conjunction with official duties. Intentional misconduct by the expert may void any of these coverages or protections, similar to the rules on awarding punitive damages.

16.4 Case Law

Numerous specific cases concerning expert dental testimony in the field of bitemarks were discussed in Chapter 14. The foundation case establishing bitemarks in American jurisprudence is a 1954 Texas criminal case, *Doyle v. State*,[6] wherein the court accepted the testimony of a firearms examiner who had made plaster casts of a piece of cheese found at a crime scene that bore several bitemarks and another plaster cast of a piece of cheese bitten by the suspect in the case. The firearms examiner, using "caliper measurements," testified that both pieces of cheese had "been bitten by the same set of teeth." A local dentist who examined "plaster casts and photographs" testified that "all were made by the same set of teeth." In the case *Niehaus v. State of Indiana*, the court accepted the testimony of a dentist in his first bitemark case based upon his years of practice and teaching experience in conjunction with his training in the field, which consisted of attendance at lectures and advanced reading. The court stated, "The determination of whether or not a witness is qualified to testify as an expert lies in the sole discretion of the trial court and may not be set aside unless there is manifest error or abuse of discretion."[7]

Cases that focus on dental identification are much fewer in number in the reported appellate decisions. While many appeals mention the fact that dental identification was utilized to establish the identity of a victim, the issue itself is not part of the appeal argument and is only mentioned in passing. The following cases are among those that do comment upon dental identification. In *Boyle v. Brigano* a dentist who was also the coroner testified that the dental records and the teeth he examined in the body "matched … perfectly." The court concluded that the evidence presented was sufficient to support the identification, based in part on the thorough cross-examination of the dentist by defense counsel.[8] A federal case in Pennsylvania, in which a dentist identified a defendant by verifying that a missing tooth reported by a witness was in fact missing, was appealed as a violation of the defendant's Fourth and Fifth Amendment rights.[9] Citing two previous Supreme Court cases the court struck down the Fifth Amendment claim because the dental examination was neither testimonial nor communicative in nature. The Fourth Amendment claim was denied due to the lack of any reasonable expectation of privacy by a defendant over what a person knowingly exhibits (in this case his smile) to the public coupled with the fact that a dental examination does not constitute a "search" (quotation marks in original).

In the Alabama case *Chafin v. State* a dentist identified a skull without the use of mathematics/percentages, but rather by comparison and based on the outline of a single filling, the bone pattern, and the outline of another tooth. The court was satisfied that even without specific forensic training, the dentist was qualified to make the identification and that any objections by the defense should go to the weight given the testimony by the jury rather than to its admissibility.[10] In a 1968 Illinois case a defendant raised questions about the ability of the prosecution to trace the path and possession of a body discovered in a burning car to the autopsy table; the court was satisfied by the testimony of the funeral director who transported the body as well as testimony from other witnesses to the fire that the body examined was in fact that from the crime scene. Although there was minor variation in two antemortem dental charts used in the process, the court accepted the expert dentist's opinion without the use of probabilities or without any declaration that the body was in fact the victim to the exclusion of all other individuals in the world.[11] In another Alabama case[12] later affirmed by the Alabama Supreme Court,[13] a forensic dentist in a difficult identification case based upon comparison to antemortem photographs, the expert was admitted to testify based on his training, skill, and experience.[10]

In the Florida case, *Tompkins v. Singletary*,[14] and later affirmed on appeal,[15] a dental identification using casts, records, x-rays, and an unusual dental feature cited by the forensic dentist was sufficient to establish the identity. The California case *People v. Mayens*[16] confirms the business record foundation for admission of dental records used in identifications. A New York case affirmed the granting of an order to exhume cremains and deliver them to Drs. Michael Baden and Lowell Levine for possible dental identification.[17] The admission of the skull of the victim into evidence over the objection of the defense that it was gross and prejudicial was allowed in Oklahoma. Because the body was not recovered for over two years, dental identification was the sole means of positive identification.[18]

In Pennsylvania in 1971 a defendant was recognized and identified at a lineup due to a dental anomaly. The case is particularly interesting because no dentist was involved.[19] In *Nash v. NYC*,[20] the court affirmed that due to a delay in obtaining dental records, during which time the decedent's unidentified body was buried in "Potter's Field," the medical examiner's office had no special duty to the family to make the identification in a timely manner.

16.5 Conclusion

Forensic dentistry is both an interesting and serious (sometimes deadly serious) field of dentistry. The forensic dentist is afforded an opportunity to interact with individuals and systems outside the normal realm of dental practice. Both general dentists and dental specialists enter with equal footing in the field. However, success requires dedication and a willingness to learn and become comfortable with the legal system, the legal profession, law enforcement, and the world of the coroner/medical examiner. The forensic dentist must be dedicated to the pursuit of the truth and must adhere to the highest ethical standard. A good forensic dentist can, without breaching ethical standards, be a good witness—one that advances the cause of justice by presenting the truth on the stand and fulfilling the expert's role to educate the attorneys, the judge, and the jury about the dental facts at issue. In today's legal arena, where some juries have come to expect the razzle-dazzle spectacle of modern CSI as depicted on television, the dental expert must be cognizant of the fact that those very same CSI techniques can disprove expert opinions not soundly grounded in science and fact—as the reversal of over a dozen criminal convictions based at least in part on bitemark evidence proclaims.

References

1. *Frye v. United States*, 293 F. 1013 (D.C. Cir. 1923).
2. *Daubert v. Merrell Dow Pharmaceuticals*, 509 U.S. 579, 113 S.Ct. 2786, 125, L.Ed2d 469 (1993).
3. *General Electric Co. v. Joiner*, 522 U.S. 136, 118 S.Ct. 512, 139 L.Ed.2d 508 (1997).
4. *Kumho Tire v. Carmichael*, 526 U.S. 137, 119 S.Ct. 1167, 143 L.Ed.2d 238 (1999).
5. Federal Rules of Evidence, Rule 702 (as amended Apr. 17, 2000, eff. Dec. 1, 2000).
6. *Doyle v. State*, 159 Tex. C.R. 310, 263 S.W.2d 779 (1954).
7. *Niehaus v. State*, 359 N.E.2d 513, 265 Ind. 655 (Ind., 1977).
8. *Boyle v. Brigano*, 25 F.3d 1047 (6th Cir., 1994, *Cert. Denied*).
9. *United States v. Holland*, 378 F.Supp. 144 (E.D. Pa., 1974).
10. *Chafin v. State*, 330 So.2d 599 (1976, Ala. Cir. App.).
11. *People v. Mattox*, 96 Ill.App.2d 148, 237 N.E.2d 845 (Ill. App. 1 Dist., 1968).
12. *Dolvin v. State*, 391 So.2d 666 (Ala. Cr. App., 1979).
13. *Ex Parte Suis Dolvin*, 391 So.2d 677 (Ala. Sup. Crt.).
14. *Tompkins v. Singletary*, (1998 U.S. Dist. Lexis 22582, Fla. Mid. Dist.).
15. *Tompkins v. Moore*, 193 F.3d 1327 (11th Cir., 1999).
16. *People v. Mayens*, (2008 Cal. App. Unpub. Lexis 4375).
17. *Buntin v. Guardian Life Ins. Co. of America*, 521 N.Y.S.2d 258, 134 A.D.2d 473 (N.Y.A.D. 2 Dept., 1987).
18. *Sheker v. State*, 644 P.2d 560 (Okla. Crim. App., 1982).
19. *Com. v. Spencer*, 275 A.2d 299, 442 Pa. 328 (Pa., 1971).
20. *Nash v. NYC*, 806 N.Y.S.2d 446 (Sup. Crt. App. Team, 1st Dept, 2005).

Evidence Management   17

SCOTT HAHN

Contents

17.1	Introduction	395
17.2	Preparation	396
17.3	Approach the Scene	396
17.4	Secure and Protect the Scene	396
17.5	Initiate a Preliminary Survey	397
17.6	Evaluate the Odontological Evidence Possibilities	398
17.7	Prepare a Narrative Description	398
17.8	Depict the Scene Photographically	399
	17.8.1 Photography Basics	399
17.9	Prepare a Diagram/Sketch	399
17.10	Conduct a Detailed Search for Forensic Dental Evidence	402
17.11	Record and Collect the Odontological Evidence	402
17.12	Conduct a Final Survey	403
17.13	Release the Scene	404
17.14	Summary	404

17.1 Introduction

Evidence management is perhaps the single most overlooked forensic requirement by odontologists pursuing a case. Failure to comply with appropriate evidence management procedures can lead to significant adverse results during litigation. It is, therefore, incumbent upon all forensic dentists to become knowledgeable in evidence handling, not only for the ultimate courtroom success, but also for the profession as a whole. This knowledge base can be established via the clear and concise evidence management guidelines presented in this chapter.

The Federal Bureau of Investigation Evidence Response Teams (ERTs) are charged with responding to crime scenes and managing all aspects of evidence collection from those scenes. Evidence management protocols for forensic dentists can be directly derived from these well-established, courtroom-tested FBI ERT evidence handling procedures with only minor modifications to ensure the applicability to odontology. Following are twelve evidence management guidelines, adapted from FBI ERT procedures, which can be utilized by odontologists as a protocol set for dealing with forensic dental evidence.

17.2 Preparation

The forensic dentist should ensure that his or her professional involvement with any evidence collection event is appropriately sanctioned by the convening legal authority. All supplies and equipment required for the collection of forensic dental evidence should be on hand and ready for deployment to a remote crime scene if warranted. Professional credentials, business cards, and other forms of professional identification should be immediately available for presentation to law enforcement as required. The most important preparatory step in evidence management for forensic dentists is the establishment of excellent liaison with other forensic professionals and all law enforcement agencies within their jurisdictions well in advance of any evidence collection/interpretation event.

17.3 Approach the Scene

For the majority of forensic dental evidence collection events, the "scene" typically comes to the dentist. Odontologists operating out of a medical examiner's facilities (or consulting from their private offices) receive evidence from law enforcement or legal consultants that may have already been entered into some type of evidence management system. The mandate is to document the arrival of this physical evidence into the odontologist's custody. Should the dentist travel to a scene containing forensic dental evidence, his or her foremost responsibility is to ensure safe entry into the scene. In some instances the odontologist will be required to provide appropriate identification to on-scene law enforcement or emergency responder personnel, to sign into an entry log, and to don appropriate personal protective gear prior to entering the scene.

17.4 Secure and Protect the Scene

Given that forensic dental evidence is more frequently subject to odontologic analysis away from the scene of the original incident, securing and protecting the scene really comprises securing and protecting the evidence in the dentist's possession. This may simply entail the forensic dentist's compliance with the security protocols of the local medical examiner's office (i.e., signing in, utilizing restricted access card entry systems, completing chain-of-custody/examination certifications, etc.).

In cases where forensic dental evidence is delivered to the dentist's office, an office protocol should be established logging the evidence into the facility. Once in the facility, the evidence should be secured from public access. Reference as to how evidence is secured within the facility, as well as log entries detailing acceptance of evidence into and departure out of the facility, should be included in the final analytical report.

17.5 Initiate a Preliminary Survey

When evidence is delivered to the odontologist, the preliminary survey should begin by assessing the composition of the evidence. The dentist should ensure that any packing list, narrative description, or telephonic conversation describing the received evidence accurately depicts the items delivered. Extensive notes should be initiated at this time. Any discrepancies in the contents of the evidence delivery from the description provided to the forensic dentist should be resolved prior to initiating any further analysis. Similarly, this survey technique should be utilized at the medical examiner's office: Are the remains within the body bag consistent with the medical examiner's report with regard to sex, stature, race, cause and manner of death? Are all loose teeth accounted for in the body bag?

The preliminary survey at a remote crime disaster scene provides an organizational baseline for the odontologic examination. While the medical examiner's office or the dentist's private facility provides a controlled environment for forensic dental evidence examination, the remote crime disaster scene is subject to external environmental influences that may degrade the quality of forensic evidence available for analysis.

To that end, any transient evidence (i.e., avulsed teeth about to be washed down a storm sewer during a rain storm) should be identified at the preliminary survey stage, photographed in place, and collected as evidence in an expeditious manner. Once transient evidence is secured, the odontologist's role in the preliminary survey stage would entail cautiously walking through the scene, with the investigator in charge, in order to assess the forensic dental implications of the available physical evidence and the impact of environmental factors thereon.

Extensive notes should be taken documenting the scene's physical and environmental conditions. Notes can be written or captured via electronic dictation. Based upon this preliminary survey, the odontologist should determine what additional personnel, equipment, and appropriate methodology would be required to preserve and collect the forensic dental-specific evidence. It is important that the odontologist limit his or her participation at a crime scene to forensic dental evidence only.

17.6 Evaluate the Odontological Evidence Possibilities

This evaluation begins upon the odontologist's arrival at the scene (or the arrival of the evidence to the examination facility) and becomes detailed in the preliminary survey stage. This step allows the forensic odontologist to assess the type of odontological evidence encountered (if at all). Odontological evidence possibilities vary greatly depending upon the nature of the incident in question. The forensic dental evidence expected to be gleaned from skeletal remains for identification purposes would constitute a wholly different subset of odontological findings versus those expected to be present for a bitemark analysis case.

At this assessment stage, the forensic dentist can begin to "piece together" evidentiary relationships regarding not only the odontological evidence but also related physical evidence. Examples would include inadvertent movement of evidence prior to odontological evaluation, inappropriate or incorrect repatriation of skeletal remains (a female mandible with an obviously male skull), intentional (contrived) movement or positioning of odontological evidence, and the interaction of physical evidence with odontological evidence (Dr. Souviron's example of a bitemark victim's clothing, visible on initial crime scene photographs, impacting the bitemark analysis performed on the unclothed body of the victim in the morgue—see Chapter 14). The ability of the odontologist to provide this assessment is directly proportional to the access granted to all physical evidence or any remote crime scenes or body recovery sites.

17.7 Prepare a Narrative Description

This narrative is a running description of the conditions and evidence encountered by the odontologist and can be written, an audio recording, or a video recording. Use of a systematic approach in recording the narrative description and maintenance of the same system or methodology throughout the entire odontological assessment will produce the most consistent results. While no item is too insignificant to record, should it catch the odontologist's attention, the narrative should not be permitted to degenerate into a sporadic and unorganized attempt to record an entire crime scene or to speculate as to motive or psychological states of perpetrators. Stay in your own lane, and that lane is as a forensic dental consultant.

17.8 Depict the Scene Photographically

Photography should begin as soon as possible in order to preserve the odontological evidence in as near a pristine state as initially observed by the forensic dentist. Even though it is highly probable that the forensic dental evidence will have been handled multiple times (by either first responders, investigators, the general public, or scavengers) prior to initial examination by the forensic dentist, the initial photographs document the dental evidence at first presentation for forensic odontological analysis. Whether in the field or at a medical examiner's office, initial photography by the odontologist upon responding to a request for forensic dental analysis corroborates the status of dental evidence upon it entering into the forensic dentist's custody. Evidence should be photographed prior to being packaged or immediately upon receipt and the opening of a package (as in the case of remains being delivered to a medical examiner's facility).

17.8.1 Photography Basics

1. All exposures should be recorded on a photographic log (Table 17.1).
2. Exposures should be made from three basic aspects:

    Long range—a view depicting dental and physical evidence in the original environment
    Mid range—a view depicting the entire body or skeletal remains
    Closeup—a view depicting a specific body part, bitemark, oral cavity, or tooth

3. Scene photographs should be taken at eye level as observed from a "normal" view.
4. Closeup photographs should be taken with and without scale.
5. Digital photography
    A. If possible, secure the original media card as evidence.
    B. If original media card cannot be secured as evidence, perform a download of the exposures to a CD or DVD. The initial download should be secured as an evidentiary copy while subsequent downloads can be utilized as work copies.

17.9 Prepare a Diagram/Sketch

The diagram/sketch supplements photographs that are two-dimensional and can tend to distort distances and relationships between objects. A

Table 17.1 Photographic Log

| ROLL# _____ | PHOTOGRAPHIC LOG | PAGE _____ OF _____ |

GENERAL INFORMATION	CAMERA/FILM INFORMATION
DATE _____	CAMERA _____
CASE ID _____	TYPE OF FILM _____
LOCATION _____	RATING _____
PREPARER/ASSISTANTS _____	REMARKS _____

PHOTO#	USE OF SCALE	DESCRIPTION OF PHOTOGRAPHIC SUBJECT / MISCELLANEOUS COMMENTS
1		
2		
3		
4		
5		
6		
7		
8		
9		
10		
11		
12		
13		
14		
15		
16		
17		
18		
19		
20		
21		
22		
23		
24		
25		
26		
27		
28		
29		
30		
31		
32		
33		
34		
35		
36		

diagram/sketch can provide additional detail as to size of and distance between objects. Examples would include radius measurements of pattern injuries suspected to be bitemarks, distances between pattern injuries on a body, and distances between skeletal remains subject to intentional dismemberment or predation. A diagram/sketch can be utilized to emphasize targets of interest or eliminate unnecessary detail. An example of this would be a faint bitemark on the body of a deceased individual exhibiting advanced postmortem mottling of the skin. The bitemark could be enhanced on the

Table 17.2 Diagram/Sketch

DIAGRAM/SKETCH		PAGE ____ OF ____
GENERAL INFORMATION		**REFERENCE**
DATE _____ CASE ID _____ LOCATION _____ PREPARER/ASSISTANTS _____		SCALE or DISCLAIMER COMPASS ORIENTATION EVIDENCE FIXED OBJECTS MEASUREMENTS KEY/LEGEND

diagram/sketch as to position and location while the mottled skin could be eliminated.

The following should be included in any diagram/sketch (Table 17.2):

1. Specific location
2. Date
3. Time
4. Case identifier
5. Preparer/assistants
6. Scale or scale disclaimer (recommended to utilize "not to scale")
7. Compass orientation "in the field"; utilize the north arrow indicator
8. Anatomical orientation in the morgue setting
9. Identification of evidence in the sketch by name or item number
10. Table of relational measurements regarding evidence
11. Key or legend for specific measurement units if needed
12. Evidence item numbers correlated with evidence recovery log

17.10 Conduct a Detailed Search for Forensic Dental Evidence

In cases where the odontologist is requested to respond to a field event, the expertise inherent in being able to recognize dental structures and differentiate them from background debris and nonhuman remains becomes an invaluable search tool. This discriminatory eye for dental-specific evidence is also a key component of differentiating pattern injuries from bitemarks during a "search" of a human body. In either case, some basic tenets to conducting an effective search are applicable:

1. Utilize a repeatable search pattern, avoiding purposeless, meandering movements.
2. Search in layers recognizing odontological evidence may be under other evidence.
3. Photograph all evidence items prior to collection (utilize a photographic log).
4. Sketch all evidence items prior to collection.
5. All items collected by the odontologist should be entered on an evidence recovery log, witnessed by at least two individuals, and noted on a chain-of-custody document.
6. The most successful searches are typically the most difficult and time-consuming.
7. You cannot overdocument the collection of physical/odontological evidence.

17.11 Record and Collect the Odontological Evidence

Once all forensic dental evidence is identified via a detailed search, it should be collected for further odontological analysis. This again refers more to field operations and less to a box of skeletal remains delivered to your office. It is important to note that appropriate evidence collection and packaging conventions, as listed below, may not have been followed by certain entities dropping off the box of dental models, patient records, or skeletal remains for forensic dental analysis. The odontologist would be well served to make notes of any unusual or patently unacceptable packaging techniques, lack of substantiating documentation (evidence log, chain-of-custody document) regarding delivered evidence, or omissions of evidence with regard to any lists or logs provided to the examiner.

Once evidence is photographed, sketched in place, and entered onto an evidence recovery log by item number and detailed description (Table 17.3);

Table 17.3 Evidence Recovery Log

ITEM #	DESCRIPTION	WHERE FOUND	RECOVERED BY / OBSERVED BY (Last name)	PACKAGING METHOD	COMMENTS (if needed)

General Information: Date, Case ID, Location, Preparer/Assistants. Personnel (Include Initials).

it can be collected. Generally, biological evidence should not be placed in airtight plastic containers which can degrade DNA sampling and lead to advancing decomposition. Paper bags, cardboard boxes, or newer type Tyvek evidence bags are suitable for packaging odontological human remains. Follow American Board of Forensic Odontology (ABFO) protocols for collection and packaging of excised bitemarks.

Odontological evidence packaging should be marked as follows (indirectly—not on the evidence itself):

1. Case identifier
2. Date of collection
3. Item number
4. Item description
5. Location where item was found
6. Full name and initials of primary collector of item
7. Full name and initials of witness to primary collector

17.12 Conduct a Final Survey

This survey is a critical review of the ten previously covered evidence management protocols. By providing an actual step to review the evidence collection procedures and protocols employed by the forensic dentist relative to a specific case, it is hoped that crucial mistakes or omissions, potentially affecting any legal outcome, can be avoided or detected and corrected. The odontologist should ensure that all dental evidence was photographed and sketched prior to collection, and that all collected evidence is accounted for. Any instruments, tools, and work products utilized at the scene should be gathered and inventoried for completeness. Critical issues should be addressed, such as ensuring that the evidentiary search has encompassed all possible venues, all essential documentation has been accurately completed, and no assumptions were made that would prove to be incorrect in the future. If the above criteria are satisfied at the final survey, it is recommended that exit photographs be taken (and logged) depicting the final status of the search site, human remains, etc. These images could prove to be invaluable in future claims against the dentist for damages to persons or property purportedly as a result of the odontological examination.

17.13 Release the Scene

In the case of forensic odontology, the odontologist is released from the scene (or releases evidence previously turned in for dental analysis). Upon completing all necessary odontological evidence collection and analysis procedures, the forensic dentist's role in the processing of said evidence is terminated. The odontologist should ensure that lead investigators or referring professionals are aware of the completed collection/examination of the dental evidence, giving them one final chance to revisit any omissions on their part. There are not many legal alibis should something be overlooked during the execution of a search warrant; thus, it's best to get it right the first time.

17.14 Summary

The aforementioned evidence management guidelines, while not being all-inclusive, should provide an easy-to-follow framework for the forensic dentist during most aspects of odontological examinations and analyses. In practice, the twelve guidelines may occasionally be combined or adjusted for field expediency, resulting in a user-friendly protocol set that will serve as an adjunct to the existing evidence collection standards and practices in use by most forensic odontologists.

Future of Forensic Dentistry

18

DAVID R. SENN
PAUL G. STIMSON

Contents

18.1 A Look to the Past 405
18.2 Research and Technology 406
18.3 Certifying Organizations and Certification 407
18.4 The Way Forward 409
References 410

18.1 A Look to the Past

> If you want to understand the causes that existed in the past, look at the results as they are manifested in the present. And if you want to understand what results will be manifested in the future, look at the causes that exist in the present.
>
> —Buddhist aphorism in Dockett et al.[1]

Forensic dentistry's history reveals a checkered past featuring logical scientific and professional activities, less scientific and sometimes bizarre activity, and forthright abuse of scientific principles. The growth from ancient, old, and recent history into its current twenty-first century status has not been uniform. Similar to what is seen in human growth there have been great spurts and periods of relative indolence. The last thirty years has not been one of the quiet periods.

During this period the role of the forensic odontologist has transformed from one primarily involving occasional dental identifications in single and multiple fatalities into a more varied role. Forensic odontologists have become valuable members of teams involved in scientific investigations. Forensic dentists today routinely perform dental identifications, estimate age from oral and dental structures, analyze and compare patterned injuries or other patterned marks to the teeth that may have created them, and teach and participate in the recognition of and intervention in cases of abuse. Many participate in the investigations into illegal civil and criminal activity involving the practice of dentistry, and they testify in court about their expert findings in those cases. Many also testify in criminal cases involving identification, age estimation, bitemarks, and abuse. The opportunities for continued and expanded activity in the future for forensic dentistry are at hand. However, those opportunities are not without cost. They arrive with increased responsibilities that must be accepted by modern forensic odontologists that include a commitment to the highest standards.

18.2 Research and Technology

Although research into the bases for and techniques used by forensic dentists has been accomplished, except for in the area of age estimation, the volume of research, and especially the volume of significant research, is seriously lacking. The path to the future for forensic dentistry must include a more rigorous program of research. Forensic odontologists must become better scientists. The technological advances in dentistry offer opportunity for contemporaneous advances in forensic dentistry. The use of three-dimensional radiography has grown exponentially and is rich with possibilities for expanded modes of identification. More discriminate and reliable methods of age estimation are currently being researched and developed. More are needed. Significant efforts are under way to develop and implement improved methods to decrease the numbers of unidentified bodies and to simultaneously lower the number of names on lists of missing persons, as many of these are the same people. These activities are long overdue but thankfully, at last, under way.

In no area of forensic dentistry is research more greatly needed than in the area of bitemark recognition, analysis, and comparison. All serious forensic odontologists must dedicate themselves to encouraging, supporting, and performing research into the bases for bitemark evidence analysis and comparison. The mistakes made in bitemark cases in recent history must be a wakeup call for all forensic dentists. The reluctance to consult, to utilize other forensic odontologists in the review of bitemark cases, must be discarded in an effort to prevent the problems that have occurred in past cases. Individuals convicted of serious crimes have endured prolonged prison time after erroneous and unsupported "scientific" opinions were reviewed and brought back to courts with drastic results. Science, realistic protocol, and peer review must replace the confirmation of others' theories of crimes using unsupported and questionable analyses. Odontologists must be advocates for the truth and use the most current scientific methods available. The art and science of bitemark analysis must be performed in a more scientific manner. To prevent repeating the problems seen in some previous bitemark trials, forensic dentists must take steps that lead to the accomplishment of three goals:

1. The uniqueness of the biting surfaces of the anterior teeth must be clearly established by scientific research.
2. A searchable database of the features of the biting surfaces of the anterior human teeth must be developed that will allow the critical science-based comparison of the teeth of suspected biters to the patterned injuries *and* the teeth of a suspected biter to all other possible biters.
3. Methods of imaging, analyzing, and comparing the patterns created by teeth on human skin and accounting for the distortion involved must be dependable and repeatable.

Until these tasks are completed, forensic odontologists must limit the degree of certainty of their bitemark analysis comparison opinions. The ABFO has already stated in their Standards for Bitemark Terminology that "Terms assuring unconditional identification of a perpetrator, or without doubt, are not sanctioned as a final conclusion."[2]

The authors/editors agree with and emphasize the principle stated earlier in this book: "The association of one individual, *in an open population*, to a bite pattern on human skin to a reasonable dental, medical, or scientific certainty based on pattern analysis *alone* cannot be scientifically supported" (see Chapter 14).

Bitemark evidence is too potentially valuable to criminal investigations and prosecutions to be lost to the system of justice. Forensic odontologists must work to accomplish whatever tasks are required to ensure that this significant resource is not lost. Forensic dentistry must commit to performing the research that is needed and to continue to develop new techniques and refine existing methodology based upon sound scientific principles.

18.3 Certifying Organizations and Certification

In 1976 the American Board of Forensic Odontology (ABFO) began certifying dentists who had developed specified levels of knowledge, experience, and skill as board-certified forensic odontologists. Currently the ABFO is the only organization in North America that is accredited by the Forensic Specialties Accreditation Board (FSAB) to certify dentists in forensic odontology. From that 1976 date through 2009, 145 dentists have been certified and 97 remain active, including 6 of the 12 original founding members. The ABFO is a dynamic and maturing certifying organization.

As stated in the ABFO diplomates manual, "certification is based upon the candidate's personal and professional record of education, training, experience and achievement, as well as the results of a formal examination."[2] The purposes of the ABFO listed in the manual include in part:

a. To encourage the study of, improve the practice of, establish and enhance standards for, and advance the specialty of forensic odontology.
b. To encourage and promote adherence to high standards of ethics, conduct, and professional practice by forensic odontologists.
c. To grant and issue certification certificates, and/or other recognition, in cognizance of special qualifications in forensic odontology, to voluntary applicants who conform to the standards established by the Board and who have established their fitness and competence therefore [sic].[2]

These are appropriately lofty goals and purposes that have served the discipline well in its first thirty-two years. The requirements for applying to the board seem daunting to dentists seeking to become board-certified forensic odontologists, but may, in fact, be not as demanding as is needed. The current requirements include the following summarized from the ABFO manual:

Applicants must:

- Possess a dental degree, DDS, DMD, or equivalent
- Be persons of good moral character, high integrity, good repute
- Possess high ethical and professional standards
- Have attended four annual meetings of national forensic organizations
- Have participated by presenting, moderating, or chairing committees at those national forensic meetings
- Have been formally affiliated with a medical/legal agency for at least two years and be currently active
- Have observed five medicolegal autopsies

In addition to those general requirements the forensic dentist applicant must have experience in actual forensic odontological casework. Recently revised casework requirements are summarized below (bold italics added):

- A minimum of 25 forensic dental cases
- A minimum of 20 person identification cases
  - 15 of which resulted in positive dental identification
  - 10 in which the candidate personally performed the postmortem radiography
  - 5 in which the candidate personally resected or surgically exposed the jaws
- A minimum of ***2 bitemark cases*** in which the marks were bitemarks of human origin and were compared to a suspect's dentition

- In a minimum of *one of these cases* the dentist must have been the primary investigator of the bitemark and have performed the documentation of the evidence
- The remaining cases can be identification, bitemark, malpractice, personal injury, human abuse, peer review cases, or other cases of forensic dental interest
- In a *minimum of 2* of the 25 cases the applicant must have provided sworn testimony
- 350 points for other designated forensic dental activities

The requirements for human identification cases are logical and provide sufficient experience for dentists to become proficient in managing those cases. However, the current requirements allow a forensic dentist to become ABFO board certified and independently participate in criminal bitemark cases, make analyses, provide reports, and testify in trials as an expert witness after having been the primary investigator in as few as one previous bitemark case. Additionally, the requirements for giving sworn testimony do not mandate that even one of the two required be given in a bitemark case. Potentially, a newly board-certified forensic odontologist can appear in court to testify in only the second bitemark case for which he or she is the primary investigator. He or she would be testifying to material that may influence a jury that has the ability to sentence a person to punishments that include loss of liberty and, in federal courts and some state courts, death. As stated in Chapter 14, the authors and editors are of the opinion that the current requirements do not provide or ensure a level of experience in bitemark analysis, bitemark case management, and expert witness testimony to justify this level of responsibility. The ABFO certification requirements for bitemark cases should be made more rigorous, and there should be a probation period for new diplomates during which their bitemark case management should be in conjunction with one or more forensic odontologists with more experience.

Currently boarded forensic odontologists must be tested for proficiency in all phases of forensic dentistry in a manner that realistically tests their knowledge and skill and periodically be retested to ensure that they are remaining current and proficient.

18.4 The Way Forward

In this book the editors and contributing authors have attempted to critically and objectively examine the disciplines of forensic identification science, especially forensic dentistry. The above-mentioned "causes" of the past have manifested a present forensic odontology that can only be classified as a "mixed bag." Forensic odontologists have unquestionably provided valuable services to forensic science, the justice system, and the society they collectively serve. Many individuals and families have expressed their gratitude to forensic dentists who have helped with difficult problems at difficult times. However, forensic odontologists cannot deny that there have been mistakes. They cannot disregard that those mistakes have severely and negatively impacted fellow humans' lives. The wrongful conviction of a person for any crime not only deprives the innocent person of freedom or life but also leaves the actual criminal free to commit similar or worse mayhem—again. Giving an expert opinion that connects a person in a cause-effect relationship to criminal activity is a tremendous responsibility that all forensic odontologists must consider very seriously.

The challenge to forensic odontologists is to embrace the positive accomplishments of the past, study and understand the errors that have occurred, and make the needed changes now, in the present, to optimize the future of forensic dentistry. Those changes include ensuring that scientific research is encouraged, supported, and performed to substantiate the procedures promoted and approved by forensic dentists and forensic odontology organizations. The forensic odontology certifying organizations must ascertain that their policies and procedures promote and ensure the highest levels of proficiency possible.

If these difficult but necessary steps are taken, the future of forensic dentistry should remain deservedly bright. With the concerted effort of those involved in the forensic sciences and forensic odontology, the editors are confident in the bright future of forensic odontology. We welcome you to join us in this confidence.

References

1. Dockett, K.H., G.R. Dudley-Grant, and C.P. Bankart. 2003. *Psychology and Buddhism: From individual to global community*, xv. International and Cultural Psychology Series. New York: Kluwer Academic/Plenum Publishers.
2. American Board of Forensic Odontology. 2008. *Diplomates reference manual*. http://www.abfo.org

Appendix
U.S. Federal and State Court Cases of Interest in Forensic Odontology

HASKELL M. PITLUCK
ROBERT E. BARSLEY

1. *Doyle v. State*, 159 Tex. Crim. 310, 263 S.W.2d 779 (Jan. 20, 1954).
2. *People v. Johnson*, 8 Ill.App.3d 457, 289 N.E.2d 722 (Nov. 16, 1972).
3. *Patterson v. State*, 509 S.W.2d 857 (Tex. Crim. App.) (Mar. 13, 1974).
4. *People v. Allah*, 84 Misc.2d 500, 376 N.Y.S.2d 399 (Nov. 20, 1975).
5. *People v. Marx*, 54 Cal.App.3d 100, 126 Cal.Rptr. 350 (Dec. 29, 1975).
6. *People v. Johnson*, 37 Ill.App.3d 328, 345 N.E.2d 531 (Apr. 7, 1976).
7. *People v. Milone*, 43 Ill.App.3d 385, 356 N.E.2d 1350 (Nov. 12, 1976). Same defendant as cases 181 and 207.
8. *Niehaus v. State*, 265 Ind. 655, 359 N.E.2d 513 (Jan. 25, 1977).
9. *State v. Routh*, 30 Ore.App. 901, 568 P.2d 704 (Sep. 12, 1977).
10. *People v. Watson*, 75 Cal.App.3d 384, 142 Cal.Rptr. 134 (Nov. 28, 1977).
11. *State v. Kendrick*, 31 Ore.App. 1195, 572 P.2d 354 (Dec. 12, 1977).
12. *People v. Slone*, 76 Cal.App.3d 611, 143 Cal.Rptr. 61 (Jan. 6, 1978).
13. *State v. Howe*, 136 Vt. 53, 386 A.2d 1125 (Mar. 15, 1978).
14. *State v. Garrison*, 120 Ariz. 255, 585 P.2d 563 (Sep. 20, 1978).
15. *State v. Bridges*, 123 Ariz. 255, 600 P.2d 756 (Aug. 2, 1979).
16. *U.S. v. Martin*, 9 M.J. 731 (NCMR 1979) (Aug. 7, 1979). Same defendant as case 31. Martin is also known as Monk in case 142.
17. *State v. Jones*, 273 S.C. 723, 259 S.E.2d 120 (Oct. 11, 1979).
18. *Deutscher v. State*, 95 Nev. 669, 601 P.2d 407 (Oct. 18 1979). Appellant is defendant in case 164.
19. *State v. Peoples*, 227 Kan. 127, 605 P.2d 135 (Jan. 19, 1980).
20. *State v. Sager*, 600 S.W.2d 541 (Mo. App.) (May 5, 1980).
21. *People v. Middleton*, 428 N.Y.S.2d 688, 76 A.D.2d 762 (Jun. 10, 1980).
22. *State v. Kleypas*, 602 S.W.2d 863 (Mo. App.) (Jul. 10, 1980).
23. *Exparte* Sue Dolvin, 391 So.2d 677 (Ala. Sup. Ct.) (Sep. 12, 1980).
24. *State v. Temple*, 302 N.C. 1, 273 S.E.2d 273 (Jan. 6, 1981).
25. *People v. Smith*, 443 N.Y.S.2d 551, 110 Misc.2d 118 (Jul. 24, 1981).
26. *People v. Geer*, 624, S.W.2d 143 (Mo. App.) (Sep. 22, 1981).
27. *People v. Middleton*, 54 N.Y.2d 42, 429 N.E.2d 100 (Oct. 27, 1981).
28. *Aguilar v. State,* 98 Nev. 18, 639 P.2d 533 (Jan. 28, 1982).
29. *Kennedy v. State*, 640 P.2d 971 (Okla. Crim. App.) (Feb. 3, 1982).
30. *State v. Turner*, 633 S.W.2d 421 (Mo. App.) (Mar. 2, 1982).

31. *U.S. v. Martin*, 13 M.J. 66 (CMA) (Apr. 19, 1982). Same defendant as in case 16. Martin is also known as Monk in case 142.
32. *State v. Green*, 305 N.C. 463, 290 S.E.2d 625 (May 4, 1982).
33. *Bludsworth v. State*, 98 Nev. 289, 646 P.2d 558 (Jun. 18, 1982).
34. *People v. Queen*, 108 Ill.App.3d 1088, 440 N.E.2d 126 (Jul. 13, 1982).
35. *Commonwealth v. Maltais*, 387 Mass. 79, 438 N.E.2d 847 (Aug. 4, 1982).
36. *Commonwealth v. Graves*, 310 Pa.Super 184; 456 A.2d 561 (Feb. 4, 1983).
37. *State of Kansas v. Galloway*, unpublished opinion (Mar. 26, 1983).
38. *People v. Jordan*, 114 Ill.App.3d 16, 448 N.E.2d 237 (Apr. 14, 1983).
39. *Miller v. State*, 448 N.E.2d 293 (Ind. Sup. Ct.) (May 6, 1983).
40. *State v. Stokes*, 433 So.2d 96 (La. Sup. Ct.) (May 23, 1983).
41. *People v. Dixon*, 191 Cal.Rptr. 917 (Cal. App. 4th Dist.) (Jun. 7, 1983).
42. *People v. Columbo*, 118 Ill.App.3d 882, 455 N.E.2d 733 (Jun. 24, 1983).
43. *State v. Sapsford*, 22 Ohio App.3d 1, 488 N.E.2d 218 (Nov. 9, 1983).
44. *Chase v. State*, 678 P.2d 1347 (Alaska App.) (Mar. 9, 1984).
45. *Marbley v. State*, 461 N.E.2d 1102 (Ind. Sup. Ct.) (Apr. 19, 1984).
46. *State v. Welker*, 683 P.2d 1110 (Wash. App.) (May 21, 1984).
47. *Bundy v. State*, 455 So.2d 330 (Fla. Sup. Ct.) (Jun. 21, 1984).
48. *People v. Smith*, 63 N.Y.2d 41, 468 N.E.2d 879 (Jul. 2, 1984).
49. *State v. Asherman*, 193 Conn. 695, 478 A.2d 227 (Jul. 17, 1984).
50. *People v. Schuning*, 125 Ill.App.3d 808, 466 N.E.2d 673 (Jul. 19, 1984).
51. *State v. Adams*, 481 A.2d 718 (R.I. Sup. Ct.) (Aug. 21, 1984).
52. *Southard v. State*, unpublished opinion, 1984 Ark. App. Lexis 1635 (Aug. 29, 1984). Southard is the appellant in case 66.
53. *Graves v. State*, 1984 Tex. App. Lexis 6097 (1st Div.–Houston) (Aug. 30, 1984).
54. *Maynard v. State*, 455 So.2d 632 (Fla. App.) (Sep. 13, 1984).
55. *People v. Jordan*, 103 Ill.2d 192, 469 N.E.2d 569 (Ill. Sup. Ct.) (Sep. 20, 1984).
56. *People v. Williams*, 128 Ill.App.3d 384, 470 N.E.2d 1140 (Oct. 22, 1984).
57. *State v. Perea*, 142 Ariz. 352, 690 P.2d 71 (Nov. 1, 1984).
58. *State v. Bullard*, 312 N.C. 129, 322 S.E.2d 370 (Nov. 6, 1984).
59. *Smith v. State*, 253 Ga. 536, 322 S.E.2d 492 (Nov. 16, 1984).
60. *People v. McDonald*, 37 Cal.3d 351, 690 P.2d 709 (Nov. 21, 1984).
61. *State v. Thornton*, 253 Ga. 524, 322 S.E.2d 711 (Nov. 21, 1984).
62. *Bradford v. State*, 460 So.2d 926 (Fla. App. 2nd Dist.) (Nov. 30, 1984).
63. *Tuggle v. Commonwealth*, 228 Va. 493, 323 S.E.2d 539 (Nov. 30, 1984). Appellant is the same defendant in cases 71, 224, and 233.
64. *People v. Bethune*, 484 N.Y.S.2d 577, 105 A.D.2d 262 (Dec. 31, 1984).
65. *People v. Queen*, 130 Ill.App.3d 523, 474 N.E.2d 786 (Jan. 11, 1985).
66. *Southard v. State*, 1985 Ark. Lexis 1911 (Ark. Sup. Ct.) (Apr. 1, 1985). Southard is the appellant in case 52.
67. *State v. Dickson*, 691 S.W.2d 334 (Mo. App.) (April 2, 1985).
68. *State v. Carter*, 74 N.C.App. 437, 328 S.E.2d 607 (May 7, 1985).
69. *Clemons v. State*, 470 So.2d 653 (Miss.) (May 29, 1985).
70. *Standridge v. State*, 701 P.2d 761 (Okla. Cr. 1985) (Jun. 6, 1985).
71. *Tuggle v. Commonwealth*, 230 Va. 99, 334 S.E.2d 838 (Sep. 6, 1985). Same defendant in cases 63, 224, and 233.
72. *State v. Ortiz*, 198 Conn. 220, 502 A.2d 400 (Dec. 31, 1985).

Appendix 413

73. *People v. Walkey*, 177 Cal.App.3d 268, 223 Cal.Rptr. 132 (Cal. App. 4th Dist.) (Jan. 23, 1986).
74. *Thornton v. State*, 255 Ga. 434, 339 S.E.2d 240 (Feb. 13, 1986).
75. *Wade v. State*, 490 N.E.2d 1097 (Ind.) (Apr. 3, 1986).
76. *People v. Vigil*, 718 P.2d 496 (Colo.) (Apr. 4, 1986).
77. *Commonwealth v. Cifizzari*, 397 Mass. 560, 492 N.E.2d 357 (May 14, 1986).
78. *State v. Bingham*, 105 Wash.2d 820, 719 P.2d 109 (May 15, 1986).
79. *Rogers v. State*, 256 Ga. 140, 344 S.E.2d 644 (Jun. 25, 1986).
80. *State v. Johnson*, 317 N.C. 343, 346 S.E.2d 596 (Aug. 12, 1986).
81. *People v. Prante*, 147 Ill.App.3d 1039, 498 N.E.2d 889 (Oct. 3, 1986).
82. *Smith v. State*, unpublished opinion, Texas Court of Appeals (Oct. 9, 1986).
83. *State v. Johnson*, 721 S.W.2d 23 (Mo. App.) (Oct. 14, 1986).
84. *State v. Stinson*, 134 Wis.2d 224, 397 N.W.2d 136 (Oct. 28, 1986).
85. *In re* the Marriage of Rimer, 395 N.W.2d 390 (Minn. App.) (Nov. 4, 1986).
86. *McCrory v. State*, 505 So.2d 1272 (Ala. Crim. App.) (Dec. 9, 1986).
87. *Marquez v. State*, 725 S.W.2d 217 (Tex. Cr. App.) (Jan. 14, 1987).
88. *Bundy v. Wainwright*, 808 F.2d 1410 (11th Cir.) (Jan. 15, 1987).
89. *DuBoise v. State*, 1987 Fla. Lexis 2701 (Fla. Sup. Ct.) (Feb. 19, 1987). Superseded by case 110.
90. *People v. Davis*, 189 Cal.App.3d 1177, 234 Cal.Rptr. 859 (Feb. 26, 1987).
91. *People v. Dace*, 153 Ill.App.3d 891, 506 N.E.2d 332 (Mar. 23 1987). Same defendant as case 213.
92. *People v. Drake*, 514 N.Y.S.2d 280, 129 A.D.2d 963 (Apr. 3, 1987). Drake is appellant in cases 302 and 364.
93. *State v. Vital*, 505 So.2d 1006 (La. App. 3rd Cir.) (Apr. 9, 1987).
94. *State v. Kendrick*, 47 Wash.App. 620, 736 P.2d 1079 (May 11, 1987).
95. *Ngoc Van Le v. State*, 733 S.W.2d 280 (Tex. App.–Houston) (May 14, 1987).
96. *People v. Wachal*, 156 Ill.App.3d 331, 509 N.E.2d 648 (May 29, 1987).
97. *State v. Moen*, 86 Ore.App. 87, 738 P.2d 228 (Jun. 24, 1987).
98. *Handley v. State*, 515 So.2d 121 (Ala. Crim. App.) (Jun. 30, 1987).
99. *Jackson v. State*, 511 So.2d 1047 (Fla. App.) (Aug. 7, 1987).
100. *State v. Crump*, 1987 Ohio App. Lexis 8380 (3rd Dist.) (Aug. 11, 1987).
101. *People v. Perez*, 194 Cal.App.3d 525, 239 Cal.Rptr. 569 (Aug. 26, 1987).
102. *Strickland v. State*, 184 Ga.App. 185, 361 S.E.2d 207 (Sep. 11, 1987).
103. *State v. McDaniel*, 515 So.2d 572 (La. App. 1st Cir.) (Oct. 14, 1987).
104. *Inman v. State*, 515 So.2d 1150 (Miss. Sup. Ct.) (Nov. 18, 1987).
105. *People v. Watson*, 521 N.Y.S.2d 548, 134 A.D.2d 729 (Nov. 19, 1987).
106. *People v. Hampton*, 746 P.2d 947 (Colo. Sup. Ct.) (Nov. 30, 1987).
107. *Busby v. State*, 741 S.W.2d 109 (Mo. App.) (Dec. 8, 1987).
108. *State v. Hasan*, 205 Conn. 485, 534 A.2d 877 (Dec. 15, 1987).
109. *Harward v. Commonwealth*, 5 Va.App. 468, 364 S.E.2d 511 (Jan. 19, 1988).
110. *DuBoise v. State*, 520 So.2d 260 (Fla. Sup. Ct.) (Feb. 4, 1988). Supersedes case 89.
111. *State v. Pierce*, unpublished opinion, 763 P.2d 16 (Kan. Sup. Ct.) (Feb. 19, 1988).
112. *Valenti v. Akron Police Dept. et al.*, 1988 Ohio App. Lexis 730 (Ohio App. 9th Dist.) (Mar. 2, 1988).
113. *People v. Howard*, 529 N.Y.S.2d 51, 139 A.2d 927 (Apr. 8, 1988). Same defendant as case 148.
114. *State v. Armstrong*, 179 W. Va. 435, 369 S.E.2d 870 (Apr. 22, 1988).

115. *Mitchell v. State*, 527 So.2d 179 (Fla. Sup. Ct.) (May 19, 1988). Same defendant as case 171.
116. *State v. Jamison*, unpublished opinion, 1988 Kan. Lexis 141 (Kan. Sup. Ct.) (Jun. 3, 1988).
117. *People v. Ferguson*, 172 Ill.App.3d 1, 526 N.E.2d 525 (Jun. 30, 1988).
118. *People v. Rich*, 755 P.2d 960 (Cal. Sup. Ct.) (Jun. 30, 1988).
119. *People v. Randt*, 530 N.Y.S.2d 266, 142 A.D.2d 611 (Jul. 5, 1988).
120. *State v. Kirsch*, unpublished opinion, Wisconsin Court of Appeals (Jul. 20, 1988).
121. *Commonwealth v. Jones*, 403 Mass. 279, 526 N.E.2d 1288 (Aug. 18, 1988).
122. *Andrews v. State*, 533 So.2d 841 (Fla. App. 5th Dist.) (Oct. 20, 1988).
123. *People v. Hernandez*, 253 Cal.Rptr. 199, 763 P.2d 1289 (Nov. 28, 1988).
124. *State v. Combs*, slip opinion, Ohio App. (Dec. 2, 1988).
125. *Commonwealth v. Edwards*, 521 Pa. 134, 555 A.2d 818 (Mar. 6, 1989).
126. *State v. Turner*, 1989 Tenn. Crim. App. Lexis 206 (Mar. 20, 1989).
127. *People v. Marsh*, 177 Mich.App. 161, 441 N.W.2d 33 (May 15, 1989).
128. *Commonwealth v. Thomas*, 522 Pa. 256, 561 A.2d 699 (Jun. 27, 1989). Thomas is the defendant in case 277 and the appellant in case 329.
129. *Bromley v. State*, 380 S.E.2d 694 (Ga. Sup. Ct.) (Jun. 30, 1989).
130. *Chaney v. State*, 775 S.W.2d 722 (Texas App.–Dallas) (Jul. 5, 1989).
131. *U.S. v. Covington*, ACM 27337 (Aug. 2, 1989).
132. *Green v. State*, 542 N.E.2d 977 (Ind.) (Aug. 30, 1989).
133. *Fox v. State*, 779 P.2d 562 (Okla. Crim. App.) (Aug. 30, 1989).
134. *State v. Mebane*, 19 Conn.App. 618, 563 A.2d 1026 (Sep. 9, 1989).
135. *State v. Worthen*, 550 So.2d 399 (La. App. 3rd Cir.) (Oct. 4, 1989).
136. *State v. Hill*, slip opinion, Ohio App. (Nov. 27, 1989). Same defendant in case 177.
137. *Cox v. State*, 555 So.2d 352 (Fla. Sup. Ct.) (Dec. 21, 1989).
138. *Commonwealth v. Henry*, 524 Pa. 135, 569 A.2d 929 (Feb. 8, 1990). Henry is same defendant as case 254.
139. *Litaker v. State*, 784 S.W.2d 739 (Tex. App.–San Antonio) (Feb. 21, 1990).
140. *People v. Bass*, 553 N.Y.S.2d 794, 160 A.D.2d 715 (Apr. 2, 1990). Bass is appellant in case 219.
141. *Bouie v. State*, 559 So.2d 1113 (Fla. Sup. Ct.) (Apr. 5, 1009).
142. *Monk v. Zelez*, 901 F.2d 885 (U.S. Ct. App. 10th Cir.–KS) (Apr. 25, 1990). Monk also known as Martin in cases 16 and 31.
143. *State v. Ford*, 301 S.C. 485, 392 S.E.2d 781 (May 7, 1990).
144. *People v. Calabro*, 555 N.Y.S.2d 321, 161 A.D.2d 375 (May 15, 1990).
145. *Williams v. State*, 790 S.W.2d 643 (Tex. Crim. App.) (Jun. 6, 1990). Williams is same defendant as cases 158, 173, and 182. Codefendant in case 165.
146. *Spence v. State*, 795 S.W.2d 743 (Tex. Crim. App.) (Jun. 13, 1990). Spence is same appellant as case 232.
147. *State v. Richards*, 166 Ariz. 576, 804 P.2d 109 (Aug. 7, 1990).
148. *Howard v. Kelly*, slip opinion (U.S. Dist. Ct. W.D. New York) (Sep. 18, 1990). Howard is same defendant as in case 113.
149. *Baker v. State*, 797 S.W.2d 406 (Tex. App.–Ft. Worth) (Oct. 19, 1990).
150. *State v. Gardner*, slip opinion (Tenn. Crim. App.) (Oct. 25, 1990).
151. *State v. Jackson*, 570 So.2d 227 (La. App. 5th Cir.) (Nov. 14, 1990). Jackson is appellant in cases 230 and 234.

152. *Mallory v. State*, 563 N.E.2d 640 (Ind. App.) (Dec. 10, 1990).
153. *Salazar v. State*, slip opinion (Tex. App.–Houston) (Jan. 10, 1991).
154. *People v. Cardenas*, 209 Ill.App.3d 217, 568 N.E.2d 102 (Jan. 16, 1991).
155. *Harris v. State*, 260 Ga. 860, 401 S.E.2d 263 (Feb. 28, 1991).
156. *State v. Wimberly*, 467 N.W.2d 499 (So. Dakota Sup. Ct.) (Mar. 20, 1991).
157. *Wilhoit v. State*, 809 P.2d 1322 (Okla. Crim. App.) (Apr. 16, 1991). Also reported same opinion at 816 P.2d 545 with appendix.
158. *Williams v. State*, 815 S.W.2d 743 (Tex. App.–Waco) (May 30, 1991). William is same defendant as cases 145, 173, and 182. Co-defendant in case 165.
159. *People v. Perkins*, 216 Ill.App.3d 389, 576 N.E.2d 355 (Jun. 28, 1991).
160. *State v. Edwards*, 1991 Ohio App. Lexis 3305 (8th Dist.) (Jul. 3, 1991).
161. *Adams v. Peterson*, 939 F.2d 1369 (U.S. Ct. App. 9th Cir.–OR) (Jul. 30, 1991). Opinion withdrawn Mar. 27, 1992. Same defendant as case 175.
162. *People v. Case*, 218 Ill.App.3d 146, 577 N.E.2d 1291 (Jul. 30, 1991).
163. *State v. Thomas*, 329 N.C. 423, 407 S.E.2d 141 (Aug. 14, 1991). Same defendant as case 242.
164. *Deutscher v. Whitley*, 946 F.2d 1443 (9th Cir.) (Oct. 15, 1991). Appellant is same as case 18.
165. *Washington v. State*, 822 S.W.2d 110 (Tex. App.–Waco) (Nov. 20, 1991). Codefendant in cases 145, 158, 173, and 182.
166. *State v. Correia*, 600 A.2d 279 (R.I. Sup. Ct.) (Dec. 5, 1991).
167. *People v. Stanciel*, 225 Ill.App.3d 1082, 589 N.E.2d 557 (Dec. 11, 1991). Same defendant as case 184.
168. *State v. Ukofia*, unpublished opinion, 1991 Minn. App. Lexis 1178 (Dec. 11, 1991).
169. *State v. Pearson*, 479 N.W.2d 401 (Minn. App.) (Dec. 31, 1991).
170. *Davasher v. State*, 308 Ark. 154, 823 S.W.2d 863 (Jan. 27, 1992).
171. *Mitchell v. State*, 595 So.2d 938 (Fla. Sup. Ct.) (Feb. 6, 1992). Appellant is same as case 115.
172. *State v. Joubert*, 603 A.2d 861 (Maine Sup. Ct.) (Feb. 21, 1992).
173. *Williams v. State*, 829 S.W.2d 216 (Tex. Crim. App. en banc) (Apr. 15, 1992). Williams is same defendant as cases 145, 158, and 182. Codefendant in case 165.
174. *State v. Williams*, 80 Ohio App.3d 648, 610 N.E.2d 545 (May 20, 1992).
175. *Adams v. Peterson*, 968 F.2d 835 (U.S. Ct. App. 9th Cir.–OR) (Jun. 24, 1992). Same defendant as case 161.
176. *Washington v. State*, 836 P.2d 673 (Okla. Crim. App.) (Jun. 29, 1992).
177. *State v. Hill*, 64 Ohio St.3d 313, 595 N.E.2d 884 (Aug. 12, 1992). Same defendant as case 136.
178. *People v. Dunsworth*, 233 Ill.App.3d 258, 599 N.E.2d 29 (Aug. 19, 1992).
179. *People v. Holmes*, 234 Ill.App.3d 931, 601 N.E.2d 985 (Sep. 8, 1992). Holmes is appellant as case 363.
180. *Freeman v. State*, 651 So.2d 573 (Ala. Crim. App.) (Sep. 18, 1992; released for publication Mar. 25, 1995).
181. *U.S. ex rel Milone v. Camp*, slip opinion (U.S. Dist. Ct. N.D. IL) (Sep. 29, 1992). Same defendant as cases 7 and 207.
182. *Williams v. State*, 838 S.W.2d 952 (Tex. App.–Waco) (Oct. 14, 1992). Williams is same defendant as cases 145, 158, and 173 above. Codefendant in case 165.
183. *Harris v. State*, unpublished opinion, 1992 Ark. App. Lexis 728 (Nov. 18, 1992).

184. *People v. Stanciel*, 153 Ill.2d 218, 606 N.E.2d 1201 (Nov. 19, 1992). Same defendant as case 167.
185. *People v. Blommaert*, 237 Ill.App.3d 811, 604 N.E.2d 1054 (Nov. 30, 1992).
186. *State v. Jones*, 83 Ohio App.3d 723, 615 N.E.2d 713 (Dec. 2, 1992).
187. *Davis v. State*, 611 So.2d 906 (Miss. Sup. Ct.) (Dec. 17, 1992).
188. *People v. Noguera*, 4 Cal. 4th 599, 842 P.2d 1160 (Dec. 28, 1992).
189. *R.M. v. Dept. of Health & Rehabilitation Services*, 617 So.2d 810 (Fla. App.) (Apr. 30, 1993).
190. *State v. Bennett*, 503 N.W.2d 42 (Iowa App.) (May 4, 1993).
191. *State v. Schaefer*, 855 S.W.2d 504 (Mo. App.) (Jun. 22, 1993).
192. *Spindle v. Berrong*, unpublished opinion, 1993 U.S. App. Lexis 15362 (10th Cir.–KS) (Jun. 24, 1993).
193. *U.S. v. Dia*, 826 F.Supp. 1237 (U.S. Dist. Ct. Ariz.) (Jul. 8, 1993).
194. *Murphy v. State*, slip opinion, not designated for publication (Tex. App.–Dallas) (Jul. 20, 1993).
195. *State v. Donnell*, 826 S.W.2d 445 (Mo. App.) (Sep. 21, 1993).
196. *State v. Williams*, 865 S.W.2d 794 (Mo. App.) (Oct. 13, 1993).
197. *Rodoussakis v. Hosey*, 8 F.3d 820 (U.S. Ct. App. 4th Cir.–WV) (Oct. 20, 1993).
198. *State v. Lyons*, 124 Ore.App. 598; 863 P.2d 1303 (Nov. 17, 1993). Same defendant as case 239.
199. *State v. Welburn*, slip opinion (Ohio App.) (Nov. 17, 1993).
200. *Verdict v. State*, 315 Ark. 436, 868 S.W.2d 443 (Dec. 20, 1993).
201. *Kinney v. State*, 315 Ark. 481, 868 S.W.2d 463 (Jan. 10, 1994).
202. *State v. Hodgson*, 512 N.W.2d 95 (Minn. Sup. Ct.) (Feb. 11, 1994). Same defendant as case 229.
203. *State v. Cazes*, 875 S.W.2d 253 (Tenn. Sup. Ct.) (Feb. 14, 1994).
204. *Mobley v. State*, 212 Ga.App. 293, 441 S.E.2d 780 (Feb. 26, 1994).
205. *People v. Gallo*, 260 Ill.App.3d 1032, 632 N.E.2d 99 (Mar. 18, 1994).
206. *Harrison v. State*, 635 So.2d 894 (Miss. Sup. Ct.) (Apr. 14, 1994).
207. *Milone v. Camp*, 22 F.3d 693 (7th Cir.–IL) (Apr. 21, 1994). Milone is the same defendant as case 7 and the appellant in case 181.
208. *Morgan v. State*, 639 So.2d 6 (Fla. Sup. Ct.) (Jun. 2, 1994).
209. *Commonwealth v. Alvarado*, 36 Mass.App.Ct. 604, 634 N.E.2d 132 (Jun. 3, 1994).
210. *State v. Hummert*, 183 Ariz. 484, 905 P.2d 493 (Ariz. App. Div. 1) (Jul. 26, 1994). Same defendant as case 246.
211. *People v. Brown*, 162 Misc.2d 555, 618 N.Y.S.2d 188 (N.Y. Co. Ct.) (Oct. 6, 1994).
212. *State v. Martin*, 645 So.2d 190 (La. Sup. Ct.) (Oct. 18, 1994). Martin is the appellant in case 281.
213. *U.S. ex rel Dace v. Welborn*, 1994 U.S. Dist. Lexis 15225 (U.S. Dist. Ct. N. D. IL) (Oct. 25, 1994). Dace is the same defendant as case 91.
214. *State v. Carpentier*, unpublished opinion, 1994 Minn. App. Lexis 1196 (Dec. 6, 1994).
215. *People v. Tripp*, 271 Ill.App.3d 194, 648 N.E.2d 241 (Mar. 10, 1995).
216. *Brim v. State*, 654 So.2d 184 (Fla. App.–2nd Dist.) (Apr. 12, 1995).
217. *State v. Warness*, 77 Wash.App. 636, 893 P.2d 665 (May 1, 1995).
218. *Chaplin v. McGrath and Donohue*, 626 N.Y.S.2d 294, 215 A.D.2d 842 (May 4, 1995).

Appendix 417

219. *Bass v. Scully*, 1995 U.S. Dist. Lexis 22115 (U.S. Dist. Ct. E. D. NY) (May 25, 1995). Bass is the same defendant as case 140.
220. *People v. Rush*, 630 N.Y.S.2d 631 (N.Y. Sup. Ct.–Kings Cty.) (Jun. 7, 1995).
221. *State v. Mann*, 1995 Ohio App. Lexis 2513 (8th Dist.) (Jun. 15, 1995).
222. *Purser v. State*, 902 S.W. 641 (Tex. App.–El Paso) (Jun. 15, 1995).
223. *State v. Krone*, 182 Ariz. 319, 897 P.2d 621 (Jun. 22, 1995).
224. *Tuggle v. Thompson*, 57 F.3d 1356 (U.S. Ct. App. 4th Cir.–VA) (Jun. 29, 1995). Same appellant as cases 63, 71, and 233.
225. *State v. Boles*, 183 Ariz. 563, 905 P.2d 572 (Ariz. App. Di. 1) (Aug. 3, 1995).
226. *State v. Teasley*, 1995 Ohio App. Lexis 3372 (8th Dist.) (Aug. 17, 1995).
227. *People v. Cumbee*, unpublished, not precedential opinion (Ill. App. 2nd Dist.) (Nov. 15, 1995). Same defendant as case 342.
228. *Franks v. State*, 666 So.2d 763 (Miss. Sup. Ct.) (Nov. 30, 1995).
229. *Hodgson v. State*, 540 N.W.2d 515 (Minn. Sup. Ct.) (Dec. 15, 1995). Hodgson is the same defendant as case 202.
230. *Jackson v. Day*, slip opinion (U.S. Dist. Ct. E.D.–LA) (Jan 9, 1996). Jackson is the same defendant as case 151 and the appellant as case 234.
231. *People v. Shaw*, 278 Ill.App.3d 939, 664 N.E.2d 97 (Mar. 26, 1996).
232. *Spence v. Scott*, 80 F.3d 989 (U.S. Ct. App. 5th Cir.–TX) (Mar. 29, 1996). Spence is the same defendant as case 146.
233. *Tuggle v. Netherland*, 79 F.3d 989 (4th Cir.–VA) (Mar. 29, 1996). Tuggle is the same appellantant as cases 63, 71, and 224.
234. *Jackson V. Day*, slip opinion, U.S. Dist. Ct. E.D. LA.) (May 2, 1996). Jackson is same defendant as in case 151 and the appellant in case 230.
235. *People v. Payne*, 282 Ill.App.3d 307, 667 N.E.2d 643 (Jun. 19, 1996).
236. *State v. Wilkinson*, 344 N.C. 198, 474 S.E.2d 375 (Sep. 6, 1996).
237. *People v. Ege*, 1996 Mich. App. Lexis 1805 (Sep. 17, 1996). Ege is appellant in cases 328, 341, and 352.
238. *Government of the Virgin Islands v. Byers*, 941 F.Supp. 513 (U.S. Dist. Ct. V.I.) (Oct. 11, 1996).
239. *State v. Lyons*, 324 Ore. 256, 924 P.2d 802 (Oct. 11, 1996). Same defendant as case 198.
240. *State v. Hamilton*, unpublished slip opinion (Ohio App. 2nd Dist.) (Oct. 25, 1996).
241. *Johnson v. State*, 326 Ark. 430, 934 S.W.2d 179 (Oct. 28, 1996). Johnson is appellant as case 291.
242. *State v. Thomas*, 344 N.C. 639, 477 S.E.2d 450 (Nov. 8, 1996). Same defendant as case 163.
243. *Al-Mosawi v. State*, 929 P.2d 270 (Okla. Crim. App.) (Nov. 21, 1996).
244. *McGrew v. State*, 673 N.E.2d 787 (Ind. App.) (Nov. 27, 1996).
245. *Brown v. State*, 690 So.2d 276 (Miss. Sup. Ct.) (Dec. 12, 1996).
246. *State v. Hummert*, 188 Ariz. 119, P.2d 1187 (Ariz. Sup. Ct) (Mar. 11, 1997). Same defendant as case 210.
247. *Moye v. State*, unpublished opinion, 1997 Tex. App. Lexis 1952 (Dallas) (Apr. 16, 1997).
248. *Howard v. State*, 697 So.2d 415 (Miss. Sup. Ct.) (Jun. 26, 1997).
249. *Brown v. Commonwealth*, 25 Va.App. 171, 487 S.E.2d 248 (Jul. 8, 1997).
250. *Marquez v. State*, unpublished opinion, 1997 Tex. App. Lexis 3655 (Dallas) (Jul. 15, 1997).

251. *State v. Kiser*, 87 Wash.App. 126, 940 P.2d 308 (Jul. 28, 1997).
252. *Rios v. State*, unpublished opinion, 1997 Tex. App. Lexis 5691 (San Antonio) (Oct. 31, 1997).
253. *Banks v. State*, 725 So.2d 711 (Miss. Sup. Ct.) (Dec. 8, 1997).
254. *Commonwealth v. Henry*, 550 Pa. 346, 706 A.2d 313 (Dec. 23, 1997). Same defendant as case 138.
255. *State v. Tankersley*, 191 Ariz. 359, 956 P.2d 486 (Mar. 12, 1998). Same defendant as case 335.
256. *People v. Daniels*, 73 Cal.Rptr.2d 399 (Cal. App.) (Mar. 12, 1998).
257. *People v. Steward*, 295 Ill.App.3d 735, 693 N.E.2d 436 (Mar. 31, 1998).
258. *State v. Butler*, slip opinion, Mo. App. (Mar. 31, 1998). Same defendant as case 282.
259. *State v. Landers*, 969 S.W.2d 808 (Mo. App.) (May 26, 1998).
260. *Brewer v. State*, 725 So.2d 106 (Miss. Sup. Ct.) (Jul. 23, 1998).
261. *Walters v. State*, 720 So.2d 856 (Miss. Sup. Ct.) (Aug. 20, 1998).
262. *Middleton v. State*, 114 Nev. 1089, 968 P.2d 296 (Nov. 25, 1998).
263. *Waltman v. State*, 734 So.2d 324 (Miss. App.) (Feb. 23, 1999).
264. *State v. Fortin*, 318 N.J. Super. 577, 724 A.2d 818 (Mar. 1, 1999). Same defendant as cases 280, 314, and 351.
265. *State v. Mataya*, 226 Wis.2d 159, 594 N.W.2d 418 (Mar. 17, 1999). Mataya is appellant in case 317.
266. *State v. Anderson*, 350 N.C. 153, 513 S.E.2d 296 (Apr. 9, 1999).
267. *People v. Koberstein*, 639 N.Y.S.2d 366, 262 A.2d 1032 (Jun. 18, 1999).
268. *State v. Davidson*, 267 Kan. 667, 987 P.2d 335 (Jul. 9, 1999).
269. *State v. Timmendequas*, 161 N.J. 515, 737 A.2d 55 (Aug. 11, 1999).
270. *Brooks v. State*, 748 So.2d 736 (Miss. Sup. Ct.) (Oct. 7, 1999).
271. *People v. Wright*, 461 Mich. 905, 605 N.W.2d 314 (Nov. 2, 1999).
272. *State v. Kunze*, 97 Wash.App. 832, 988 P.2d 977 (Nov. 10, 1999).
273. *Cazes v. State*, 1999 Tenn. Crim. App. Lexis 1194 (Dec. 9, 1999).
274. *State v. Best*, 56 Conn.App. 742, 745 A.2d 223 (Dec. 14, 1999).
275. *State v. Calloway*, 528 S.E.2d 490 (Sup. Ct. of App. W. Va.) (Dec. 16, 1999).
276. *People v. May*, 702 N.Y.S.2d 393, 263 A.D.2d 215 (Jan. 6, 2000).
277. *Commonwealth v. Thomas*, 560 Pa. 249, 744 A.2d 713 (Jan. 18, 2000). Thomas is defendant in case 128. Appellant in case 329.
278. *Wilson v. Commonwealth*, 31 Va.App. 495, 525 S.E.2d 1 (Feb. 15, 2000).
279. *People v. Culuko*, 78 Cal. App. 4th 307 (Feb. 17, 2000).
280. *State v. Fortin*, 162 N.J. 517, 745 A.2d 509 (Feb. 23, 2000). Same defendant as in cases 264, 314, and 351.
281. *Martin v. Cain*, 206 F.3d 450 (U.S. Ct. App. 5th Cir.–LA) (Mar. 8, 2000). Martin is defendant in case 212.
282. *State v. Butler*, 24 S.W.3d 21 (Mo. App. W. Dist.) (Mar. 21, 2000). Same defendant as case 258.
283. *Carr v. State*, 728 N.E.2d 125 (Ind. Sup. Ct.) (Apr. 18, 2000).
284. *Otero v. Warnick*, 241 Mich.App. 143, 614 N.W.2d 177 (May 12, 2000).
285. *State v. Krider*, 530 S.E.2d 569 (N.C. App.) (May 16, 2000).
286. *Seivewright v. State*, 7 P.3d 24 (Wyo. Sup. Ct.) (May 31, 2000).
287. *State v. Wommack*, 770 So.2d 365 (La. App. 3rd Cir.) (Jun. 7, 2000).
288. *Parks v. Commonwealth*, unpublished opinion, 2000 Va. App. Lexis 597 (Aug. 15, 2000).

289. *State v. Prade*, 139 Ohio App.3d 676, 745 N.E.2d 475 (Aug. 23, 2000).
290. *Collman v. State*, 116 Nev. 687, 7 P.3d 426 (Aug. 23, 2000).
291. *Johnson v. State*, 342 Ark. 186, 27 S.W.3d 405 (Oct. 5, 2000). Johnson is appellant in case 241.
292. *State v. Blamer*, 2001 Ohio App. Lexis 444 (Feb. 6, 2001).
293. *In re* Pederson, 187 Misc.2d 486, 723 N.Y.S.2d 344 (Fam. Ct.) (Mar. 13, 2001).
294. *Brown v. State*, 798 So.2d 481 (Miss. Sup. Ct.) (Aug. 9, 2001).
295. *State v. Duncan*, 802 So.2d 533 (La. Sup. Ct.) (Oct. 16, 2001).
296. *Furtado v. State*, unpublished opinion, 2001 WL 959437 (Tex. App. El Paso) (Aug. 23, 2001).
297. *Carter v. State*, 766 N.E.2d 377 (Ind. Sup. Ct.) (Apr. 22, 2002).
298. *People v. Moldowan*, 466 Mich. 862; 643 N.W.2d 570 (May 15, 2002). Moldowan is appellant in case 347.
299. *Brewer v. State*, 819 So.2d 1169 (Miss. Sup. Ct.) (Jun. 6, 2002).
300. *U.S. v. Randolph Valentino Kills in Water*, 293 F.3d 432 (U.S. Ct. App. 8th Cir.) (Jun. 7, 2002).
301. *Keko v. Hingle et al.*, 318 F.3d 639 (U.S. Ct. App. 5th Cir) (Jan. 8, 2003).
302. *Drake v. Portuondo*, 321 F.3d 338 (U.S. Ct. App. 2nd Cir.–NY) (Jan. 31, 2003). Drake is the same defendant as case 92 and the appellant in case 364.
303. *People v. Mostrong*, 2003 Cal. App. unpublished Lexis 1179 (Cal. App. 4th Dist.) (Feb. 4, 2003).
304. *Stubbs v. State*, 845 So.2d 656 (Miss. Sup. Ct.) (Mar. 20, 2003).
305. *Howard v. State*, 853 So.2d 781 (Miss. Sup. Ct.) (Jul. 24, 2003).
306. *State v. McBride*, 666 N.W.2d 351 (Minn. Sup. Ct.) (Jul. 31, 2003).
307. *Wheeler v. Commonwealth*, 121 S.W.3d 173 (Ky. Sup. Ct.) (Aug. 21, 2003).
308. *Hernandez v. State*, 118 S.W.3d 469 (Tex. App. 11th Dist.–Eastland) (Aug. 29, 2003).
309. *People v. Moreno*, 2003 Cal. App. unpublished Lexis 8937 (Cal. App. 2nd Dist.) (Sep. 16, 2003).
310. *State v. Smith*, 2003 Ohio 5524, 2003 Ohio App. Lexis 4952 (Ohio Ct. App. 4th Dist.) (Oct. 8, 2003).
311. *People v. Quaderer*, unpublished opinion, 2003 Mich. App. Lexis 2958 (Nov. 25, 2003).
312. *State v. Arredondo*, 269 Wis.2d 369, 674 N.W.2d 647 (Wis. App. 1st Dist.) (Dec 23, 2003).
313. *Kunco v. A. G. of Pennsylvania*, unpublished opinion, 85 Fed.Appx. 819 (U.S. 3rd Cir.) (Dec. 29, 2003).
314. *State v. Fortin*, 178 N.J. 540, 843 A.2d 974 (Feb. 3, 2004). Same defendant as cases 264, 280, and 351.
315. *People v. Pre*, 11 Cal.Rptr.3d 739, 117 Cal. App. 4th 413 (Apr. 1, 2004).
316. *State v. Swinton*, 268 Conn. 781, 847 A.2d 921 (May 11, 2004).
317. *Mataya v. Kingston*, 371 F.3d 353 (U.S. Ct. App. 7th Cir. WI) (Jun. 3, 2004). Mataya is same defendant as case 265.
318. *Pardo and McLemore v. Simons et al.*, 148 S.W.3d 181 (Tex. App. 10th Dist.–Waco) (Jul. 28, 2004).
319. *O'Donnell v. State*, 5 Misc.3d 604, 782 N.Y.S.2d 603 (Ct. of Claims N.Y.) (Sep. 13, 2004).

320. *Gary v. Schofield*, 336 F.Supp.2d 1337 (U.S. Dist. Ct. M.D. Ga.) (Sep. 28, 2004). Same appellant as cases 337 and 353.
321. *Leal v. Dretke*, 2004 U.S. Dist. Lexis 22919 (U.S. Dist. Ct. W.D. TX) (Oct. 20, 2004). Same appellant as case 332.
322. *Moses v. State*, 893 So.2d 258 (Miss. App.) (Nov. 16, 2004).
323. *Garrison v. State*, 2004 OK CR 35, 103 P.3d 590 (Okla. App.) (Nov. 30, 2004).
324. *Boyd v. State*, 910 So.2d 167 (Fla. Sup. Ct.) (Feb. 10, 2005).
325. *Calhoun v. State*, 932 So.2d 923 (Ala. Crim. App.) (Apr. 29, 2005; corrected Jan. 9, 2007).
326. *Burke v. Town of Walpole et al.*, 405 F.3d 66 (U.S. Ct. App. 1st Cir.–MA) (Apr. 26, 2005).
327. *Meadows v. Commonwealth*, 178 S.W.3d 527 (Ky. App.) (Jun. 3, 2005; modified Jul. 8, 2005; released for publication Dec. 22, 2005). Same appellant as case 361.
328. *Ege v. Yukins*, 380 F.Supp.2d 852 (U.S. Dist. E.D. MI) (Jul. 22, 2005). Ege is defendant in case 237 and appellant in cases 341 and 352.
329. *Thomas v. Beard*, 388 F.Supp.2d 489 (U.S. Dist. E.D. PA) (Aug. 19, 2005). Thomas is defendant as cases 128 and 277.
330. *State v. McDermott*, 2005 Ohio 5233 (Ohio App. 2nd Dist.) (Sep. 30, 2005).
331. *State v. Neal*, 34 Kan.App.2d 485, 120 P.3d 366 (Sep. 30, 2005).
332. *Leal v. Dretke*, 428 F.3d 543 (U.S. 5th Cir.–TX) (Oct. 13, 2005). Same appellant as case 321.
333. *Dross v. State*, 915 So.2d 203 (Fla. App. 5th Dist.) (Oct. 21, 2005).
334. *State v. Marshall*, 2005 Ohio 5947 (Ohio App. 9th Dist.) (Nov. 9, 2005).
335. *State v. Tankersley*, 211 Ariz. 323, 121 P.3d 829 (Oct. 25, 2005). Same defendant as case 255.
336. *Patrick v. State*, 11 Misc.3d 296, 806 N.Y.S.2d 849 (Ct. of Claims N.Y.) (Dec. 5, 2005).
337. *Gary v. Terry*, 2005 U.S. Dist. Lexis 37996 (U.S. Dist. M.D. GA) (Dec. 23, 2005). Same appellant as cases 320 and 353.
338. *Gonzales v. State*, unpublished opinion, no. 08-04-00153-CR (Tex. App. 8th Dist.–El Paso) (Dec. 29, 2005).
339. *Simmons v. Runnels*, 2006 U.S. Dist. Lexis 2887 (U.S. Dist. Ct. E.D. Cal.) (Jan. 24, 2006).
340. *People v. Starks*, 365 Ill.App.3d 592, 850 N.E.2d 206 (Mar. 23, 2006).
341. *Ege v. Yukins*, 2006 U.S. Dist. Lexis 18986 (U.S. Dist. Ct. E.D. MI) (Apr. 12, 2006). Ege is the defendant as case 237 and the appellant as cases 328 and 352.
342. *People v. Cumbee*, 366 Ill.App.3d 476, 851 N.E.2d 934 (Jun. 30, 2006). Same defendant as case 227.
343. *Drayton v. State*, 935 So.2d 1290 (Fla. App. 2nd Dist.) (Aug. 25, 2006).
344. *State v. Anderson*, 2006 Ohio 4618, 2006 Ohio App. Lexis 4581 (Sep. 1, 2006).
345. *Howard v. State*, 945 So.2d 326 (Miss. Sup. Ct.) (Sep. 28, 2006).
346. *U.S. v. Studnicka*, 450 F.Supp.2d 680 (U.S. Dist. E. D. Tex.) (Oct. 6, 2006).
347. *Moldowan v. City of Warren et al.*, 2006 U.S. Dist. Lexis 82161 (Oct. 31, 2006). Moldowan is defendant as case 298.
348. *Bufkin v. State*, 207 S.W.3d 779; 2006 Tex. Crim. App. Lexis 2111 (Nov. 1, 2006).
349. *State v. Lester*, 2006 Mich. App. Lexis 3520 (Nov. 28, 2006).
350. *State v. Charley*, 2007 Ohio 1108; 2007 Ohio App. Lexis 1029 (Mar. 6, 2007).

Appendix

351. *State v. Fortin*, 189 N.J. 579, 917 A.2d 746 (Mar. 28, 2007). Same defendant as cases 264, 280, and 314.
352. *Ege v. Yukins*, 485 F.3d 364 (U.S. Ct. App. 6th Cir.–MI) (Apr. 24, 2007). Ege is the defendant in case 237 and the appellant as cases 328 and 341.
353. *Gary v. Schofield*, 493 F.Supp. 1255 (U.S. Dist. M.D. GA) (May 30, 2007). Same appellant as cases 320 and 337.
354. *Williams v. State*, 967 So.2d 735 (Fla. Sup. Ct.) (Jun. 21, 2007).
355. *Blackwelder v. State*, unpublished opinion, 2007 Tex. App. Lexis 6051 (Jul. 31, 2007).
356. *People v. Madrigal*, 2008 Cal. App. unpublished Lexis 612 (Jan. 24, 2008).
357. *Villa v. State*, 2008 Tex. App. Lexis 1025 (4th Dist.–San Antonio) (Feb. 13, 2008; released for publication Sep. 18, 2008).
358. *U.S. v. Lizarraga-Tirado*, 2008 U.S. Dist. Lexis 30013 (U.S. Dist. Ct. KS) (Apr. 10, 2008).
359. *State v. Kaufman*, 2008 Minn. App. unpublished Lexis 388 (Apr. 15, 2008).
360. *U.S. v. Cerno*, 529 F.3d 926 (U.S. Ct. App. 10th Cir.–NM) (Jun. 24, 2008).
361. *Meadows v. Commonwealth*, unpublished opinion, no. 2007-CA-000155-MR (Ky. App.) (Feb. 22, 2008). Same appellant as case 327.
362. *State v. Finley*, 2008 Ohio 4904, 2008 Ohio App. Lexis 4119 (Sep. 26, 2008).
363. *Holmes v. Pierce*, 2009 U.S. Dist. Lexis 681 (U.S. Dist. Ct. N.D. IL) (Jan. 7, 2009). Holmes is defendant as case 179.
364. *Drake v. Portuondo*, 553 F.3d 230 (U.S. Ct. App. 2nd Cir.–NY) (Jan. 23, 2009). Drake is defendant in case 92 and appellant as case 302.

Index

A

Abdomen, bitemarks, 337, 347, 372
ABFO. *See* American Board of Forensic Odontology (ABFO)
ABFO 2˚ scale, 211, 212
ABFO 3 scale, 211
Abrasion, 93, 287, 302
Abuse, 28–29
    child, 370–371
    dating, 372
    elder, 374–376, 377
    during pregnancy, 371–372
    spouse, 372–374
Acute intoxication, 48
Adams v. Peterson, 415
Administration Simplification (AS), 170
Adobe Photoshop®, 215
AFDIL. *See* Armed Forces DNA Identification Laboratory (AFDIL)
AFIS. *See* Automated Fingerprint Identification System (AFIS)
Age determination, 144–146
    dental *vs.* skeletal, 143
    dentin and, 291
    diaphyseal lengths and, 144
    histological, 145
    osteoarthritic changes and, 145
    radiographic, 145
Age estimation, 29, 263–295
    for adolescents and adults, 279–292
    aspartic acid racemization and, 291–292
    cementum annulations and, 284–285
    composite, 264
    enamel uptake of ^{14}C and, 292
    Gustafson method, 281–284
    root transparency and, 284
    third molar formation and, 288–291
    tooth eruption/tooth emergence and, 264–269
    tooth mineralization and, 269–279
    tooth wear and, 286–288
Aggripina, the younger, 12
Aguilar v. State, 411
Al-Mosawi v. State, 417
ALI. *See* Alternate light imaging (ALI)
Alleleles, 112
    defined, 127
    in population genetics, 142
Alternate light imaging (ALI), 219, 239
Amalgam filling, 182
American Academy of Forensic Sciences, 210, 330, 332
American Board of Forensic Anthropology, 253
American Board of Forensic Odontology (ABFO), 8, 288, 290, 332, 351
    bitemark analysis guidelines, 350–351
    bitemark scoring guide, 344
    bitemark workshop, 355–358
    certification, 409
    guidelines and standards, 177–178, 258, 331
    guides and terminology, 258
    protocols for collection and packaging of excise bitemarks, 403
    reference manual, 172
    website, 172, 177
American Dental Association, 177, 184, 194
American Society of Crime Laboratory Directors, 114
Amino acid racemization, 294, 303
Amoëdo, Oscar, 17–18
Amylase, in saliva, 121
Andrews v. State, 414
Anthropologists, 54
Arm injury, 238
Armed Forces DNA Identification Laboratory (AFDIL), 107, 109, 110, 112, 127, 156
    collection team, 117
Armed Forces Institute of Pathology, 103, 332
Aspartic acid racemization, 291–292
Attorney(s), 380
    expert's role om education, 392
    responsibility, 361

Attrition
    age progression and, 288
    crown changes and, 283
    extensive, 283
    occlusal, 281, 286
Automated Fingerprint Identification
        System (AFIS), 75, 97–100
Autopsy(ies), 3, 50, 70
    anatomic changes discovered at, 47
    bitemark evidence at, 313
    bitemarks identified at, 7
    evisceration of organs at, 54
    examination, 53
    forensic performance standards, 57
    of individuals dying in hospital settings,
        50
    invasive procedures, 211
    medical facility for, 49
    medicolegal, 57
    in MFI, 246
    payment for, 50–51
    postmortem team, 255
    rate in U. S., 50
    room, 53
    technicians, 54
    third-party payers of, 51
Avulsion, 345, 370

B

Badge systems, for mass disaster sites, 250
Baker v. State, 414
Bandpass filter, 223
Banks v. State, 417
"Bar code" effect, 152
Bass v. Scully, 416
Bayer model, 216
Bayesian statistical model, 147
Bitemark(s), 331, 344, 363
    abdomen, 337, 347, 372
    ALI image, 221
    analysis, 331, 344, 363
        future of, 364
        guidelines, 350–351
        methods of, 346–348
        reporting conclusions of, 350–351
        statistical and mathematical, 354–355
        unique of dentition and, 351–353
        variables and, 346–348
    arm, 337
    back, 337
    breast, 309, 310, 316, 318, 372

cases, 306–332
    chronology of, 306–308
    management of, 337–351
        evidence analysis in, 344–348
        evidence collection in, 337–344
        injury *vs.* dental evidence in,
            348–351
    problem, 316–332
    significant, 308–316
characteristics, 332–337
    class, 334
    frequency and distribution, 336–337
    individual, 336
color image, 221
defined, 332
evidence, 54
    at autopsy, 313
    collection and analysis, 28, 337–348
    fingerprint *vs.*, 8
    forensic value of, 359–360
    injury *vs.* dental, 348–351
    as "junk" science, 362
    management of, 395–404
        approaching scene for, 396
        conducting final survey in,
            403–404
        diagram or sketch, 399–401
        evaluation process in, 398
        initiating preliminary survey in,
            397–398
        narrative descriptions in, 398
        photographic depicting of scene
            in, 399
        preparing for, 396
        recording and collecting in,
            402–403
        releasing scene in, 404
        securing and protecting for,
            396–397
    salivary DNA in, 121–123
face, 337
foot, 337
genitalia, 337
hand, 337
head, 337
hemorrhage and, 329, 331, 347
with high forensic/evidentiary value, 333
individual characteristics, 336
leg, 337
with limited forensic/evidentiary value,
    335
locations, 335

Index

neck, 337
nose, 311
scoring guide, 344
shoulder, 221, 230, 337, 347, 374
skin as medium for recording of, 353–354
teeth mark *vs.*, 332
terminology, 350, 407
testimony, 308
workshop, 355–358
Bitemark Severity and Significance Scale, 345
Bitemark Terminology, 350
standards, 407
Blackwelder v. State, 421
Blade injury, 151–153
Blood
detection, 233
latent, 206
light absorption, 222
samples, 121
Bludsworth v. State, 412
Blunt force injury, 153
Blunt trauma, 374
Bone. *See* Skeletal remains
Booth, John Wilkes, 17
Bouie v. State, 414
Boyd v. State, 420
Boyle v. Brigano, 391, 393
Bracketing, photographic, 215
Bradford v. State, 412
Breast
bitemarks, 309, 310, 316, 318, 372
excised, 237, 238
removal of, 319
Brewer, Kennedy, 327–332
Brewer v. State, 327, 418, 419
Bridges. *See* Restorations
Brim v. State, 416
Bromley v. State, 414
Brooks v. State, 418
Brown v. Commonwealth, 417
Brown v. State, 417, 419
Bruises, 206, 219, 221, 345, 370
multicolored, 376
photographing, 208
Buffkin v. State, 420
Bundy v. State, 412
Bundy v. Wainwright, 413
Buntin v. Guardian Life Ins. Co. of America, 393
Burke v. Town of Walpole et al, 420
Burns, 69, 193, 370
Busby v. State, 413

C

Calhoun v. State, 420
California v. Marx, 311–312
Cameras
35 mm, 212
automatic point and shoot, 212, 216
digital, 204
imaging software, 215
nonvisible light photography for, 228
saving images with, 217
use of light, 204
visible light spectrum, 212
size limitations, 212
Cancer, 70
CAPMI. *See* Computer-Assisted Postmortem Identification (CAPMI)
CAPTA. *See* Child Abuse Prevention and Treatment Act (CAPTA)
^{14}Carbon uptake of enamel, age estimation, 292
Caries, 28, 283, 293
interproximal, 190
tertiary dentin and, 282
Carr v. State, 418
Carter v. State, 419
Cazes v. State, 418
CBCT. *See* Cone beam computed tomography (CBCT)
Cementum
annulations, 284–285
analysis, 294
apposition, 282–283
Chafin v. State, 391, 393
Chain of evidence, 237, 249, 258
Chaney v. State, 414
Chaplin v. McGrath and Donohue, 416
Charlemagne, 33
Charles the Bold, 13
Chase v. State, 412
ChemFinder, 125
Child abuse, 370
Child Abuse Prevention and Treatment Act (CAPTA), 370–371
Cholelithiasis, 70
Chromatographic analysis, 55, 292
Civil litigation, 29, 385–388
CJIS. *See* Criminal Justice Information Services (CJIS)
Clemons v. State, 412

CMOS. *See* Complementary meta-oxide semiconductor (CMOS)
CODIS. *See* Combined DNA Index System (CODIS)
Collman v. State, 418
Com. v. Spencer, 393
Combined DNA Index System (CODIS), 75, 126–127, 156, 157
    Convicted Offender Index, 126
    Forensic Index, 126–127
    Quality Assurance Standards, 113
Commonwealth v. Alvarado, 416
Commonwealth v. Cifizzari, 413
Commonwealth v. Edwards, 414
Commonwealth v. Graves, 412
Commonwealth v. Henry, 414, 418
Commonwealth v. Jones, 414
Commonwealth v. Maltais, 412
Commonwealth v. Thomas, 414, 418
Complementary meta-oxide semiconductor (CMOS), 216
Computed tomography (CT), 50
Computer(s), 172
    for comparing dental radiographs, 178
    digital fingerprint images, 85
    genetic analyzer, 112
    photography and, 216
    security and, 250
    software programs, 125
    use in identification, 74
Computer-Assisted Postmortem Identification (CAPMI), 125, 170, 258
Cone beam computed tomography (CBCT), 196–198
    duplication in, 199
    imaging possibilities, 197, 198
    of jaw fragments, 200
    panoramic, 200
    pitfalls and tips, 199
    of skeletal and carbonized remains, 199
    star artifacts, 197
    victim identification and, 198, 199
Constitutional rights, 308, 380, 383, 391
Contamination
    of DNA, 109, 122, 124–125
    at mass disaster site, 250
Contract for services, 49
Contusion, 334
Convicted Offender Index, 126
Coroner(s)
    medical examiners *vs.,* 40
    requirements for position of, 39
    system
        historical perspectives, 35–37
        in England, 35–37
        in Germany, 37
        in Scotland, 37
        shortcomings of, 40
    organizational facets in common with ME offices, 56
Cox v. State, 414
Cremains, 180, 183, 392
Crime scene, 123, 314, 327, 338, 396
    DNA profiles recovered at, 126
    evidence, 126
        processing of, 113
    grid, 250
    latent prints at, 2
    management, 249–251
    material left at, 342
    photograph, 211, 309, 317, 398
    serological fluids left at, 219
Criminal Justice Information Services (CJIS), 75
Criminal litigation, 29
Criminal Master File, 97
Criminal offenders, 126
Cristini, Michael, 325–326
Crown v. Hay, 309
Crowns, 147, 177, 182, 270

D

Dating, physical injuries during, 372
Daubert v. Merrell Dow Pharmaceuticals, 4, 384–385
Davasher v. State, 415
Davis v. State, 416
Death
    ancient understanding of, 34
    natural, 53
    timing of, 149
    trauma resulting in, 45, 50, 55
Death certificate, 43–44, 49, 62, 166, 254
Death certification, 42–49
    cause of death, 45–47
    manner of death, 47–49
    rationale for, 44–45
    requirements, 43
Death investigation, 31–34
    historical perspectives
        in ancient Egypt, 33
        in ancient Greece, 32

Index 427

in China, 35
in Roman empire, 32–33
modern systems, 38–42
  components, 38
office, 49–56
  funding of, 49
systems
  common denominator of all, 42
  types of, 43
  in U. S., 43
Dental arches, 197, 334
Dental casts, 28, 311, 312, 352
Dental charts, 120, 141, 147, 391
Dental exemplars, 348–350
Dental identification, 5, 17, 26–27, 54, 64,
    163–184
  ABFO guidelines and standards,
      177–178
  in Australia, 21
  cases, 22
  from dental records, 5, 19, 392
  in England, 21
  history, 12–23, 166
  in multiple fatality incidents, 245–261
  philosophy and legal basis for, 166–167
  statistical and mathematical models,
      172–177
  steps, 167–172
    antemortem examination, 168–170,
        171
    comparison, 170–172
    postmortem examination, 167–168,
        169
  technological and scientific advances,
      178–181
  threshold, 125
  tips and cautions, 181–183
  tips and cautions for, 181–183
Dental radiology, 178
Dental records, 4
  antemortem, 22, 26, 64, 168, 181
  availability, 27
  dental identification from, 5, 19, 392
  HIPAA and, 257
  mass disasters and, 259
  postmortem, 167, 253
  reconstruction of, 27
Dentin, 104
  age determination and, 291
  changes in, 284
  for DNA recovery, 120
  from maxillary molar, 107

primary teeth, 270
radiographic images, 190
sclerotic, 283
secondary, 282
tertiary, caries and, 282
transparent, 284
Dentinal tubules, 119, 284
Denture(s), 314
  identification system, 21
  marking, 20
Deutscher v. State, 411
Deutscher v. Whitley, 415
Diagnostic imaging techniques, 50
Diaphyseal lengths, 144
Digital photography, 96, 214, 216–219
Disaster Mortuary Operational Response
    (DMORT), 76
Disaster-portable morgue units (DPMUs),
    251
Disaster Squad, 82
Disaster Victim Identification (DVI), 82–83,
    251
Disasters. *See* Mass incidents; Multiple
    fatality incidents
Dismemberment, intentional, 400
Dissection
  organ, 53, 54
  postmortem, 33
  techniques, 178
DMORT. *See* Disaster Mortuary
    Operational Response (DMORT)
DNA
  amplification, 108–111
  analysis, 111–112
  contamination of, 109, 122, 124–125
  databases, 125–127
  equipment performance, 113, 396
  evidence, 165–166
  isolation, 106–108
  laboratory processing, 106–114
  laboratory quality assurance, 113–114
  in management of mass incidents,
      114–119
    communicating with laboratory in,
        116
    data management in, 119
    establishing scope in, 116
    evidence collection teams for, 117–118
    planning in, 115
    transportation and storage in, 118
  mitochondrial, 105
  molecular structure, 105

saliva and oral mucosa as sources of, 120–125
teeth as sources of, 119–120
DNA Advisory Board, 113
DNA Fingerprint Act of 2005, 127
DNA genome
   defined, 130
   mitochondrial, 105–106
   PCR amplification, 110
   site, 6
DNA identification, as gold standard for biological human identification, 5
DNA Identification Act (1994), 113, 126
Dog bite, 230
Dog tags, 82
Dolvin v. State, 393
Doyle v. State, 22, 307, 308–309, 365, 390, 391, 393, 411
Doyle v. Texas, 22
DPMUs. *See* Disaster-portable morgue units (DPMUs)
Drake v. Portuondo, 419, 421
Drayton v. State, 420
Dross v. State, 420
Drug intoxication, 55
Drug-related death, 51, 55
DuBoise v. State, 413
DVI. *See* Disaster Victim Identification (DVI)

E

Earl of Shrewbury, 12
EARR. *See* External apical root resorption (EARR)
EDS. *See* Energy dispersive spectroscopy (EDS); Energy dispersive X-ray spectroscopy (EDS)
Ege v. Yukins, 420
Elder abuse, 374–377
Energy dispersive spectroscopy (EDS), 180
Equipment. *See also specific types*
   camera, 204
   digital photographic, 212
   for mass disaster, 197, 255
   performance, 113, 396
   personal protective, 255
   radiographic, 196–199
   reliability, 113
   upgrading, 228
   validation of new, 113
Evidence
   at autopsy, 313

bitemark, 54, 313
   collection and analysis, 28
   fingerprint *vs.,* 8
   salivary DNA in, 121–123
chain of, 237, 249, 258
collection
   management of, 337–344
   teams for, 117–118
crime scene, 126
   processing of, 113
DNA, 117–118, 165–166
from forensic dental photography, 237, 241
of malpractice, 389
management, 395–404
   approaching scene for, 396
   conducting final survey in, 403–404
   diagram or sketch, 399–401
   evaluation process in, 398
   initiating preliminary survey in, 397–398
   narrative descriptions in, 398
   photographic depicting of scene in, 399
   preparing for, 396
   recording and collecting in, 402–403
   releasing scene in, 404
   securing and protecting for, 396–397
recovery log, 403
relevant, 3
response teams, 396
rules of, 3, 385
Exhibits, 120
   with fluorescence, 219
Exhumation, 148, 183
Exparte sues Dolvin, 393, 411
Expert testimony, 3, 29
Expert witness, 384–385, 409
   in civil disputes, 4–5
   defendants, 390
   earliest account of dentist as, 13
   qualified, 3
   as subject of legal action, 332
   workshop in testimony of, 332
External apical root resorption (EARR), 283

F

Facial reconstruction, 21
Facial trauma, 61, 164, 369
   from domestic abuse, 375
Family Assistance Act (1996), 248, 251

Index 429

Family Assistance Center, 247, 248
Family member(s)
 complaints, 251
 for decedent history, 52–53
 for DNA profile, 118, 123, 127
 elder abuse and, 376
 emotional trauma, 164
 identification by, 64, 164, 246
 misidentification by, 164
 release of victim to, 255
Faveon X3 CMOS sensor, 217
FBI. *See* Federal Bureau of Investigation (FBI)
Federal Bureau of Investigation (FBI)
 Civil File, 97
 Criminal Master File, 97
 Disaster Squad, 82
 Disaster Victim Identification, 82–83
 DNA Advisory Board, 113
 evidence response teams, 396
 Identification Division, 81
 Supplementary Homicide Reports, 372
Federal Emergency Management Agency (FEMA), 76
Federal Reporter, 384
Federal Rules of Evidence, 3, 385
Federation Dentaire Internationale (FDI), 20
Fellatio, forced, 28
FEMA. *See* Federal Emergency Management Agency (FEMA)
Fight bite, 332
Fillings, 20, 179–180
Film, 109
 image processing, 191–192
 transparency, 95
Fingerprint(s), 79–100. *See also* Friction ridge skin
 antemortem *vs.* postmortem, 98
 arches, 84, 85
 classification, 85
 formations, 84
 fundamentals, 83–85
 historical overview, 80–83
 identification, 164–165
  automated, 97–100
 impression, 82, 83, 84, 85
 individuality, 88–89
 known, 85
 latent, 85
 loops, 84, 85
 patent prints, 85
 pattern types, 85
 plastic impressions, 85
 postmortem recovery process, 89–96
  inspecting and cleansing friction skin, 90
  reconditioning compromised friction skin, 90–94
  recording postmortem impressions of, 94–96
 whorls, 84, 85
Flash units, 212
 IR light sources, 226
 UV light sources, 226
Florida v. Bundy, 313–314
Florida v. Stewart, 314–316
Fluorescent light, 220–221, 222
Focus shift, 226-=228
Force feeding, 28
Forensic anthropology, 71
 analysis, 138–155
  animal *vs.* human in, 139
  biological distance and, 143
  biological profile in, 141–147
   ancestry, 142–143
   biotype, 142
   features, 142
   nonmetric variations, 142
   sex, 141–142
   statistics, 142
  defined, 137
  function tests, 143
  individualization in, 147–148
  limb proportion indices, 143
  medicolegal significance of human remains in, 140
  minimum number of individuals in, 140
  nonmetric traits, 143
  postmortem interval in, 148–149
  skeletal and dental age, 143–146
  stature and physique, 146–147
  trauma and, 149–155
   antemortem, 150
   perimortem, 150–154
   postmortem, 154–155
  typical case progression, 138–155
  unique identifiers, 147
Forensic dental consultant, 398
Forensic dental identification, 166–184. *See also* Dental identification
Forensic dental photography, 203–244
 alternate light imaging, 219
 application and use, 234–236
 digital, 216–219

film-based, 214–215
fluorescent imaging, 219–222
focus shift, 226–229
infrared, 226, 232–234
injured skin properties and, 207–211
management of evidence from, 237, 241
nonvisible light, 222–226
properties of illumination, 205–206
   absorption, 205
   fluorescence, 205
   reflection, 205
   transmission, 205
reflective long-wave ultraviolet, 229–232
spectrum of light, 204, 205
standard techniques, 211
types and techniques, 211–212
ultra light, 226, 229–232
visible light in, 212–213
Forensic dental radiography, 187–201
   biological effects, 192–194
   equipment, 196–198 (*See also* Cone beam computed tomography (CBCT))
   film image processing, 191–192
   intraoral techniques, 192
   radiation protection, 194–196
      handheld X-ray devices and, 196
      maximum permissible dose for, 194–195
      personnel monitoring for, 195–196
   radiographic image and image receptors, 189–191
   X-ray generators, 188–189
Forensic dentist(s)
   civil litigation and, 385–388
   as defendants, 388
   as expert witness defendants, 390
   as expert witnesses, 388
   as fact witnesses, 388
   malpractice, 381, 386, 388–389
   personal injury litigation, 389–390
Forensic dentistry
   certifying organizations and certification, 407–409
   dry fingered, 25
   forensic identification and, 4–8
   history, 12–23
   legal issues in, 384–392
      case law, 390–392
      civil litigation, 385–388
      expert witnesses, 384–385
      personal care litigation, 389–390
      standard of care litigation, 388–389
   research and technology, 406–407
Forensic identification
   forensic dentistry and, 4–8
   in history, 12–23
   steps in, 2–3
Forensic Index, 126–127
Forensic Information System for Handwriting, 125–126
Forensic investigation requirement, 3
Forensic odontology, 104, 352, 353, 404
   bitemark testimony and, 308
   Bundy case and, 314
   contributions to, 175
   education programs in, 331, 363
   egocentricity and, 358–359
   experts, 320
   Gustafson method, 281
   imaging, 180
   research in age estimation, 293
   scope, 25–30
   U. S. federal and state court cases in, 411–421
Forensic pathologist, 41
Forensic pathology, 40
   criteria for assessment, 57
   as medical subspecialty, 41
Forensic science, law *vs.*, 1–2
Forensic Specialties Accreditation Board (FSAB), 407
Fox v. State, 414
Franks v. State, 417
Freeman v. State, 415
Friction ridge skin, 83–84
   bifurcations, 84
   comparison and identification of impressions of, 86–89
   detail and structure, 84
   dividing ridges, 84
   ending ridges, 84
   individuality, 88–89
   inspecting and cleansing, 90
   methodology in examination of, 86–87
      analysis, 86
      comparison, 86–87
      evaluation, 87
      exclusion, 87
      individualization, 87
      verification, 87
   persistency, 89
   reconditioning compromised, 90–94

recording postmortem impressions, 94–96
ridge arrangements, 84–85
substructure, 84
Frye v. Merrell Dow Pharmaceuticals, 4
Frye v. United States, 393
Furtado v. State, 419

G

Garrison v. State, 420
Gary v. Schofield, 419, 420
Gary v. Terry, 420
General Electric Co. v. Joiner, 393
General Electric v. Joiner, 385
Genitalia, bitemarks, 337
Global positioning system (GPS), 250
Gonzales v. State, 420
Gorgas, F. J. S., 16
Government of the Virgin Islands v. Byers, 417
GPS. *See* Global positioning system (GPS)
Graves v. State, 412
Green v. State, 414
Gross, Winfield, 16–17
Guerin, 14
Gunshot injury, 150–151
Gustafson method, 281–284

H

Handley v. State, 413
Harris v. State, 415
Harrison v. State, 416
Harward v. Commonwealth, 413
Hazardous materials (HAZMAT), 250
HAZMAT, 250
Health Insurance Portability and Accountability Act (HIPPA), 170
Hernandez v. State, 419
High performance liquid chromatography (HPLC), 292
HIPPA. *See* Health Insurance Portability and Accountability Act (HIPPA)
His Majesty's Advocate v. Pattison et al., 13–14
Hitler, Adolf, 21–22
Hodgson v. State, 417
Holmes v. Pierce, 421
Homicide, 52
Howard v. Kelly, 414
Howard v. State, 417, 419, 420
HPLC. *See* High performance liquid chromatography (HPLC)

Human identification. *See* Identification of remains
Human remains
adolescent, 144
adults, 145–146
age determination, 144–146
of children, 144
fetal, 144
medicolegal significance, 140
of young adults, 144
Hurricane Katrina, 22
Hurricane Rita, 22

I

IAFIS. *See* Integrated Automated Fingerprint Identification System (IAFIS)
IBIS. *See* Integrated Ballistics Identification System (IBIS)
Ident-A-Drug, 125
Identification of remains
age determination in, 144–146
birthmarks in, 67
circumstances of death as aid to, 64–65
establishing positive, 63–64
external characteristics in, 66–70
fingerprints and, 79–100 (*See also* Fingerprint(s))
human *vs.* nonhuman, 62–63
internal characteristics in, 70
jewelry as aid in, 67
in massive head trauma, 67
occupation stigma in, 67–69
presumptive, 63
radiographs in, 70–71
scars in, 68
sources of comparison, 75
tattoos in, 68
visual, 64
Illinois v. Johnson, 310
Illinois v. Milone, 312
Impression, 28
dental, 340, 342, 354
dermal, 99
epidermal, 99
fingerprint, 82, 83, 84, 85
military use, 82
powder, 96, 97
In re Pederson, 419
In re the Marriage of Rimer, 413

Incident Command System (ICS)
  chain of command, 252
Incised wound, 345
Injured skin, 207–211
Injury patterns, 203–204
  in blunt trauma, 374
  dental exemplars and, 348
  photographing, 221, 233
  thin skin, 210
Inman v. State, 413
Integrated Automated Fingerprint
  Identification System (IAFIS), 97
Integrated Ballistics Identification System
  (IBIS), 125
Iroquois Theatre fire, 19

J

Jackson v. Day, 417
Jackson v. State, 413
Jai Chand, 12
Jewelry, 67
Johnson v. State, 418
Joint Commission on Accreditation of
  Health Care Organizations, 50
Jonestown, Guyana, 22

K

Keep, Nathan Cooley, 15
Keko v. Hingle, 419
Kennedy v. State, 411
Kinney v. State, 416
Knife wound, 241
Krone, Ray, 323–325
Kumho Tire v. Carmichael, 393
Kunco v. A. G. of Pennsylvania, 419

L

Laboratory Accreditation Board, 114
Laceration, 90
  elliptical, 311
Law, 3–4
Leal v. Dretke, 419, 420
Legal issues, 379–392. *See also* Forensic
  dentistry, legal issues
  civil litigation, 381, 385–390
  courts
    American, 380–383
    decision reporting systems, 383–384
    federal, 382
    state, 381
  U.S. Constitution and, 382
  criminal litigation, 380
Liquid chromatography mass spectrometry
  analyzer, 55
Litaker v. State, 414
Litigation. *See* Legal issues
Louis XVII, 14–15

M

Magnetic resonance imaging, 50
Mallory v. State, 414
Malpractice, 381, 386, 388–389
  preponderance of evidence of, 389
Mandible, 265, 268, 270, 289, 290
  bony structures, 168
  crown-root development, 288
  resecting, 192
  rocker, 143
Marbley v. State, 412
Marquez v. State, 413, 417
Martin v. Cain, 418
Mass incidents. *See also* Multiple fatality
  incident (MFI)
  DNA management, 114–119
    communicating with laboratory for,
      116
    data management for, 119
    establishing scope for, 116
    evidence collection teams for, 117–118
    planning for, 115
    transportation and storage for, 118
Mataya v. Kingston, 419
Maxilla, 268, 270, 289, 290
  bony structures, 168
  crown-root development, 288
  resecting, 192
Maynard v. State, 412
McCrory v. State, 413
McGrew v. State, 417
Meadows v. Commonwealth, 420, 421
Media, 116
  CODIS and, 126
  court decision reporting systems and, 383
  forensic anthropology and, 138
  MFI and, 254
  mistakes, 254
  role in identification, 74
Medical examiner(s)
  civil service protection, 40
  coroners *vs.,* 40

office
  accreditation, 57
  autopsy section, 53–54
  clerical section, 56
  divisions of, 51
  quality assurance, 56–58
  toxicology section, 55
 system, 41–42
  elements of, 41–42
Methylene blue, 284
MFI. *See* Multiple fatality incident (MFI)
Middleton v. State, 418
Miller v. State, 412
Milone, Richard, 320–322
Milone v. Camp, 416
Mitchell v. State, 413, 415
Mobley v. State, 416
Moldowan, Jeffrey, 325–326
Moldowan v. City of Warren et al, 420
Monk v. Zelez, 414
Morgan v. State, 416
Moses v. State, 419
Motor vehicle death, 52
Moye v. State, 417
Multiple cone beam CT images, 198
Multiple fatality incident (MFI), 27–28, 245–261
  crime scene, 249–251
  defined, 246
  identification section, 252–255
    computer-assisted programs in, 258
    digital photography and radiography in, 258–260
  incident perimeter, 249–250
  interagency relations, 251
  natural disasters, 248
  odontology section, 255–258
  preparedness training and planning for, 260
  site of, 249–251
    global positioning of, 250
    grid, 250
    security and safety of, 250–251
  terrorism, 249
  transportation accidents, 248–249
  types of, 248–249
  weapons of mass destruction, 249
Murphy v. State, 416

N

NAME. *See* National Association of Medical Examiners (NAME)
NamUS, 75
Nash v. NYC, 392, 393
National Association of Medical Examiners (NAME), 48, 56, 57
  Standards and Accreditation Committee, 57–58
National Automotive Paint File, 125
National Center for Missing and Exploited Children, 76
  Missing Person File, 75
  Unidentified Person File, 75
National Committee for Prevention of Elder Abuse, 376
National Council on Radiation Protection and Measurements (NCRP), 194
National Crime Information Center (NCIC), 75–76
National Disaster Medical System (NDMS), 251
National DNA Identification Index. *See* Combined DNA Index System (CODIS)
National Institute of Justice (NIJ), 75
National Integrated Ballistic Information Network, 125
National Missing Persons DNA Database (NMPDD), 127
National Response Plan (NRP), 251
National Transportation Safety Board (NTSB), 76, 248, 251
Natural death, 53
NCIC. *See* National Crime Information Center (NCIC)
NCRP. *See* National Council on Radiation Protection and Measurements (NCRP)
NDMS. *See* National Disaster Medical System (NDMS)
Neck injury, 238
Nephrolithiasis, 70
New York Civil Service Commission, 81
Ngoc Van Le v. State, 413
Niehaus v. State, 391, 393, 411
NIJ. *See* National Institute of Justice (NIJ)
NMPDD. *See* National Missing Persons DNA Database (NMPDD)

NRP. *See* National Response Plan (NRP)
NTSB. *See* National Transportation Safety Board (NTSB)

O

O'Donnell v. State, 419
OdontoSearch, 165–166
Oklahoma v. Wilhoit, 307, 316
Organ dissection, 54
Osteon fragments, 146
Oswald, Lee Harvey, 22
Otero v. Warnick, 418

P

Pardo and McLemore v. Simons et al, 419
Parkman, George, 15
Parks v. Commonwealth, 418
Patrick v. State, 420
Patterson v. State, 411
People v. Allah, 411
People v. Bass, 414
People v. Bethune, 412
People v. Blommaert, 415
People v. Brewer, 308
People v. Brown, 416
People v. Bundy, 307
People v. Calabro, 414
People v. Cardenas, 415
People v. Case, 415
People v. Columbo, 412
People v. Culuko, 418
People v. Cumbee, 417, 420
People v. Dace, 413
People v. Daniels, 418
People v. Davis, 413
People v. Dixon, 412
People v. Drake, 413
People v. Dunsworth, 415
People v. Ege, 366, 417
People v. Ferguson, 414
People v. Gallo, 416
People v. Geer, 411
People v. Hampton, 413
People v. Hernandez, 414
People v. Holmes, 415
People v. Howard, 413
People v. Johnson, 307, 310, 365, 411, 418
People v. Jordan, 412
People v. Koberstein, 418
People v. Krone, 308, 316
People v. Madrigal, 421
People v. Marsh, 414
People v. Marx, 307, 311–312, 411
People v. Mattox, 393
People v. May, 418
People v. Mayens, 392, 393
People v. McDonald, 412
People v. Middleton, 411
People v. Milone, 307, 312, 316, 320–322, 366, 411
People v. Moldowan, 366, 419
People v. Moldowan and Cristini, 307, 339
People v. Moreno, 419
People v. Mostrong, 419
People v. Noguera, 416
People v. Payne, 417
People v. Perez, 413
People v. Perkins, 415
People v. Prante, 413
People v. Pre, 419
People v. Quaderer, 419
People v. Queen, 412
People v. Randt, 414
People v. Rich, 414
People v. Rush, 416
People v. Schuning, 412
People v. Shaw, 417
People v. Slone, 411
People v. Smith, 411, 412
People v. Stanciel, 415
People v. Starks, 420
People v. Steward, 418
People v. Stewart, 307
People v. Tripp, 416
People v. Vigil, 413
People v. Wachal, 413
People v. Walkey, 412
People v. Watson, 411, 413
People v. Wilhort, 307
People v. Williams, 412
People v. Wright, 418
Personal effects, 165
Peter Halket, 13
Photographic log, 400
Photography. *See also* Forensic dental photography
    digital, 96, 214, 216–219, 256
        advantages, 233
    evidence management and, 399
    of images over time, 207, 208, 209, 210
Polymerase chain reaction (PCR), 108–111, 112, 130–131

Index

Pregnancy, abuse during, 371–372
Pseudoborne objects, 139
Public Prosecutor v. Torgerson, 309
Purser v. State, 417

Q

QLF. *See* Quantitative light-induced fluorescence (QLF)
Quality Assurance Standards, 113
Quality Assurance Standards for Convicted Offender DNA Databasing Laboratories, 113
Quality Assurance Standards for DNA Testing Laboratories, 113
Quantitative light-induced fluorescence (QLF), 179

R

R. M. v. Dept. of Health & Rehabilitation Services, 416
Radiography. *See* Forensic dental radiography
Restoration(s), 171, 181
    amalgam, 181
    designation, 177
    materials used, 179
    resin, 179
Revere, Paul, 13
Ridgeology, 86
Rios v. State, 417
Robinson, A. I., 16
Rodoussakis v. Hosey, 416
Rogers v. State, 413

S

Salazar v. State, 414
Saliva and oral mucosa, as sources of DNA, 120–125
Scanning electron microscopy (SEM), 180
Scars, 68, 165
Scientific Working Group on DNA Analysis Methods (SWGDAM), 113
Scientific Working Group on Friction Ridge Analysis Study and Technology (SWGFAST), 87
Seivewright v. Warnick, 418
SEM. *See* Scanning electron microscopy (SEM)
Sheker v. State, 393

Shoulder, bitemarks, 221, 230, 337, 347, 374
Simmons v. Runnels, 420
Skeletal remain(s), 6, 61, 71
    commingled, 140
    decomposition, 54
    dessicated, 106
    evaluation of, 54
    evidence of parity, 141
    first steps in examination of, 139
    radiography, 72, 73
    visibly unidentifiable, 66
Skull. *See also* Skeletal remain(s)
    contrecoup fracture, 153
    CT scan, 198
    male *vs.* female, 73
    reconstruction of shattered, 153
    ring fractures of, 153
    sutures, 29
SLICE. *See* Spectral Library Identification and Classification Explorer (SLICE)
Smith v. State, 412, 413
Southard v. State, 412
Spectral Library Identification and Classification Explorer (SLICE), 180
Spence v. Scott, 417
Spence v. State, 414
Spindle v. Berrong, 416
Spousal abuse, 372–374
Standridge v. State, 412
State of Kansas v. Galloway, 412
State v. Adams, 412
State v. Anderson, 418, 420
State v. Armstrong, 413
State v. Arredondo, 419
State v. Asherman, 412
State v. Bennett, 416
State v. Best, 418
State v. Bingham, 413
State v. Blamer, 419
State v. Boles, 417
State v. Bridges, 411
State v. Bullard, 412
State v. Butler, 418
State v. Calloway, 418
State v. Carpentier, 416
State v. Carter, 412
State v. Cazes, 416
State v. Charley, 420
State v. Combs, 414
State v. Correia, 415
State v. Crump, 413

State v. Davidson, 418
State v. Dickson, 412
State v. Donnell, 416
State v. Duncan, 419
State v. Edwards, 415
State v. Finley, 421
State v. Ford, 414
State v. Fortin, 418, 419
State v. Gardner, 414
State v. Garrison, 411, 420
State v. Green, 412
State v. Hamilton, 417
State v. Hasan, 413
State v. Hill, 414, 415
State v. Hodgson, 416
State v. Howe, 411
State v. Hummert, 416, 417
State v. Jackson, 414
State v. Jamison, 413, 414
State v. Johnson, 413
State v. Jones, 411, 416
State v. Joubert, 415
State v. Kaufman, 421
State v. Kendrick, 411, 413
State v. Kirsch, 414
State v. Kiser, 417
State v. Kleypas, 411
State v. Krider, 418
State v. Krone, 323, 366, 416
State v. Kunze, 418
State v. Landers, 418
State v. Lester, 420
State v. Lyons, 416, 417
State v. Mann, 416
State v. Marshall, 420
State v. Martin, 416
State v. Mataya, 418
State v. McBride, 419
State v. McDaniel, 413
State v. McDermott, 420
State v. Mebane, 414
State v. Moen, 413
State v. Neal, 420
State v. Ortiz, 412
State v. Pearson, 415
State v. Peoples, 411
State v. Perea, 412
State v. Pierce, 413
State v. Prade, 418
State v. Richards, 414
State v. Routh, 411
State v. Sager, 411

State v. Sapsford, 412
State v. Schaefer, 416
State v. Smith, 419
State v. Stinson, 413
State v. Stokes, 412
State v. Swinton, 419
State v. Tankersley, 418
State v. Tankersly, 420
State v. Teasley, 417
State v. Temple, 411
State v. Thomas, 415, 417
State v. Thornton, 412
State v. Timmendequas, 418
State v. Turner, 411, 414
State v. Ukofia, 415
State v. Van Winkle, 366
State v. Vital, 413
State v. Warness, 416
State v. Welburn, 416
State v. Welker, 412
State v. Wilkinson, 417
State v. Williams, 415, 416
State v. Wimberly, 415
State v. Wommack, 418
State v. Worthen, 414
Strickland v. State, 413
Stubbs v. State, 419
Suicide, 48
Supplementary Homicide Reports, 372
SWGDAM. *See* Scientific Working Group on DNA Analysis Methods (SWGDAM)
Symbionese Liberation Army, 22

T

Tattoos, 68, 165, 242
Technical Working Group on DNA Analysis Methods (TWGDAM), 113
Teeth
   age estimation, 29
   age progressive changes, 281–283
   attrition
      crown changes and, 283
      extensive, 283
      occlusal, 281, 286
   cementum apposition, 282–283
   crown root development, 271, 275, 288
   occlusal attrition, 281–282
   periodontosis, 282
   root resorption, 283

Index

root transparency, 283
secondary dentin, 282
Ten-State Nutrition Survey, 266
Terrorist attacks, 22
Texas v. Doyle, 308–309
Thomas v. Beard, 420
Thornton v. State, 412
Tompkins v. Singletary, 393
Tooth cementum. *See* Cementum
Tooth decay, 182
Tooth eruption/tooth emergence, 264–269
Tooth mineralization, 269–279
Tooth numbering system, 20
Tooth wear, 286–288
Torgerson, Frederick Fasting, 316–320
Trauma, 31, 149–155
    abrasion, 93
    antemortem, 150
    blunt, 374
    death-induced, 45, 50, 55
    detection, 55
    disfigurement due to, 61
    emotional, of family members, 164
    evidence, 53
    facial, 61, 164, 369, 375
    inflicted, 370, 372
    intraoral, 28
    massive head, 67
    perimortem, 150–154
    postmortem, 154–155
    during pregnancy, 371
Tuggle v. Commonwealth, 412
Tuggle v. Netherland, 417
Tuggle v. Thompson, 417
TWGDAM. *See* Technical Working Group on DNA Analysis Methods (TWGDAM)

U

U. S. ex rel Dace v. Welborn, 416
U. S. Standard of Certificate of Death, 44

Unidentified persons, 76
United State v. Holland, 393
U.S. ex rel Milone v Camp, 415
U.S. v. Cerno, 421
U.S. v. Covington, 414
U.S. v. Dia, 416
U.S. v. Lizarraga-Tirado, 421
U.S. v. Martin, 411
U.S. v. Randolph Valentino Kills in Water, 419
U.S. v. Studnicka, 420

V

Valenti v. Akron Police Dept. et al, 413
Vehicular accidents, 334
Verdict v. State, 416
VICTIMS, 75
Villa v. State, 421
Vinyl polysiloxane, 340

W

Wade v. State, 413
Walsh, Caroline, 14
Walters v. State, 418
Waltman v. State, 418
Warren Joseph, 13
Washington v. State, 415
Webster, John, 15
Wheeler v. Commonwealth, 419
Wilhoit, Greg, 322–323
Wilhort v. State, 415
William the Conqueror, 15–16
Williams v. State, 414, 415, 421
Wilson v. Commonwealth, 418
WinID3©, 170, 171

X

X-ray fluorescence, 179

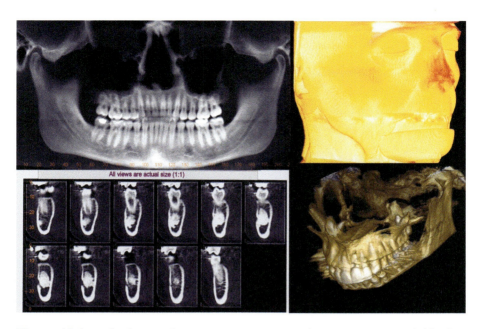

Figure 10.4 Multiple cone beam CT images. Several programs are available that provide three-dimensional rendering of the soft tissue.

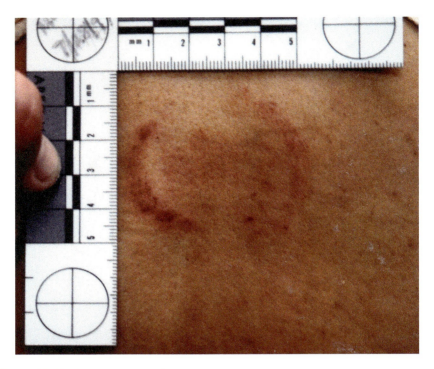

Figure 11.2 Back, color, day 1.

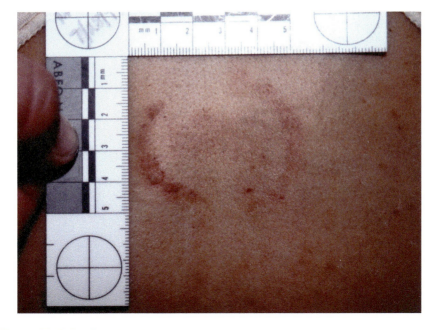

Figure 11.3 Back, color, day 8.

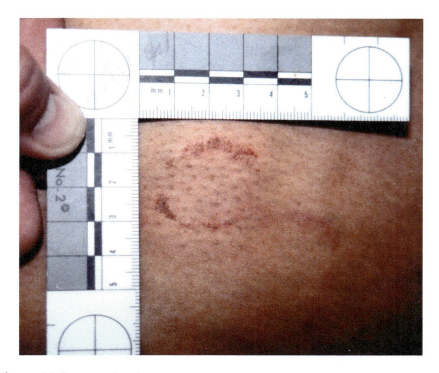

Figure 11.6 Hip, color, day 1.

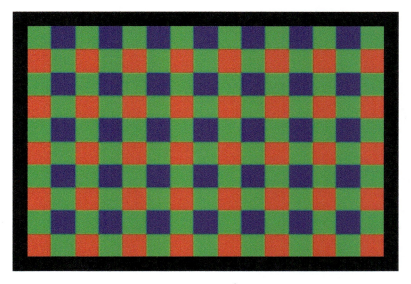

Figure 11.14 Bayer model. (Source: www.photo.net/learn/raw) RGB = red, blue, green. Each pixel in the sensor responds to either red, green, or blue light. There are two green sensitive pixels for each red and blue pixel because the human eye is more sensitive to green.

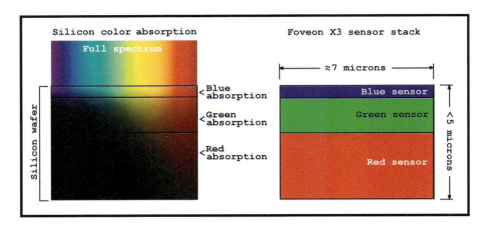

Figure 11.15 Foveon X3 CMOS sensor, method of capture and creation of a color digital image.

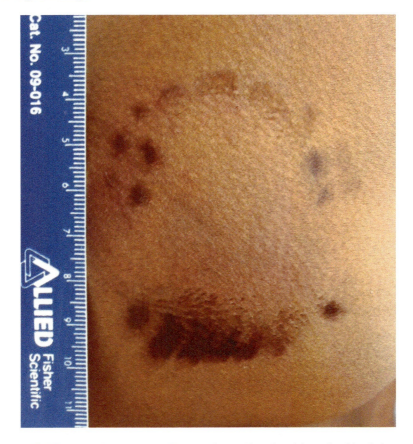

Figure 11.19 Color image of a bitemark on the shoulder of a black homicide victim.

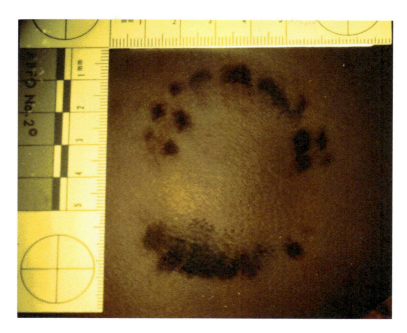

Figure 11.20 ALI image of same bitemark seen in Figure 11.19. *Note the enhancement of the bruise pattern in the fluorescent image acquired with ALI.*

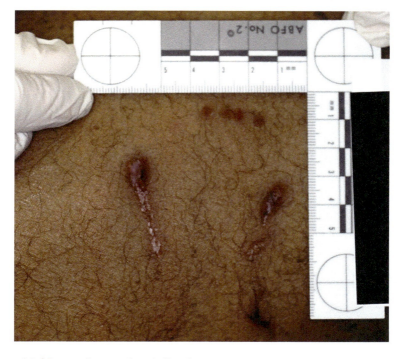

Figure 11.32 Dog bite on leg, full color spectrum.

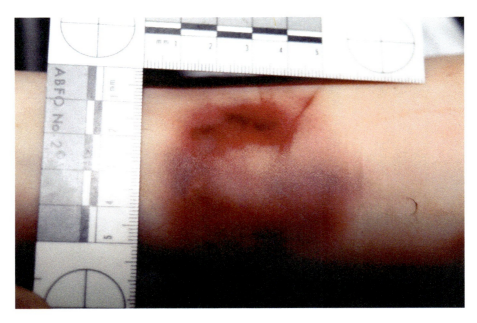

Figure 11.43 Full spectrum color image of patterned injury with very little forensic or evidentiary value.

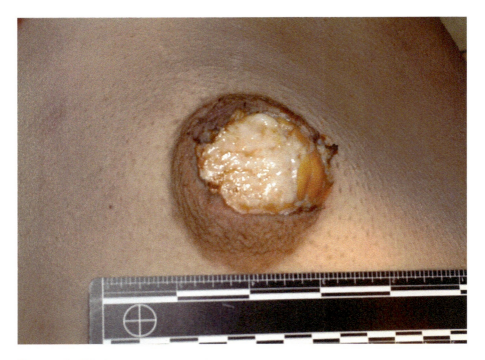

Figure 11.48 Excised breast, color.

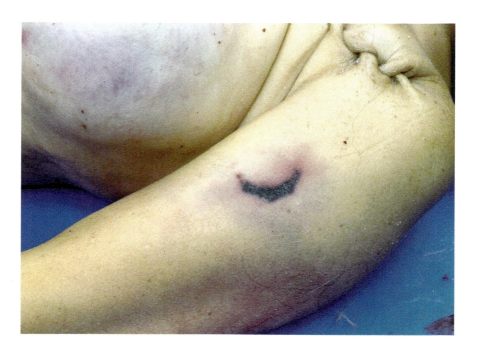

Figure 11.50 Orientation photograph of a patterned injury on the left arm that is suggestive of being a human bitemark, using ABFO terminology.

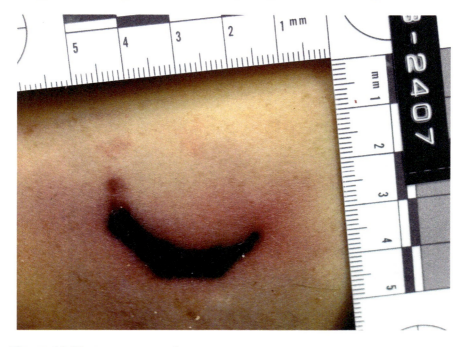

Figure 11.51 Arm injury, color.

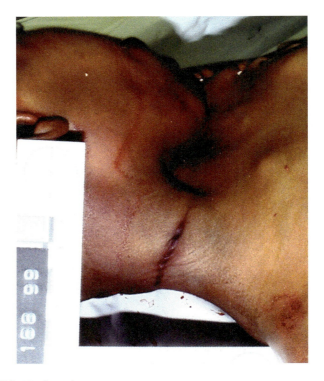

Figure 11.54 Neck, color.

Figure 11.62 Tattoo, color.

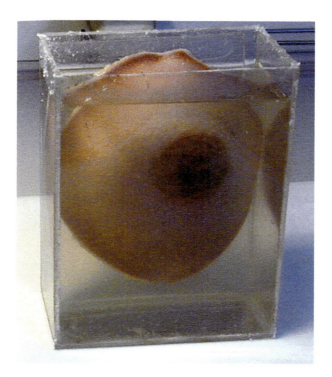

Figure 14.12 Breast removed in 1957 in the just opened case in 2001.

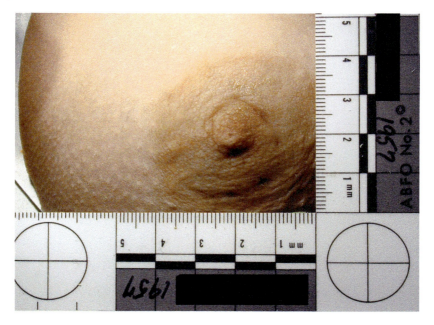

Figure 14.13 Closeup of bitemark on breast, imaged in 2001.

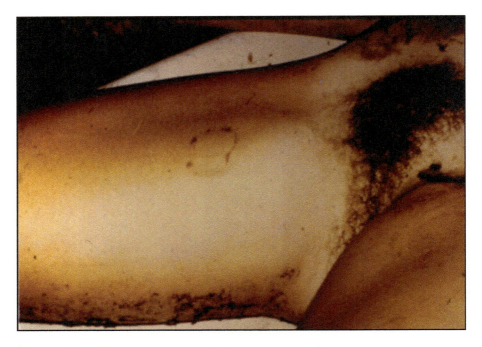

Figure 14.14 Bitemark on right thigh of Sally Kandel.

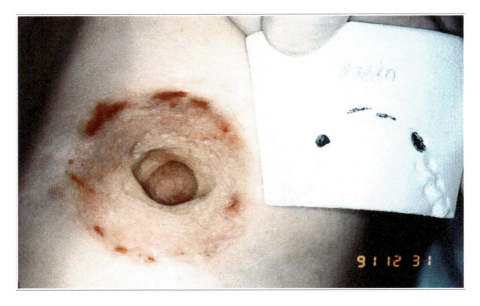

Figure 14.16 Side-by-side comparison of Ancona bitemark to Krone test bite in expanded polystyrene.

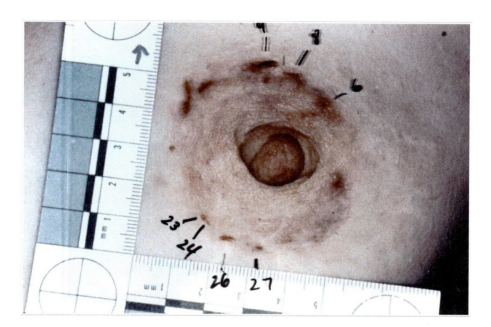

Figure 14.17 Ancona bitemark: Labels indicate one odontologist's opinion of marks made by specific teeth.

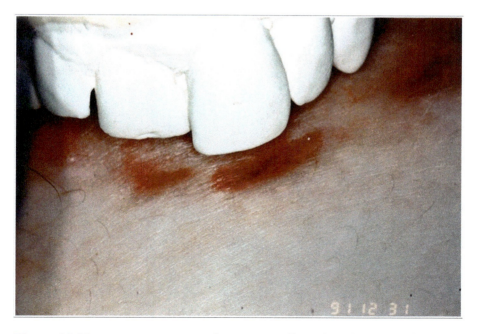

Figure 14.18 Direct comparison of Krone maxillary dental model to bitemark injury on breast of Ancona.

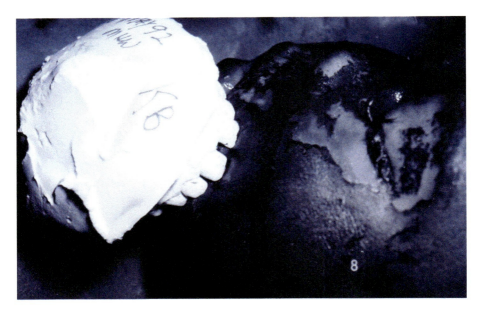

Figure 14.21 Direct comparison of Brewer's maxillary model to Jackson's face (pattern 19 on the MHW body chart).

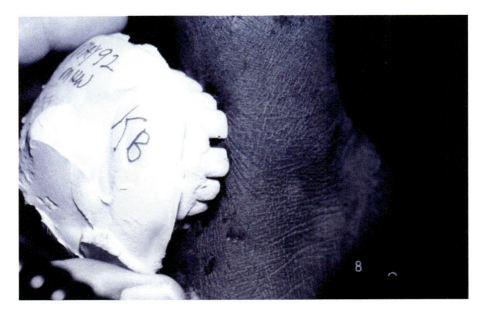

Figure 14.22 Direct comparison of Brewer's maxillary model to Jackson's foot (pattern 1 or 2 on MHW body chart).

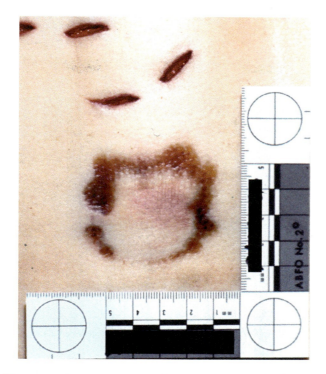

Figure 14.23 Bitemark and stab wounds made through clothing.

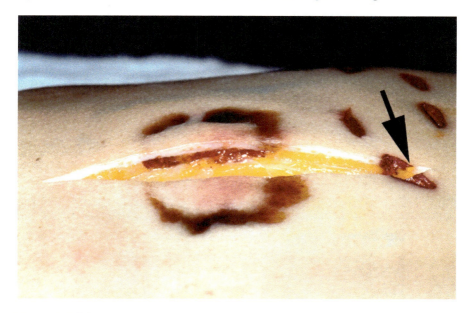

Figure 14.24 Incision through bitemark and stab wound. Note subepithelial hemorrhage in bitemark, none in stab wound (arrow).

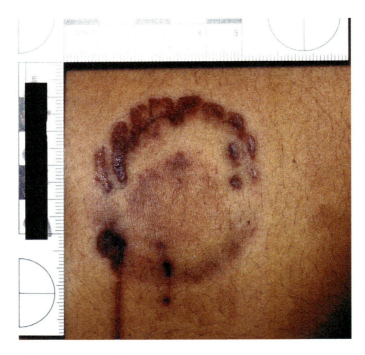

Figure 14.25 Bitemark with high forensic/evidentiary value.

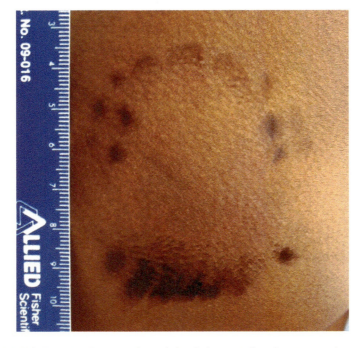

Figure 14.26 Another bitemark with high forensic/evidentiary value.

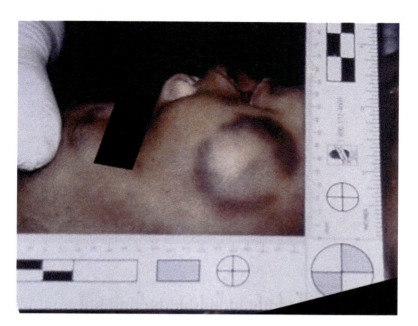

Figure 14.27 Bitemark with useful but limited forensic/evidentiary value.

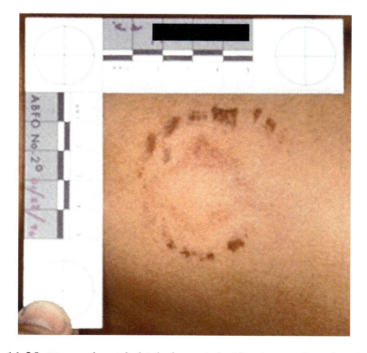

Figure 14.30 Bitemark with high forensic/evidentiary value showing class and individual characteristics. (Profile suggested that biter had a missing lower incisor and broken or malposed upper right incisors.)

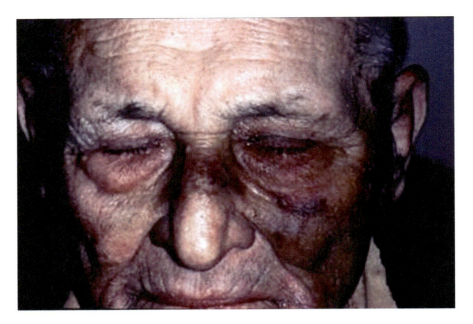

Figure 15.7 Ninety-three-year-old male with zygomaticomaxillary complex fracture.

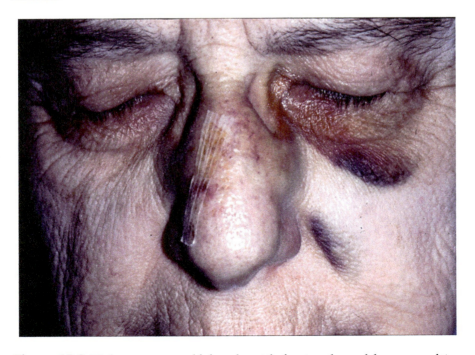

Figure 15.8 Eighty-two-year-old female with fractured nasal bones resulting from inflicted trauma.